Muscles, Nerves and M

KINESIOLOGY IN DAILY LIVING

Muscles, Nerves

KINESIOLOGY

and Movement
IN DAILY LIVING

BARBARA TYLDESLEY TDipCOT
School of Health Science,
Department of Occupational Therapy,
The University of Liverpool,
PO Box 147,
Liverpool

JUNE I GRIEVE Bsc, Msc
Formerly London School of Occupational Therapy,
Department of Health and Paramedical Studies,
West London Institute of Higher Education,
Lancaster House,
Borough Road,
Isleworth, Middlesex

OXFORD
BLACKWELL SCIENTIFIC PUBLICATIONS
LONDON EDINBURGH BOSTON
MELBOURNE PARIS BERLIN VIENNA

To the London and Liverpool students
of occupational therapy —
past, present and future

Editorial Offices:
Osney Mead, Oxford OX2 0EL
25 John Street, London WC1N 2BL
23 Ainslie Place, Edinburgh EH3 6AJ
238 Main Street, Cambridge,
MA 02142, USA
54 University Street, Carlton,
Victoria 3053, Australia

Other Editorial Offices:
Librairie Arnette SA
1, rue de Lille
75007 Paris
France

Blackwell Wissenschafts-Verlag GmbH
Kurfürstendamm 57
10707 Berlin
Germany

Blackwell MZV
Feldgasse 13
A-1238 Wien
Austria

First published, 1989
Reprinted 1991, 1993, 1994

Set by Times Graphics, Singapore
Printed and bound
in Great Britain at the Alden Press
Limited, Oxford and Northampton

DISTRIBUTORS

Marston Book Services Ltd
PO Box 87
Oxford OX2 0DT
(*Orders*: Tel: 0865-791155
Fax: 0865-791927
Telex: 837515)

USA
Blackwell Scientific
Publications, Inc.
238 Main Street
Cambridge, MA 02142
(*Orders*: Tel: 800 759-6102
617 876-7000)

Canada
Times Mirror Professional
Publishing, Ltd
130 Flaska Drive
Markham, Ontario L6G 1B8
(*Orders*: Tel: 800 268-4178
416 470-6739)

Australia
Blackwell Scientific
Publications Pty Ltd
54 University Street
Carlton, Victoria 3053
(*Orders*: Tel: 03 347-5552)

British Library
Cataloguing in Publication Data

Tyldesley, Barbara
Muscles, nerves & movement.
1. Kinesiology
I. Title II. Grieve, June I.,
612'.76

ISBN 0-632-01643-4

Contents

movement, and that it will serve as a sound basis for students who are involved in the rehabilitation of those people who have functional limitations. If the knowledge gained from this book leads a student to ask questions, and to seek the answers in more advanced reading, then our objectives have been achieved.

Acknowledgements

We would like to thank those who have given us expert help in the preparation of this book: Dr Geoffrey L Kidd, MSc, PhD, director of the Neurotech Research Institute, Department of Physiology, University of Liverpool; Laura Mitchell, MCSP, TDip, formerly teacher of Physiotherapy, St. Thomas' Hospital and tutor in movement at the London School of Occupational Therapy; June Sutherland, DipCOT, Department of Occupational Therapy, Westminster Hospital. We also remember Barbara J Barrard, BSc MIBiol, formerly of the Polytechnic of Central London, who devoted so many years of her life to teaching neurology to paramedical students. Our three editors at Blackwell Scientific Publications, each in their own way, have stimulated our endeavour.

The book would not have been written without the constant support and encouragement of our colleagues and our families.

Barbara Tyldesley
June Grieve

Section 1 Introduction to Movement

FUNCTIONAL UNITS, TERMINOLOGY, COMPONENTS OF THE NERVOUS SYSTEM

1 / Components of the Locomotor System — Functional Units

The study of movement of the body as a whole must include some understanding of its basic components. The unit of structure is the *cell*, studied in detail by the histologist, biochemist and pharmacologist. In movement studies, it is our concern to appreciate how cells and tissues are organized to produce a moveable joint, a contractile muscle, an active nerve fibre, and how some connective tissues limit movement. With this in mind, the unit of structure of the cell, and the organization of cells into the functional units of the locomotor system will be considered.

1.1 The cell and basic tissues

Every cell has an outer limiting *membrane* which supports the cell and controls what substances will enter and leave it. The cell membrane is largely composed of fat and protein molecules. The fat is usually arranged in two layers, with protein molecules peppered in between (Fig. 1.1). The fatty layers (in the form of phospholipids, glycolipids and sterol cholesterol) play a structural role, and the protein assists transport of substances across the membrane, i.e. in and out of the cell. Some substances pass through the membrane by diffusion, and small particles, e.g. ions, will pass through with greatest ease. All cell membranes are selectively permeable, and the presence of enzymes in the protein part of the membrane allows larger particles to move in or out. In addition, the presence of enzymes allows some particles to move against the tide of diffusion, i.e. against the concentration gradient, by active transport. For example, some of the protein in the cell membrane acts as an enzyme to continuously transport sodium ions out of the cell.

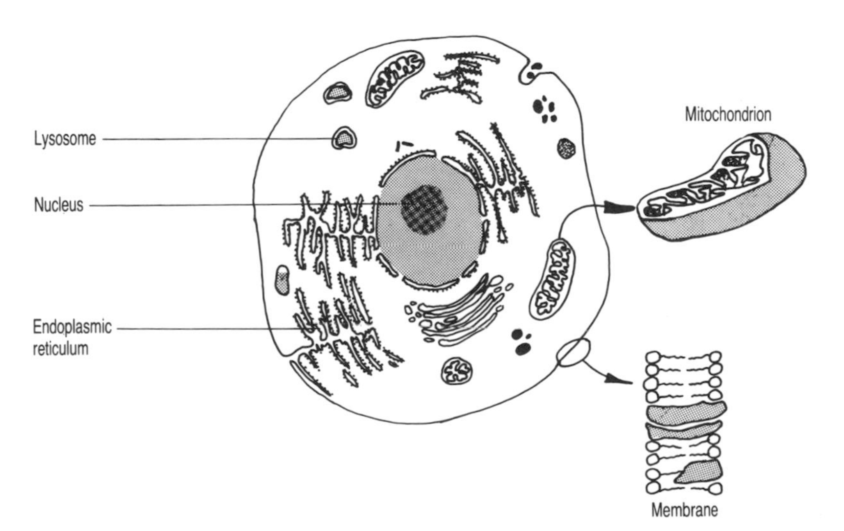

Fig. 1.1. A typical cell, membrane and mitochondrion enlarged.

This active transport, known as the 'sodium pump', removes sodium that would be harmful to the internal function of the cell.

Inside the cell lies the nucleus embedded in cytoplasm. *Organelles,* each with their own membrane, are found within the cytoplasm. Some examples of organelles are as follows.

1 **Endoplasmic reticulum** is an elaborate system of tubes which extends from the cell membrane to the nuclear membrane. It provides a conducting system for the movement of substances inside the cell and divides the cell into compartments, with sites, known as *ribosomes,* for production of protein. In muscle cells, the endoplasmic (sarcoplasmic) reticulum stores calcium ions, which play an essential role in the contractile mechanism.

2 **Mitochondria** are the 'power houses' of the cell where energy rich compounds, e.g. adenosine triphosphate (ATP), are stored and carbohydrate (glycogen) is broken down to release energy. Each mitochondrium has a double membrane, the inner one is thrown into folds, giving a large surface area for chemical reactions to occur (Fig. 1.1). Cells with high energy requirements, such as muscle cells and neurones have a large number of mitochondria. The aerobic reactions in the release of energy occur in the mitochondria. Pyruvic acid enters the mitochondria to be broken down to carbon dioxide and water, under the action of enzymes. Different parts of the process take place in specific areas of the mitochondria.

3 **Lysosomes** are organelles with a single membrane, varying in size and shape, and form the disposal units of the cell. Lysosomes are important in cell growth and repair when cell molecules are broken down before being removed from the cell. They also break down toxins, bacteria and viruses that have entered the cell. The resultant products are either excreted to the exterior of the cell or absorbed into it.

Cells are collected together to form a *tissue,* a collection of similar cells lying in an intercellular substance. Each tissue has a particular function. Within the neuromotor system, three basic tissues play important roles: *connective tissue, muscle* and *nerve.*

1.2 Connective tissues in the musculoskeletal system

The overall function of connective tissue is to unite or connect structures in the body, and to give support. Bone provides the rigid framework for support. Where bones articulate with each other dense, fibrous connective tissue rich in collagen fibres, surrounds the ends of the bones, allowing movement to occur whilst maintaining stability. Cartilage is also found associated with joints, where it forms a compressible link between two bones, or provides a low friction surface for smooth movement. Connective tissue also attaches muscles to bone, either in the form of a cord (tendon) or a flat sheet (fascia). The three connective tissues which play a major role in movement will be described below.

1.2.1 Dense fibrous tissue

Dense fibrous connective tissue has few cells and is largely made up of fibres of collagen and elastin — protein strands that give the tissue great strength. The fibres are produced by fibroblast cells that lie in between the fibres (Fig. 1.2), and the tissue has high tensile

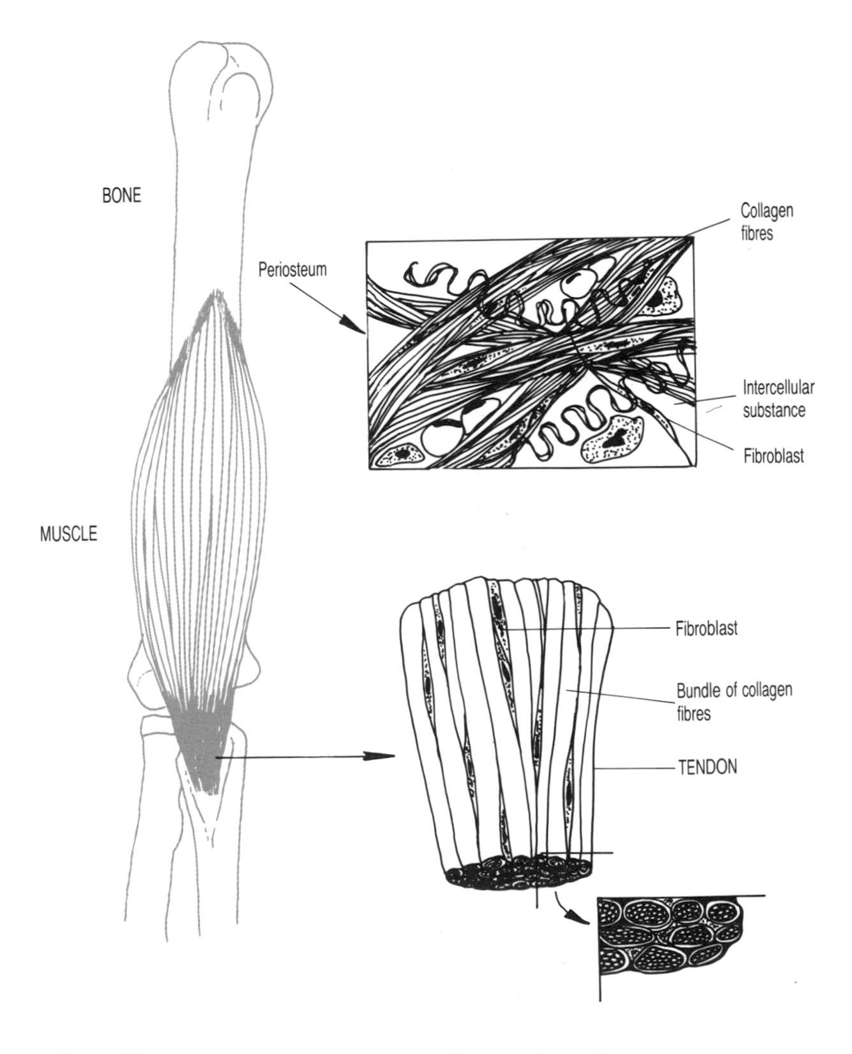

Fig. 1.2. Dense fibrous connective tissue covering bone and muscle and forming tendon.

strength to resist stretching forces. The toughness of this tissue can be felt when cutting through stewing steak with a blunt knife: the muscle fibres are easily sliced, but the covering of white connective tissue is very tough. Examples of this tissue are as follows.

1 The **capsule** surrounds the moveable (synovial) joints, binding the bones together (see Fig. 1.8).

2 **Ligaments** form strong bands to strengthen the joint capsules in particular directions and limit movement. Some ligaments contain a large proportion of elastin, which increases their elasticity, such as those between the arches of the vertebrae (ligamenta flava).

3 **Tendons** unite the contractile fibres of muscle to bone. In both tendons and ligaments the collagenous fibres lie in parallel in the direction of greatest stress.

4 **Aponeurosis** is a strong flat membrane with collagen fibres that lie in different directions to form sheets of connective tissue. Aponeuroses can form the attachment of a muscle, such as the oblique abdominal muscles, which meet in the midline of the abdomen (see Fig. 10.7, p. 220). In the palm of the hand and the sole of the foot an aponeurosis lies deep to the skin and forms a protective layer for the tendons underneath (see Figs 6.3 and 8.22, pp. 132 and 193).

5 **Retinaculum** is a band of dense fibrous tissue which binds tendons of muscles and prevents bowstring during movement. An example is the flexor retinaculum of the wrist, which holds the tendons of muscles passing into the hand in position (see Fig. 6.12, p. 141).

6 **Fascia** is a term used for the large areas of dense fibrous tissue that surround the musculature of all the body segments. Fascia is particularly developed in the limbs, where it dips down between the large groups of muscles and attaches to the bone. In some areas, fascia provides a base for the attachment of muscles, for example the thoracolumbar fascia gives attachment to the long muscles of the back (see Fig. 10.4, p. 217).

7 **Periosteum** is the protective covering of bones. Tendons and ligaments blend with the periosteum around bone (see Fig. 1.4d).

8 **Dura** is thick fibrous connective tissue protecting the brain and spinal cord (see Fig. 3.23, p. 75).

Fibrous connective tissue unites structures in the body while still allowing movement to occur. When the tissue loses its strength and elasticity, movement is affected, and if the change is prolonged, deformity may occur. For example, contraction of the palmar aponeurosis in the hand produces a disabling deformity when the ring and little fingers become curled. (Dupuytren's contracture) (see Fig. 6.13, p. 141).

1.2.2 Cartilage

Cartilage is a tissue that can be compressed and has resilience. The cells (chondrocytes) are oval and lie in a ground substance that is not rigid like bone. There is no blood supply to cartilage so there is a limit to its thickness. The tissue does have a great resistance to wear, but cannot be repaired when damaged.

Hyaline cartilage is commonly called gristle. It is smooth and glass-like forming a low friction covering to the articular surfaces of joints. In the elderly, the articular cartilage tends to become eroded or calcifies, so that joints become stiff. Hyaline cartilage forms the costal cartilages which join the anterior ends of the ribs to the sternum (Fig. 1.3). In the developing foetus, most of the bones are formed in hyaline cartilage. When the cartilaginous model of each bone reaches a critical size for the survival of the cartilage cells, ossification begins.

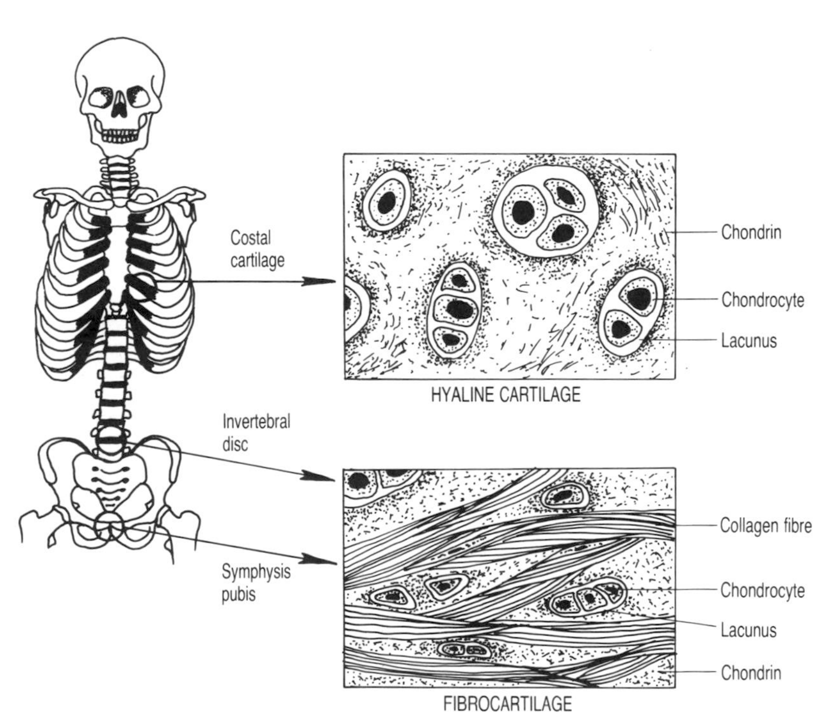

Fig. 1.3. Microscopic structure of hyaline and fibrocartilage, location in the skeleton of the trunk.

• *LOOK at some large butchers' bones to see the cartilage covering the joint surfaces at the end. Note that it is bluish and looks like glass.*

Fibrocartilage consists of cartilage cells lying in between densely packed collagen fibres (Fig. 1.3). The fibres give extra strength to the tissue whilst retaining its resilience. Examples of where fibrocartilage is found are: (i) the discs between the bones of the vertebral column; (ii) the pubic symphysis joining the two halves of the pelvis; and (iii) the menisci in the knee joint.

1.2.3 Bone

Bone is the tissue that forms the rigid supports for the body by con-

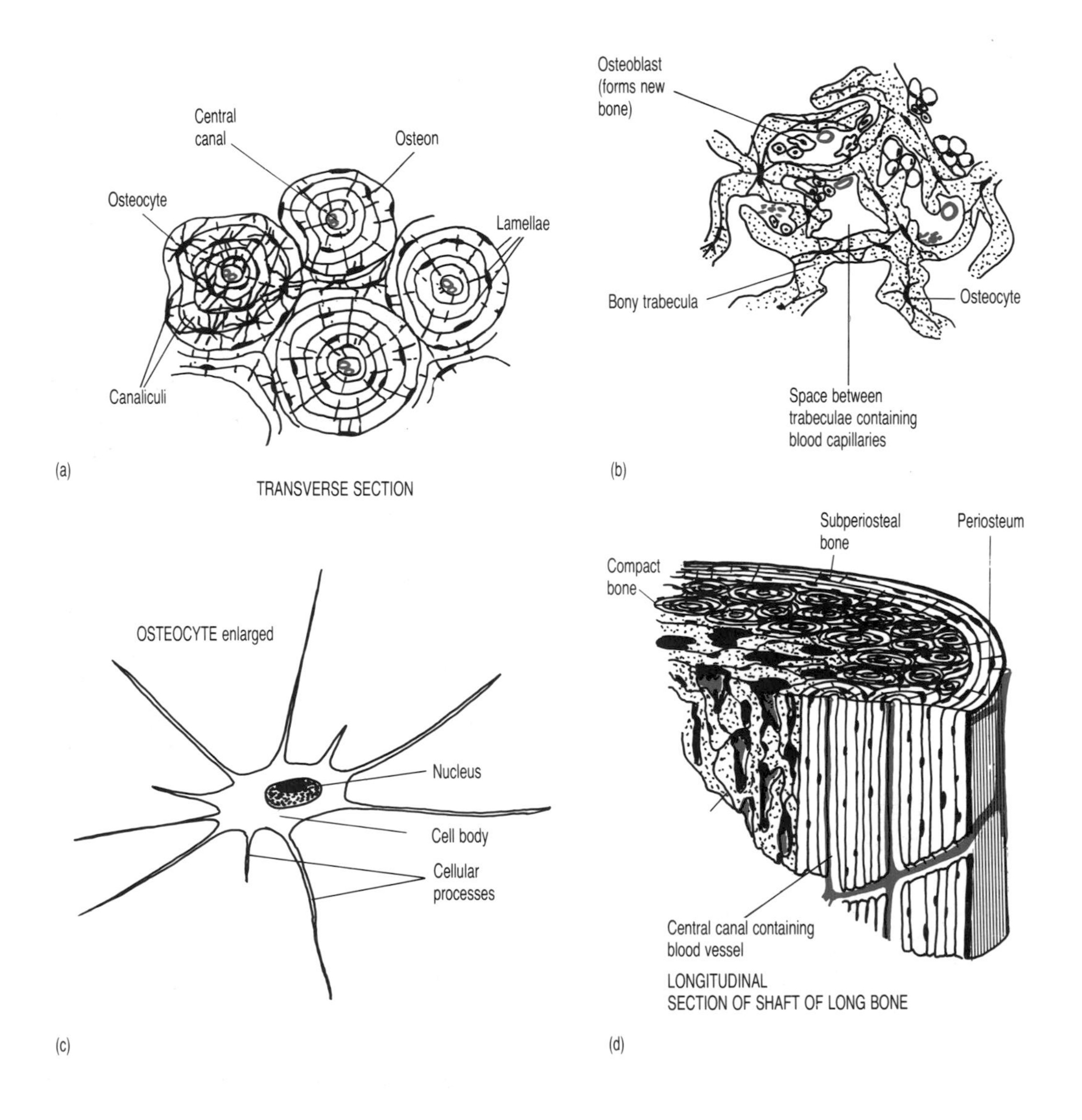

Fig. 1.4. Microscopic structure of bone; (a) arrangement of osteons (Haversian systems) in compact bone; (b) lamellae of bone containing osteocytes in cancellous bone; (c) an individual osteocyte; and (d) a section of the shaft of a long bone.

taining a large proportion of calcium salts (calcium phosphate and carbonate). It must be remembered that bone is a living tissue composed of cells and an abundant blood supply. It has a greater capacity for repair after damage than any other tissue in the body, except blood. The strength of bone lies in the thin plates (lamellae) of collagen fibres with calcium salts deposited in between. The lamellae lie in parallel, held together by fibres, and the bone cells (*osteocytes*) are found in between. Each bone cell lies in a small space or lacuna, and connects with other cells and to blood capillaries by fine channels called canaliculi.

In **compact bone**, the lamellae are laid down in concentric rings around a central canal containing blood vessels. Each system of concentric lamellae (known as a Haversian system or an osteon) lies in a longitudinal direction. Many of these systems are closely packed to form the dense compact bone found in the shaft of long bones (Fig. 1.4a).

In **cancellous** or **trabeculate bone**, the lamellae form plates arranged in different directions. The plates are known as trabeculae and the spaces in between contain blood capillaries. The bone cells lying in the trabeculae communicate with each other and with the spaces by canaliculi (Fig. 1.4b). The expanded ends of long bones are filled with cancellous bone covered with a thin layer of compact bone. The central cavity of the shaft of long bones contains bone marrow. This organization of the two types of bone produces a structure with great rigidity without excessive weight (Fig. 1.5). Bone has the capacity to remodel its shape in response to imposed stresses, so that the structure lines of the trabeculae at the ends of the bone follow the lines of force on the bone. For example, the lines of trabeculae at the ends of weight bearing bones, such as the femur, provide maximum strength to support the body weight against gravity. Remodelling of bone is achieved by the activity of 'bone forming' cells known as osteoblasts, and 'bone destroying cells' known as osteoclacts, both these types of cell are found in bone tissue. The calcium salts of bone are constantly interchanging with calcium ions in the blood, under the influence of hormones (parathormone & thyrocalcitonin). Bone is a living, constantly changing connective tissue, that provides a rigid framework on which muscles can exert forces to produce movement.

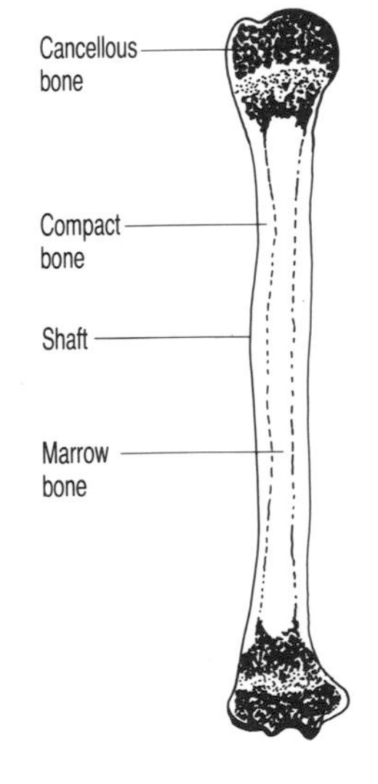

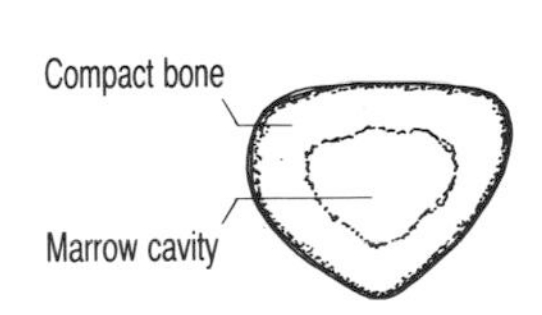

Fig. 1.5. Long bone, gross structure. Transverse and longitudinal section.

• *LOOK at any of the following examples of connective tissue that are available to you*:

1 *Microscope slides of dense fibrous tissue, cartilage and bone, noting the arrangement of the cellular and fibre content.*

2 *Dissected material of joints and muscles which include tendons, ligaments, aponeurosis and retinaculum.*
3 *Fresh butcher's bone — note the pink colour (blood supply), and central cavity in the shaft of long bones.*
4 *Fresh red meat to see fibrous connective tissue around muscle.*

1.3 Articulations

Where the rigid bones of the skeleton meet, connective tissues are organized to bind the bones together and to form *joints*. These allow movement of the segments of the body relative to each other. The joints or articulations between bones can be divided into three types based on the particular connective tissues involved. The three main classes of joint are *fibrous, cartilaginous* and *synovial*.

1.3.1 Fibrous joints

Here the bones are united by dense fibrous connective tissue.

The **sutures** of the skull are fibrous joints which allow no

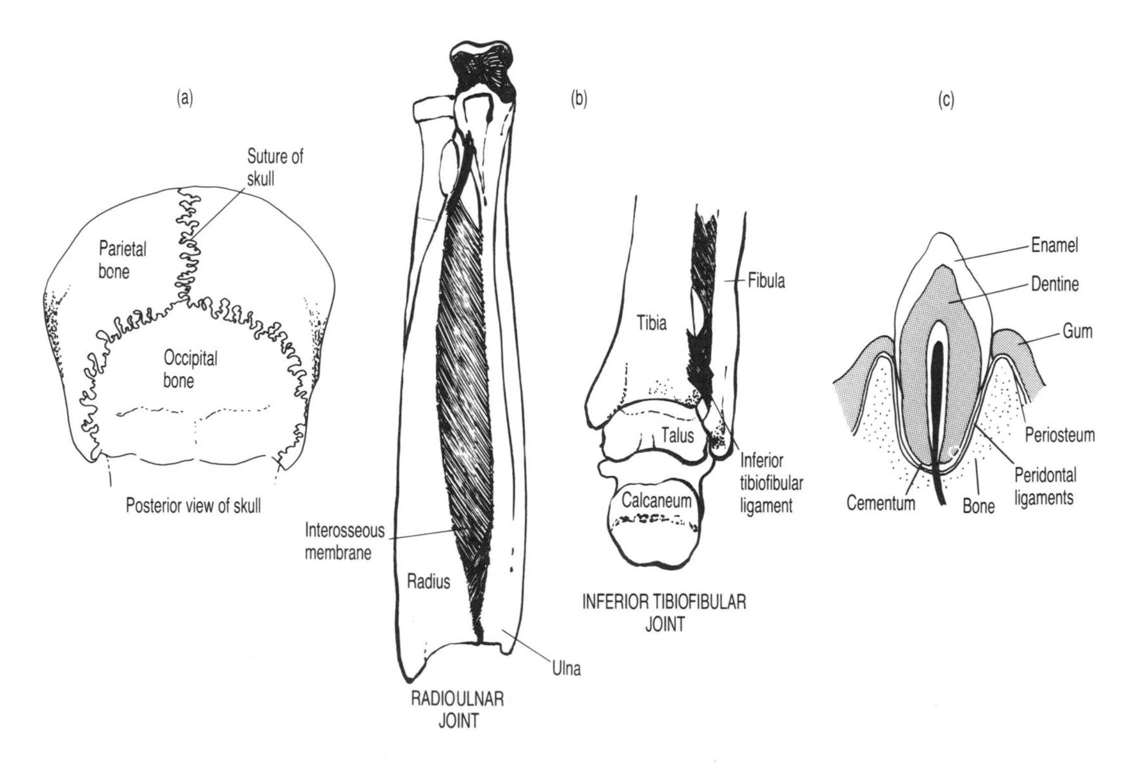

Fig. 1.6. Fibrous joints: (a) suture between bones of the skull; (b) syndesmosis between the radius and ulna, and tibia and fibula; and (c) gomphosis — tooth in socket.

movement between the bones. The edge of each bone is irregular, and interlocks with the adjacent bone, a layer of fibrous tissue linking them (Fig. 1.6a).

A **syndesmosis** is a joint where the bones are joined by a ligament which allows some movement between the bones. A syndesmosis is found between the tibia and fibula bones of the leg just above the ankle (inferior tibiofibular joint) (Fig. 1.6b). The interosseous membrane between the radius and ulna, which allows movement of the forearm, is often classified as a syndesmosis.

A **gomphosis** is a specialized fibrous joint which fixes the teeth in the jaw sockets (Fig. 1.6c).

1.3.2 Cartilaginous joints

In these joints the bones are united by cartilage.

A **synchondrosis** is a joint where the union is composed of hyaline cartilage. The articulation of the first rib with the sternum is by a synchondrosis. During growth of the long bones of the skeleton, there is a synchondrosis between the ends and the shaft of the bone, where cartilage forms the epiphyseal plate. These plates disappear when growth is completed (Fig. 1.7a).

A **symphysis** is a joint where the joint surfaces are covered by a thin layer of hyaline cartilage and united by a disc of fibrocartilage. This type of joint allows a limited amount of movement between the bones by compression of the cartilage. The bodies of the vertebrae articulate with a disc of fibrocartilage. Movement between two vertebrae is small, but when all the intervertebral discs are compressed in a particular direction, considerable movement of the vertebral column occurs (Fig. 1.7b). Little movement, however, occurs at the pubic symphysis, the joint where the right and left halves of the pelvis meet anteriorly. Movement is thought to

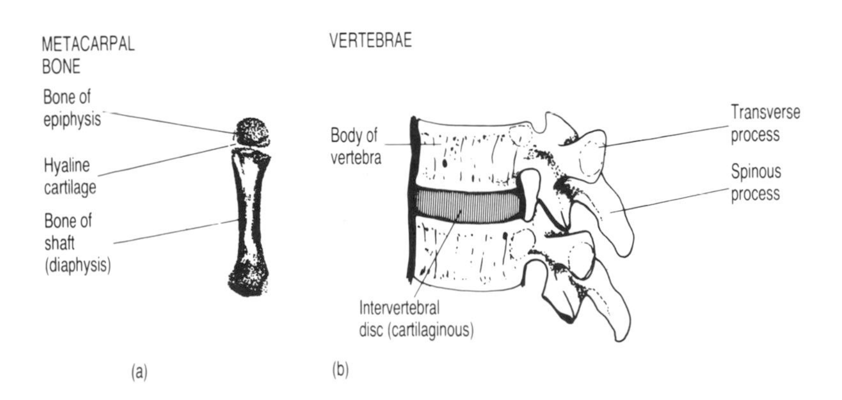

Fig. 1.7. Cartilaginous joints: (a) synchondrosis in child's metacarpal; as seen by X-ray; (b) symphysis between bodies of vertebrae.

increase at the pubic symphysis in the late stage of pregnancy and during childbirth, to increase the size of the birth canal.

1.3.3 Synovial joints

These are the mobile joints of the body, where the two ends of the bones are surrounded by a sleeve-like *capsule* of dense fibrous tissue. In the embryo, the capsule is attached to the site of the epiphyseal plates, enveloping the ends of the articulating bones. The capsule is strengthened by *ligaments*, some of which blend with the capsule, whilst others are found attached to the bones near to the joint. There is a *joint cavity* inside the capsule which allows free movement between the bones. A *synovial membrane* lines the joint capsule and all the non-articular surfaces inside the joint, i.e. any structure within a joint not covered by hyaline cartilage. A viscous lubricating fluid is secreted by the synovial membrane. Pads of fat (liquid at body temperature) are present in some joints. Figure 1.8 shows a section through a typical synovial joint.

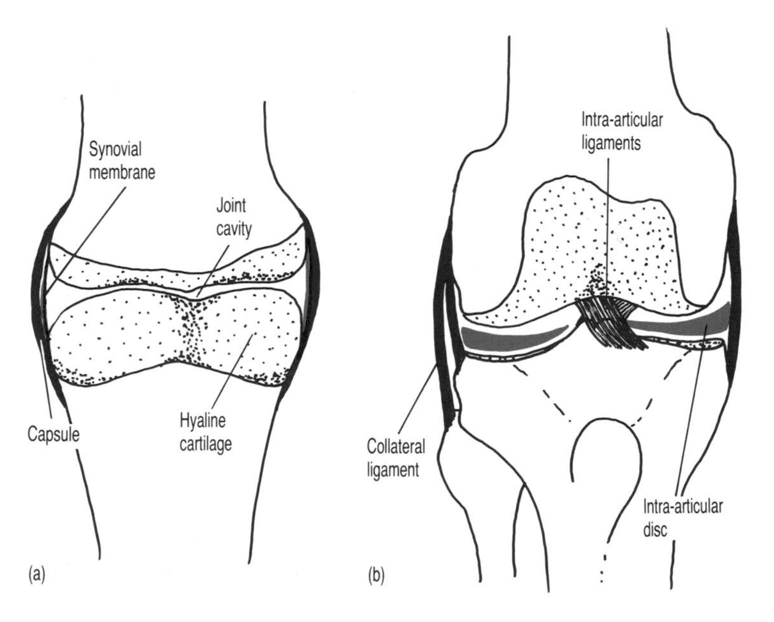

Fig. 1.8. (a) Typical synovial joint; and (b) the knee joint with anterior capsule and muscle removed.

All the large moveable joints of the body (e.g. the shoulder, elbow, wrist, hip, knee and ankle), are synovial joints. The direction and range of their movements depend on the shape of the articular surfaces and on the presence of ligaments and muscles close to the joint. The different types of synovial joint are described in Chapter 2 when the directions of joint movement are discussed. Examples of synovial joints are illustrated in Appendix 2.

1.4 Skeletal muscle

The special feature of muscle tissue is its ability to actively contract. The connective tissues already described have elasticity, they can be stretched and return to the original length. Muscles cells can also be passively stretched, but they have the additional capacity to actively shorten.

The function of skeletal muscle is to: (i) shorten to produce movement of the body at joints; and (ii) to resist active stretching by external forces acting on it. If a weight and/or gravity is acting on a body part, e.g. holding a full glass in the hand, active muscle tissue prevents the hand from falling by resisting stretching, and lifts the weight by shortening.

- *HOLD a glass of water in the hand and feel the activity in the muscles above the elbow as they resist stretching.*
- *LIFT the glass to the mouth and feel the muscle activity in the same muscles as they shorten.*

Skeletal muscle is only active when nerve impulses reach the muscle through the nerve supplying it. If the nerve supply is damaged, the muscle is unable to function. Details of the way in which nerve impulses arrive at the muscle cells will be discussed in Section 1.5.

1.4.1 Muscle fibres — gross and microscopic structure

The muscle fibres, the unit of structure of a skeletal muscle, are elongated cells with many nuclei surrounded by a strong outer membrane, the *sarcolemma*.

A muscle fibre can just be seen with the naked eye. If one fibre is viewed under a light microscope, the nuclei can be seen close to the membrane around the fibre. The chief constituent of the fibre is several hundreds of *myofibrils*, strands of protein, extending from one end of the fibre to the other (Fig. 1.9). The arrangement of the two main proteins, actin and myosin, that form each myofibril present a banded appearance. The light and dark bands in adjacent myofibrils coincide, so that the whole muscle fibre is striated.

The electron microscope reveals the detail of the cross striations in each myofibril. A repeating unit, known as the *sarcomere*, is revealed along the length of the myofibril. Each sarcomere links to the next at a disc called the Z line. The thin filaments of actin are attached to the Z line and project towards the centre of the sarcomere. The thicker myosin filaments lie in between the actin

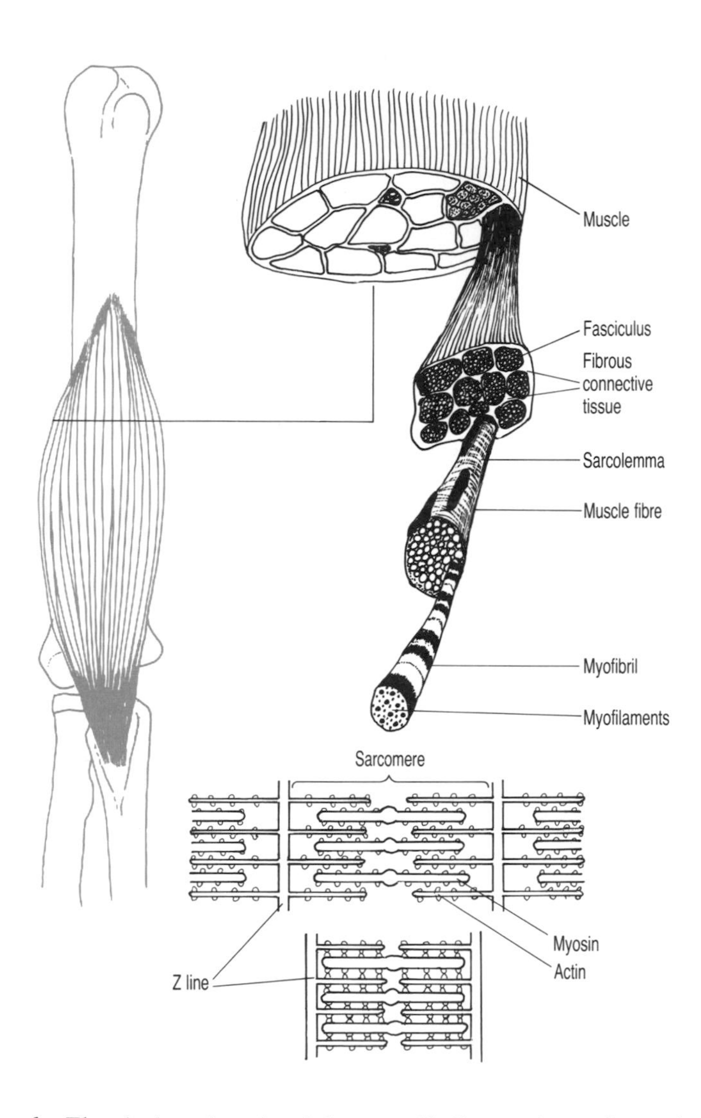

Fig. 1.9. Skeletal muscle; muscle fibre, myofibril, sarcomere.

strands. The darkest bands of the myofibril are where the actin and myosin overlap in the sarcomere.

• *LOOK closely at Fig. 1.9 to see clearly how: (i) sarcomeres lie end to end to form a myofibril; (ii) myofibrils are packed tightly together inside a muscle fibre; and (iii) the sarcolemma encloses the myofibrils in a muscle fibre.*

The arrangement of the myosin molecules in the thick myosin filaments forms cross bridges which link with special sites on the actin filaments when the muscle fibre is activated. The result of this linking is to allow the filaments to slide past one another, so that each sarcomere becomes shorter. This in turn means that the myofibril is shorter, and since all the myofibrils respond together, the muscle fibre shortens. The initiation of the active state in the myofibrils depends on the release of calcium from the endoplasmic reticulum of the muscle fibre. During relaxation of the fibre, the calcium returns to the tubules of the endoplasmic reticulum.

All cells contain high energy compounds (mainly adenosine triphosphate) in the mitochondria which provide the *energy* for cell activity. Muscle cells have a higher level and rate of energy output than other cells. This is supplied by ATP and a 'back-up' of another high energy compound, phosphocreatine. The store of ATP is replenished by chemical reactions in the mitochondria using oxygen and glucose brought by the blood in the network of capillaries surrounding muscle fibres (Fig. 1.10). In this way, the muscle fibres have a continuous supply of energy, as long as the supply of oxygen is maintained (aerobic metabolism).

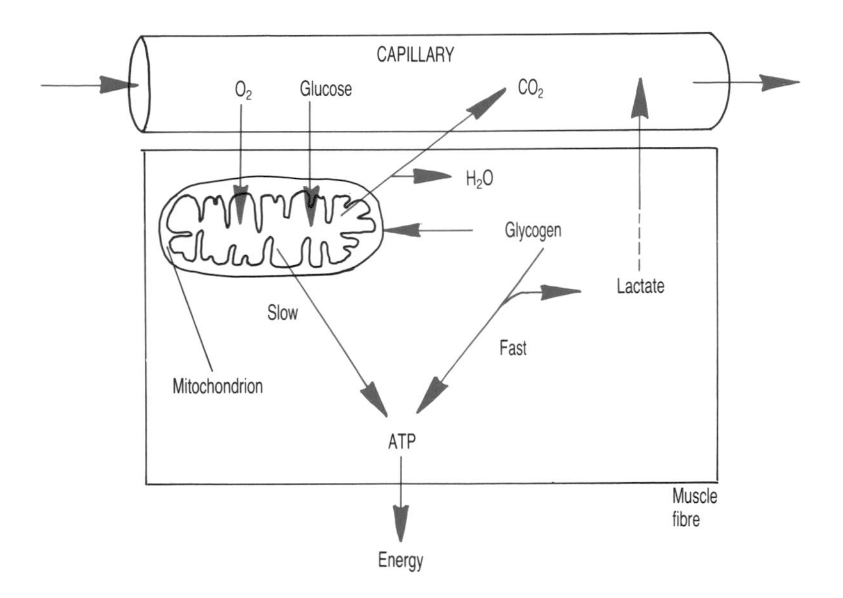

Fig. 1.10. Energy for muscle contraction (simplified). The slow process predominates in Type I fibres with myoglobin which facilitates the uptake of oxygen: The fast process predominates in Type II fibres where glycogen is the main source of energy (anaerobic metabolism).

Glycogen is another source of energy that is stored in muscle fibres. When there is insufficient oxygen to replenish ATP by oxidative reactions, energy released from breakdown of glycogen is also used to maintain ATP levels during a short burst of high level muscle activity.

Types of muscle fibre

Different types of fibres in a muscle have been identified by the relative amounts of oxidative and glycolytic reactions used to produce energy.

• *LOOK at the muscle seen in chicken meat to see the white muscles of the breast and the more vascular red muscles of the legs.*

In human muscle, the distinction is not so marked, and all muscles contain fibres of each type, but the proportion depends on the function of the muscle. Postural muscles contain more slow oxidative fibres known as Type I or SO, whilst muscles involved in rhythmic or phasic activity contain more fast glycolytic fibres known as Type II or FG.

Most muscles are involved in both postural and phasic activity at the same or at different times. There is evidence that Type II fibres can change, so that they become more like Type I, and the muscle can then perform over longer periods of time. The implications of this are important for the athlete who wants to increase his or her endurance. The energy capacity of a given muscle fibre is not fixed, but is determined by the type of activity performed by the muscle.

1.4.2 Shape and form of skeletal muscle

The structure of a whole muscle is the combination of muscle and connective tissues, both of which contribute to the function of the muscle when it is active. In a whole muscle, groups of contractile muscle fibres, of varying diameter, are bound together by fibrous connective tissue to form *fasciculi*. Further coverings of connective tissue bind the fascilculi together and an outer layer surrounds the whole muscle (Fig. 1.9). The total connective tissue element lying in between the contractile muscle fibres is known as the *parallel elastic component*. The tension that is built up in muscle when it is activated depends on the tension in the muscle fibres and in the parallel elastic component. The fibrous connective tissue which links the whole muscle to bone, e.g. the tendon, is known as the *series elastic component*. The initial tension that builds up in an active muscle tightens the series elastic component and then the muscle can shorten. A model of the elastic and contractile parts of a muscle is shown in Fig. 1.11. If the connective tissue components lose their elasticity, through lack of use in injury or disease, a muscle may go into contracture. Lively splints are used to maintain elasticity and prevent contracture while the muscle recovers.

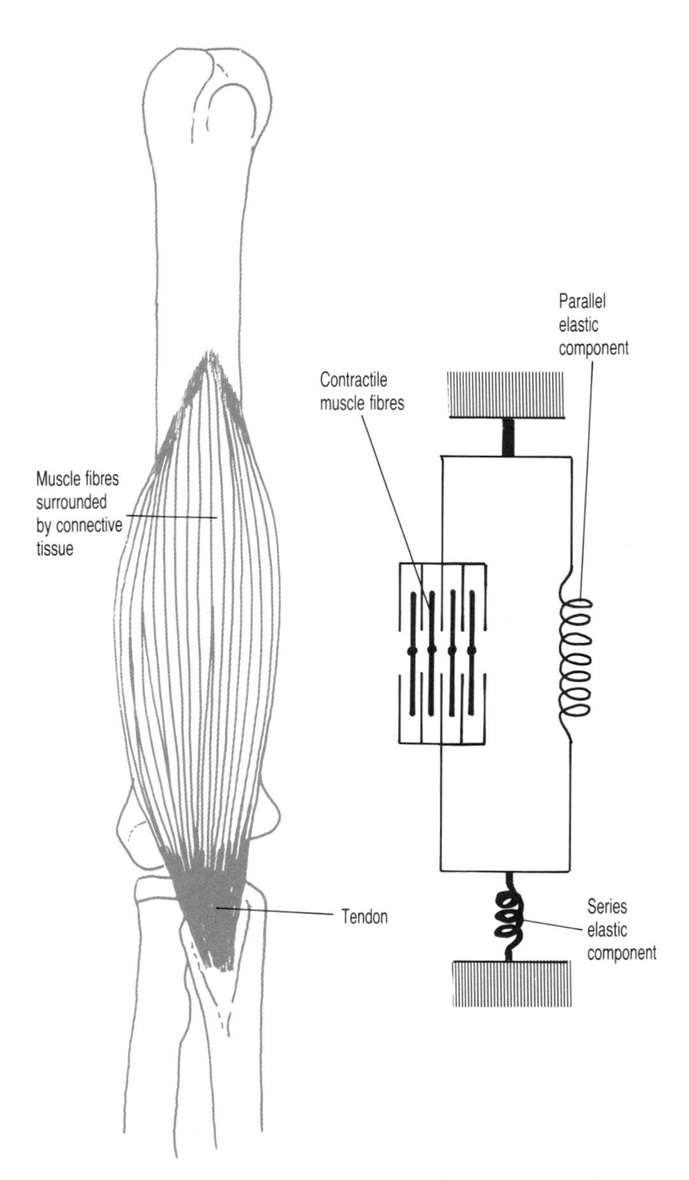

Fig. 1.11. Elastic components of muscle.

The individual fasciculi in a muscle lie in one of two ways, parallel or oblique to the line of pull of the muscle.

Parallel fibres are seen in *strap* and *fusiform* muscles illustrated in Fig. 1.12. These muscles have long fibres and are capable of a wide range of movement.

Oblique fibres are seen in *triangular* and *pennate* muscles. The tendon of the pennate muscles extends as a strip along the muscle so that a large number of short fibres can be accommodated in a given volume. This arrangement means that the muscle can develop great strength at the expense of range of movement. Figure 1.12 illustrates the variety of pennate arrangement found in different

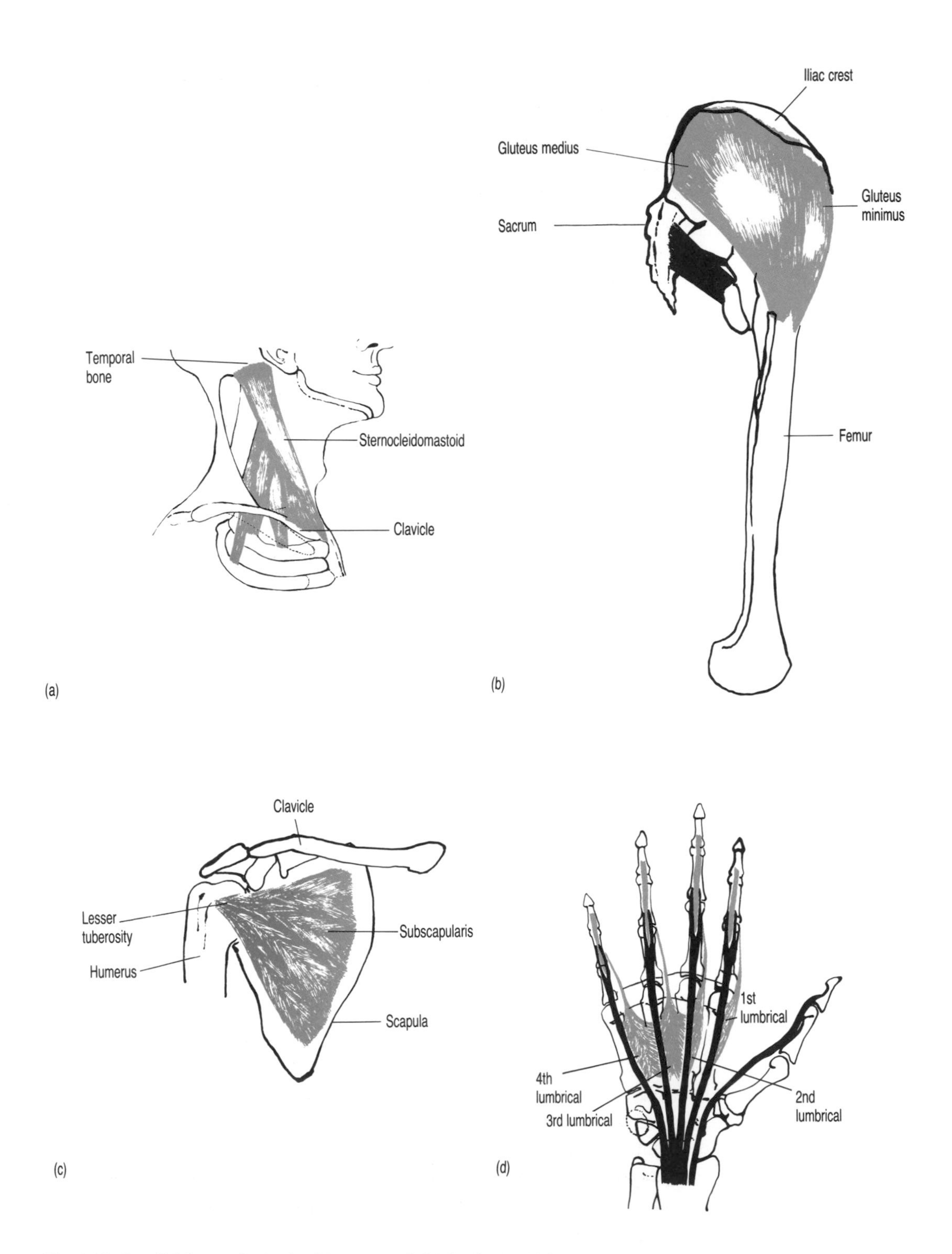

Fig. 1.12. Parallel form of muscle: (a) strap and (b) fusiform. Oblique form of muscle: (c) multipennate and (d) unipennate and bipennate.

muscles. Some of the large muscles of the body combine parallel and oblique arrangements. The deltoid muscle of the shoulder has one group of fibres that are multipennate, and two groups that are fusiform, to combine strength to lift the weight of the arm with a wide range of movement. The form of a particular muscle reflects the space available and the demands of range and strength of movement.

Strength and flexibility

The strength of a muscle is the maximum force it can develop in a particular direction. Increase in strength can be achieved by exercising the muscle against gradually increasing loads. The muscle responds by an increase in the size of individual muscle fibres, but there is no change in the total number of fibres in the muscle. Fitness programmes, and the re-education of weak muscles after injury, include weight training.

Increase in strength alone may result in shortening of the muscle and loss of range of movement at the joint. It is important for the muscle to remain flexible to allow the joint to move with the wide range and speed of movement required by a particular person. Flexibility depends on the elasticity of not only the muscle fibres, but also all the connective tissue in the muscle and the joint on which it acts. Stretching exercises are designed to elongate or extend the length of a muscle. Flexibility training reduces the incidence of torn muscles.

Some muscle stretching can be achieved by the positioning of a body part. For example, if one arm is straightened and the weight of the trunk is supported on it while the other arm is used to do an activity, the muscles of the supporting arm will be stretched.

Active stretching of muscles involves slowly moving the joint through its maximum range, holding it, and then letting go. Here one group of muscles is active to stretch the opposing group. For example, moving the leg backwards in this way actively stretches the muscles and ligaments on the front of the hip. The muscle fibres of the stretching muscles must be relaxed, then the connective tissues of the muscle and the ligaments of the joint yield to the prolonged stretch.

The muscles of each individual have unique properties of strength and flexibility related to his or her daily activities. Efforts to improve muscle function must include an increase in both strength and mobility.

Muscle is a highly specialized tissue which is adapted to the demands of the movement it performs. Muscles have limited capacity for repair, although small areas of damage to muscle fibres may regenerate. In more extensive damage, the connective tissue responds by producing more collagen fibres and a 'scar' is formed. An intact nerve and adequate blood supply is essential for muscle function; if these are interrupted the muscle may never recover. Movement can then only be restored by other muscles taking over the functions of the damaged muscles.

1.5 Nervous tissue

1.5.1 Neurone

The neurone is the structural unit of the nervous system, which is composed of the brain, spinal cord and the nerves supplying all the structures in the body. Neurones are *excitable*, they generate signals in response to stimulation, and they also *conduct* signals from one point to another in the system. Networks of neurones code or *integrate* the information before passing it on to other parts of the

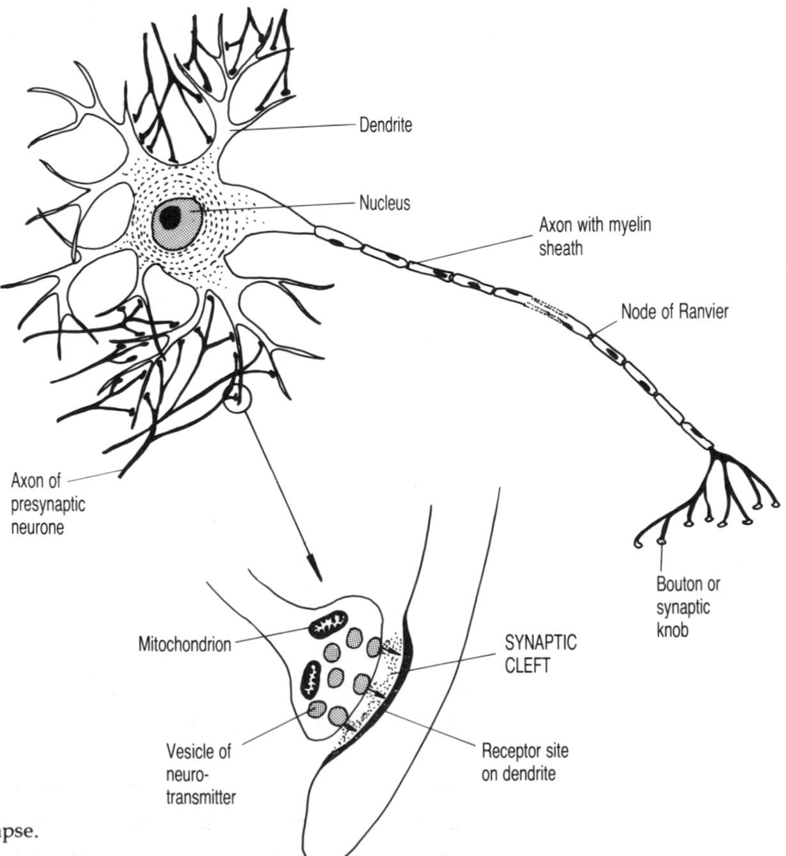

Fig. 1.13. Neurone and synapse. Synaptic cleft enlarged.

body to produce a response. Each neurone has a cell body and numerous processes extending outwards from the cell. The processes are living structures and their membrane is continuous with that of the cell body (Fig. 1.13). (Think of the cell body like a conker with spines projecting out in all directions.) The projections vary in length, some are short, known as *dendrites,* and each neurone has one long process, known as the *axon.*

The dendrites are adapted to receive information: signals or *impulses* received by the dendrites, are passed on to the cell body. Some neurones, particularly in the brain, have a large number of complex branching dendrites, so that information from many other neurones can be received and processed.

The axon is the output end of every neurone. The length of an axon varies from a few millimetres to up to 1 m long. Cell bodies of motor neurones in the spinal cord in the lower back have long axons that extend down the leg to supply the foot muscles. The axon may be surrounded by a sheath of *myelin,* a fatty material that increases the rate at which impulses are conducted. The myelin is laid down between layers of membrane of Schwann cells that wrap round the axon. Gaps in the myelin occur between successive Schwann cells forming nodes of Ranvier (Fig. 1.13). The axon branches at its end, and each branch is swollen to form a bouton or synaptic knob which contains a chemical required for the conduction of impulses to the next neurone. The chemical is called a *neurotransmitter.* The boutons lie near a dendrite or cell body of another neurone. Axons also terminate on muscle fibres at neuromuscular junctions, near blood vessels and in glands.

Impulses are localized changes in the membrane of a neurone. When a neurone is excited, the membrane over a small area allows charged particles (ions) to pass across the membrane (a process known as depolarization) and an impulse is generated. The area of depolarization then moves to the adjacent area and the impulse travels down the membrane in one direction only. Each impulse is the same size, like a morse code of dots only, but the information carried can be varied by the rate and pattern of the impulses conducted along the neurone. If the impulse is to be conducted on to another neurone, sufficient transmitter substance must be released from the boutons at the end of the axon. Each neurone has a *threshold* level of stimulation and the level of *excitation* reaching a neurone must be sufficient to depolarize the membrane, so that impulses are generated. Some impulses reaching a neurone affect the membrane in such a way that no impulses are propagated, this is known as *inhibition.* The activity in some small neurones in the central nervous system always produces inhibition by the release of an inhibitory transmitter substance from the boutons of the axon. The mechanism of inhibition will be discussed in more detail in Chapter 13.

1.5.2 Synapse

A synapse is the junction between neurones, where impulses pass from one neurone to another. Each bouton or synaptic knob at the end of the axon lies near to a special receptor site on the cell body or dendrite of the other neurone, but there is a gap between called the synaptic cleft (Fig. 1.13). A neurone may have as many as 5000 synaptic knobs lying over the cell body and dendrites, so that the variety of input to the neurone is enormous. There is a delay in the conduction of an impulse at the bouton while the neurotransmitter is released and diffuses into the synaptic cleft. The effect of the transmitter substance is to depolarize the membrane and generate impulses in the next neurone. The transmitter substance is broken down by enzymes, but can be taken up again by the bouton to be reformed into transmitter substance and stored.

A neurotransmitter released at most of the synapses in the central nervous system, and at the neuromuscular junction, is acetylcholine. Drugs which prevent the release of acetylcholine at synapses are used as relaxants for muscles, e.g. in abdominal surgery.

Impulses only travel in one direction at a synapse, i.e. from the axon of one neurone to the dendrites and cell body of the next. This ensures one way traffic in the nervous system.

Neurones that carry impulses *down* from the brain to the spinal cord, and *away* from the spinal cord to all parts of the body are known as *motor neurones.*

Neurones that carry impulses *towards* the central nervous system and *up* from the spinal cord to the brain are *sensory neurones.*

1.5.3 Motor neurones. The motor unit

The spinal cord has a central H shaped core of cell bodies called the grey matter. The motor neurones lying in the anterior (ventral) limb of the H are known as 'lower motor neurones' or anterior horn cells. The motor neurones that activate a particular group of muscles lie together and form a *motor neurone pool* (Fig. 1.14). Activity in a particular muscle is generated by impulses from its anterior horn cells, along axons lying in a particular spinal nerve which branches to form the nerve supplying the muscle. Since there are fewer

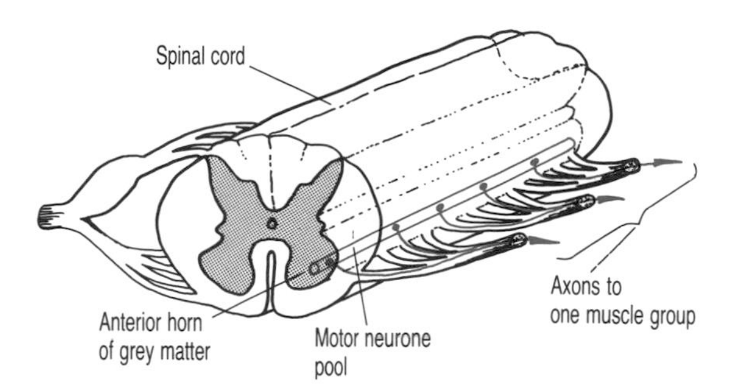

Fig. 1.14. Motor neurone pool in the spinal cord.

motor neurones in the pool than muscle fibres in the muscle, each neurone must supply a number of muscle fibres.

A **motor unit** consists of one motor neurone in the anterior horn of the spinal cord, its axon, and all the muscle fibres innervated by the branches of the axon (Fig. 1.15). The number of muscle fibres in one motor unit depends on the function of the muscle rather than its size. Muscles performing large, strong movements have motor units with a large number of muscle fibres. For example, the large muscle of the calf has approximately 1900 muscle fibres in each motor unit. In muscles that perform fine precision movements, the motor units have a small number of muscle fibres (e.g. up to 100 in muscles of the hand). The muscle fibres of one motor unit do not necessarily lie together in the muscle, but may be scattered in different fasciculi. The number of motor units that are active in a muscle at any one time determines the amount of work performed by the muscle.

In the sustained muscle activity that holds the posture of the body, low threshold motor units supplying slow Type I muscle

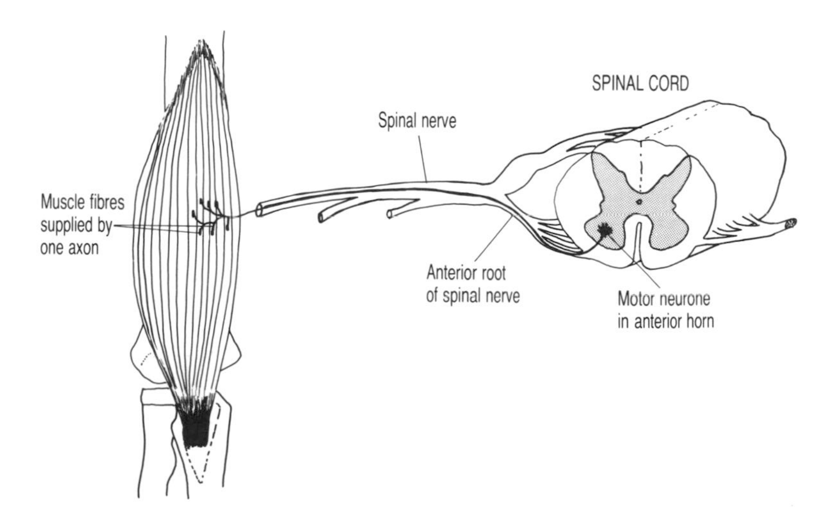

Fig. 1.15. The motor unit.

fibres are involved. The number of active motor units remains constant, but activity alternates between all the low threshold neurones. The slow Type I muscle fibres do not fatigue easily and the activity is maintained over long periods. In fast active movements which move the parts of the body from one position to another, high threshold motor units with large diameter axons supplying fast Type II muscle fibres are involved. These motor units soon fatigue, but they are adapted for fast strong movements such as running and jumping.

In a strong purposeful movement, such as pushing forwards on a door, the motor units are activated or recruited in a particular order. The slow units are active at the start of the movement, and then the fast units become active as the movement reaches its peak.

All muscle activity includes a combination of slow and fast motor units. The slow units contribute more to the background postural activity, whilst the fast units play a greater part in rapid phasic movements. In manipulative activities the shoulder muscles have sustained postural activity to hold the limb steady, so that the hand can do rapid precision movements such as writing, sewing or using a tool.

1.5.4 Sensory neurones and receptors

The basic units for conduction of nerve impulses *into* the central nervous system are the *sensory neurones*, which lie in all the nerves distributed all over the body. Sensory neurones bring information

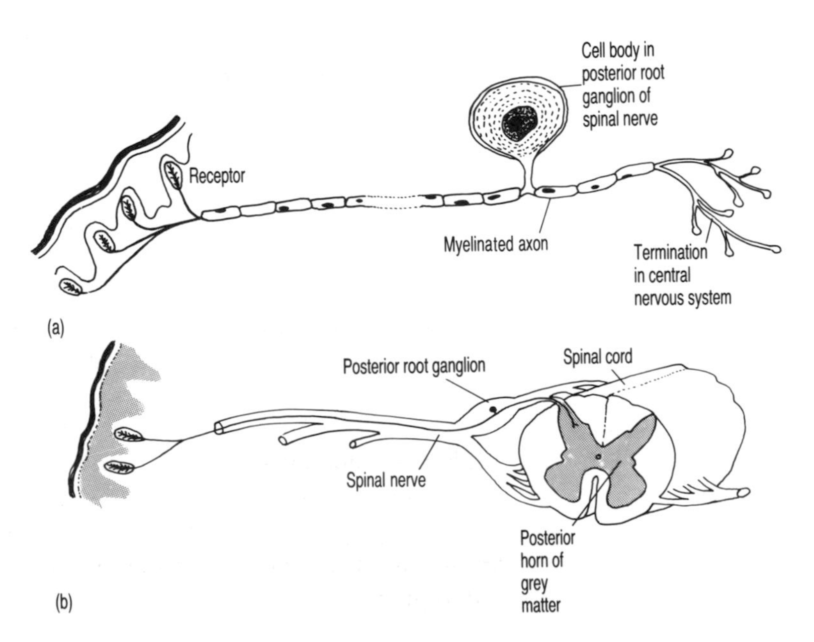

Fig. 1.16. Sensory neurones: (a) Typical sensory neurone; (b) sensory neurone in position in spinal nerve and spinal cord, showing route of entry.

from all parts of the body, including the muscles, to the central nervous system. Axons of sensory neurones are found in all the spinal nerves that emerge from the spinal cord, and many of the cranial nerves arising from the brain. The cell bodies of these neurones are found in ganglia just outside the spinal cord. There are no synaptic junctions on the cell bodies, and the axon divides into two almost immediately after it leaves the cell. The two long processes formed by this division are: (i) axon or nerve fibre in the spinal nerve and its branches which ends in a specialized sensory receptor; and (ii) nerve fibre that enters the spinal cord and terminates in the central nervous system.

Figure 1.16a shows the arrangement of a typical sensory neurone. It is sometimes called 'pseudo unipolar', since it has one axon but appears to be bipolar. Compare this with the multipolar motor neurone shown in Fig. 1.13 (p. 20).

Figure 1.16b shows the position of a sensory neurone in relation to the spinal cord, a spinal nerve and its branches. Note the cell body lying in a ganglion (swelling) and the axon entering the spinal cord.

The function of these sensory neurones is to carry information about the external environment, and the internal state of the body, into the central nervous system for processing.

Receptors

Sensory receptors are specialized structures which respond to a *stimulus*, and generate nerve impulses in sensory neurones. Many receptors are free branched endings, whilst others are encapsulated in a variety of ways. Figure 11.2 (see p. 242) shows some of the types of receptor found in the skin.

Many sensory neurones branch and end in a group of receptors lying in a small area, e.g. the skin. The area covered by all the receptors activating one sensory fibre is call a *receptive field*. There may be overlap in receptive fields, so that stimulation of one point may excite more than one sensory neurone (Fig. 1.17). In the fingertips, for example, where the receptive fields are small and there is great overlap, a stimulus such as a pin prick can be very precisely interpreted by the central nervous system.

When a receptor is stimulated, the membrane of the receptor ending is depolarized and impulses are generated. If the same stimulus continues for some time, the rate of firing of impulses falls and may stop, even though the stimulus is still present. This is known as *adaptation* of receptors. Different receptors adapt at different rates.

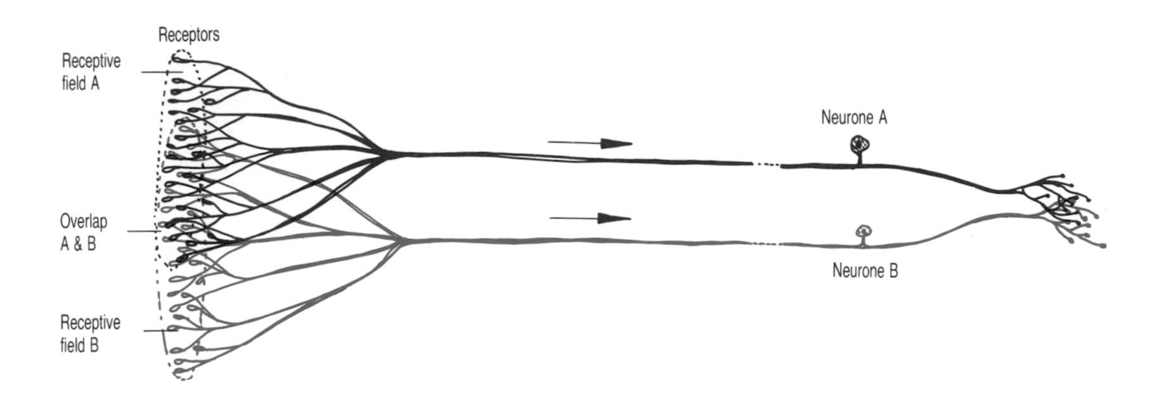

Fig. 1.17. Receptors: receptive fields of two neurones.

Slow adapting receptors continue to produce impulses at the same rate all the time the stimulus is applied. The function of these receptors is to give continuous monitoring of background sensory information. We are not aware of most of this activity, as it never reaches consciousness. An example of slowly adapting receptors are those lying in between muscle fibres (muscle spindles) which give information about the length of muscles in the body. These will be discussed in more detail in Section 1.6.

Fast adapting receptors generate a short burst of impulses in response to the stimulus, but activity ceases if the stimulus continues at the same level. Sensation from these receptors usually reaches consciousness. Touch receptors in the skin are fast adapting. When we put clothes on, we feel the clothes at first, and then we are no longer aware of them. If the strength of stimulus changes, e.g. a belt becomes tighter, another burst of impulses is generated, and we sense the change.

Adaptation of receptors allows the nervous system to process the changing features of the environment inside and outside the body, whilst information of unchanging features is reduced. More detail of the structure and function of receptors will be given in Chapter 11.

1.6 The myotatic unit

There are sensory neurones in the nerves supplying skeletal muscle, so that muscles are a source of sensation in the body, as well as a mechanism for motor action. The receptors lie in parallel with the muscle fibres and are known as *muscle spindles*. Each of the spindles consists of a capsule of connective tissue enclosing

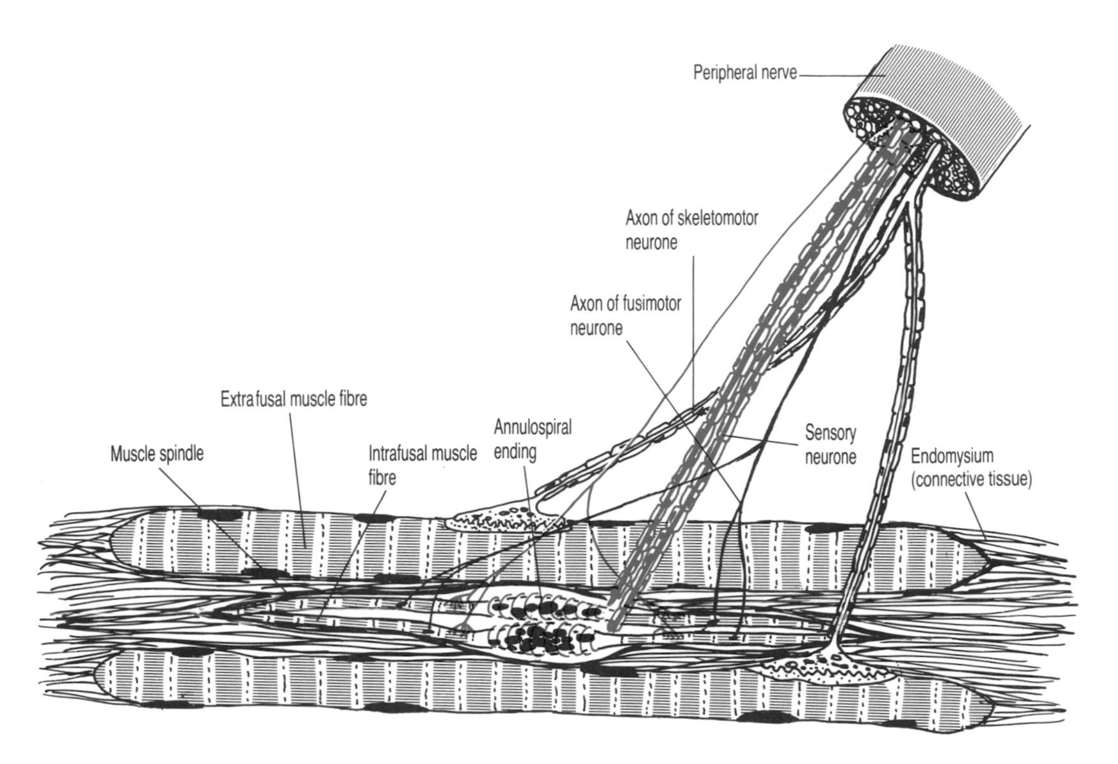

Fig. 1.18. Muscle spindle lying in between two extrafusal muscle fibres.

5–14 specialized small muscle fibres known as *intrafusal fibres*. The central part of these intrafusal fibres of the spindle contains the nuclei and is non contractile. Wound round this central area is the primary sensory ending, called the *annulospiral ending* (Fig. 1.18). Impulses from the annulospiral ending pass along the sensory neurone into the spinal cord where they excite the motor units of the same muscle. The fibres of the muscle are known as *extrafusal fibres* to distinguish them from those of the spindle. The neurones supplying the extrafusal fibres are large diameter alpha motor neurones or *skeletomotor neurones*. This complete pathway is sometimes called the 'fusimotor loop' and the reflex activity is known as the 'stretch reflex'. The use of the term 'stretch' is misleading since the spindles do not only respond to a muscle increasing in length. The activity of the spindle depends on the balance of forces developed inside the muscle, compared with the external forces acting on it: a muscle may also be responding to prolonged stretch or to a rapid change in length. In each type of stretch, the response of the spindle is different.

Looking in more detail at the muscle spindle, the intrafusal fibres are themselves contractile and are supplied by small diameter gamma motor neurones originating in the spinal cord. Stimulation of these *fusimotor neurones* makes the intrafusal fibres contract. The spindle then shortens, becomes more sensitive to distortion, and reflex activity is increased. Fusimotor neurones are under the control of descending pathways from the brain; these will be discussed in Chapter 12.

There are two types of skeletomotor neurones which supply the two types of skeletal muscle fibres described in Section 1.4.1. Large diameter skeletomotor neurones supply the fast Type II muscle fibres active in rhythmic or phasic movements. Smaller diameter skeletomotor neurones supply the slow Type I muscle fibres involved in tonic postural activity. The fusimotor neurones are even smaller in diameter and supply the intrafusal fibres of the muscle spindle. All these motor neurones lie close together in the anterior horn of the spinal cord.

Figure 1.19 shows the myotatic unit, which includes the muscle spindle, the sensory neurone, the skeletomotor and fusimotor neurones, and the extrafusal fibres of the muscle.

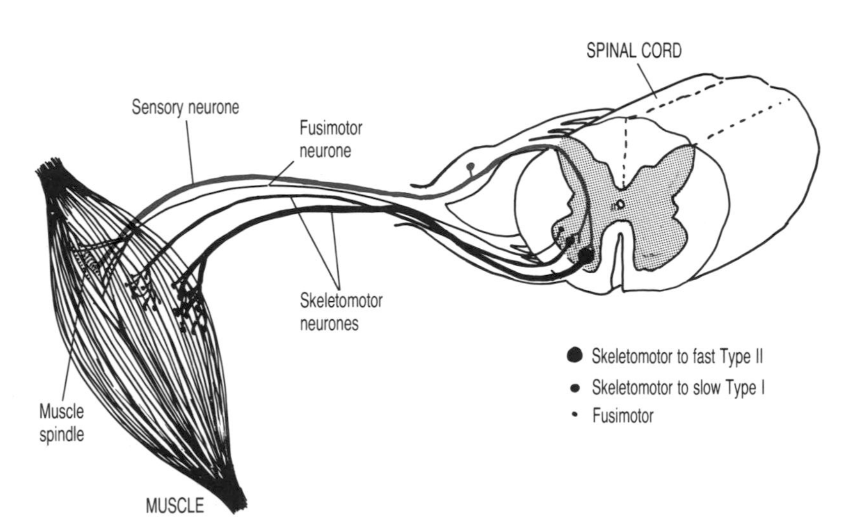

Fig. 1.19. The myotatic unit showing skeletomotor and fusimotor neurones.

Spinal reflex activity in the myotatic units in all the muscles of the body is the basis of *muscle tone*. The level of activity is greatest in muscles holding a body segment against gravity: for example in sitting upright the head is held up by activity in the muscles at the back of the neck to prevent the head from falling forwards (Fig. 1.20a). In standing, there is a tendency for the body to sway forwards and this is counteracted by activity in the muscles of the calf (Fig. 1.20b). Reflex activity in muscles never completely stops: even when the body is asleep some muscle tone is present, except in periods of deep sleep.

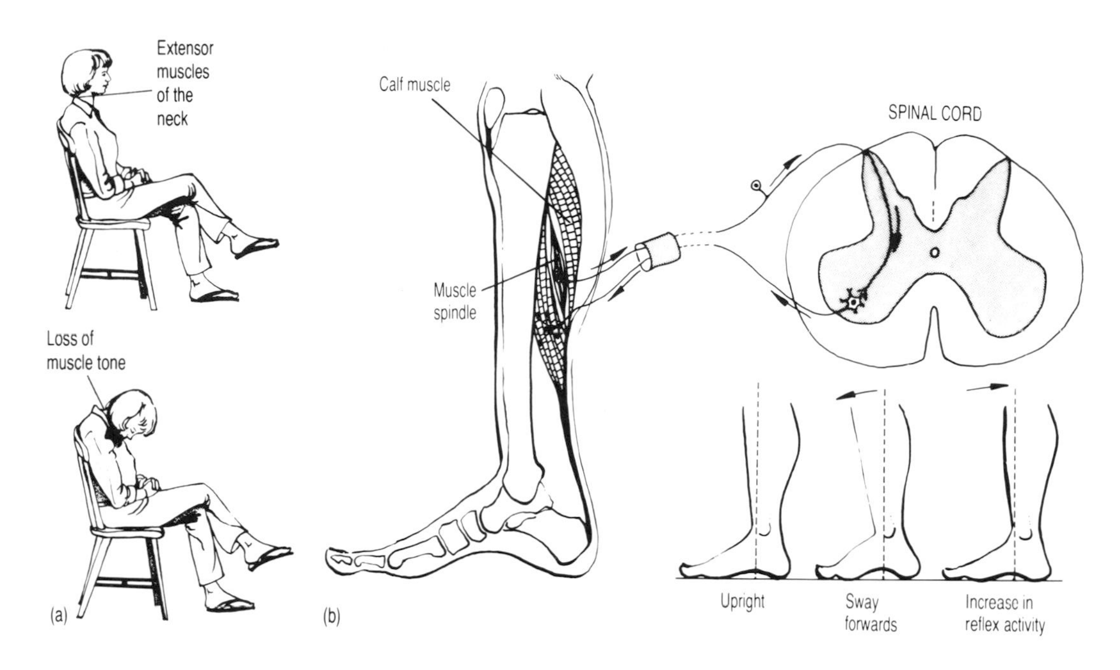

Fig. 1.20. Examples of 'holding a position': (a) the head in sitting; (b) calf muscles in standing.

- *FEEL the muscles around the shoulder of a partner while the arm is hanging by his or her side. The muscles are not limp, but are 'lively'. Now ask the partner to lift his or her arm sideways to the horizontal and hold the position. Feel the muscles again, and notice that they are more lively, the tone of the muscles has increased.*

Interruption

Interruption of the nerve supplying a muscle abolishes all activity in the myotatic unit, and both sensory and motor neurones will be involved. The muscle then becomes limp or *flaccid*.

At the end of this chapter, you should be able to:

1 Explain the functional importance of the connective tissues in providing support, strength and elasticity in the locomotor and neuromuscular systems.

2 Summarize the three main classes of joints found in the body, and describe a typical synovial joint.

3 Describe the gross and microscopic structure of skeletal muscle in relation to function.

4 Outline the generation and conduction of impulses in neurones, and the transmission of impulses at synapses.
5 Distinguish between the form and function of sensory and motor neurones in the nervous system.
6 Define a motor unit.
7 Describe the myotatic unit. Discuss its function as the basic spinal reflex for muscle tone and for maintaining body position (the reflex is also known as the 'stretch reflex').

2 / Movement Terminology and Biomechanical Concepts

A vocabulary is needed to communicate to others a useful description of the moving body. The appropriate vocabulary depends on the purpose of the description, and different disciplines have developed their own languages.

Anatomical terminology defines a reference position, and names the direction of movement at each of the joints. For example, in standing up from sitting, we can describe the direction of movement at the hip, the knee and the ankle. A description based on direction alone does not distinguish the difference between rising from a low seat or a high stool, doing it quickly or slowly, staying in balance or losing it. To discuss the dynamic aspects of movement, which involve the forces required and the weights of the body parts, we use terms taken from the vocabulary of mechanics.

The aim of this chapter is to introduce the anatomical terms used in the analysis of daily activities. The application of some simple mechanics to the moving body will also be considered.

2.1 The anatomical position

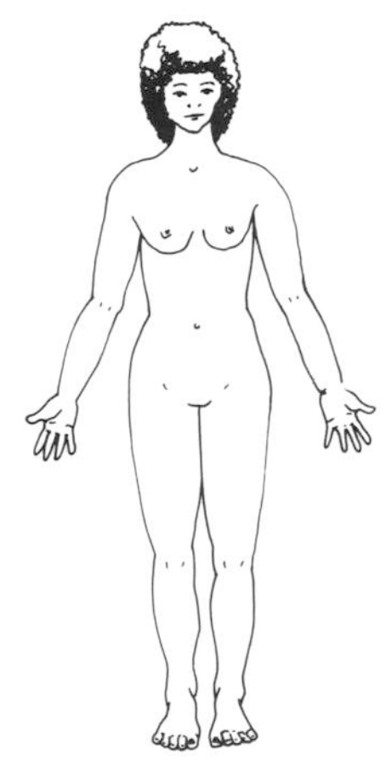

Fig. 2.1. The anatomical position. Stand upright with palms of hands facing forwards, feet parallel and facing forwards.

All movement starts from a posture or position, which must be first defined before describing the changes that follow. We need to use a common reference to describe the positions, relationships and directions of movement.

The reference is the *anatomical position*, standing upright with the palms of the hands facing forwards, the feet parallel and facing forwards (Fig. 2.1). Note that the usual standing position, with the palms of the hands facing the sides, is not used.

• *LOOK at the articulated skeleton to see the difference between the anatomical position and the natural standing position. In the anatomical position, the bones of the forearm are parallel, and the whole of the palm of the hand can be seen from the front.*
• *STAND in a natural position, and then change to the anatomical position. Note the change in position of the forearm and hand. Check that the feet are slightly apart and facing forwards.*

2.2 Planes and axes of movement

The reference anatomical position can now be divided into three planes which lie at right angles to each other. The planes are the fixed lines of reference for movement (Fig. 2.2).

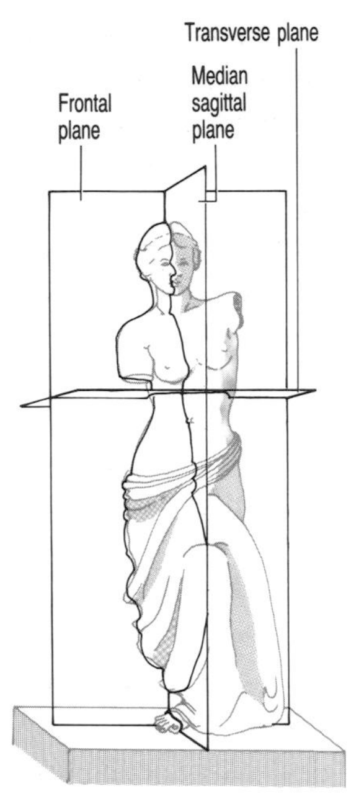

Fig. 2.2. Planes of movement. Median sagittal plane, frontal plane, transverse plane.

2.2.1 Median sagittal plane

The median sagittal plane is a vertical plane which divides the body into right and left halves. Any plane parallel to the median plane, dividing the body into unequal right and left parts is said to be a *sagittal plane*, parallel to the sagittal suture of the skull in the midline of the skull. The terms 'medial' and 'lateral' relate to this plane. A structure nearer to the median plane is medial, and one further away from the median plane is lateral. For example the medial ligament of the knee is on the inside of the joint, while the lateral ligament is on the outside.

2.2.2 Coronal or frontal plane

This divides the body into front and back halves. Frontal planes are parallel to the frontal suture of the skull across the crown of the head. The terms 'anterior' and 'posterior' relate to this plane. The anterior shaft of the femur is the front of the bone in the anatomical position, the posterior shaft is the back of the bone.

2.2.3 Transverse or horizontal plane

This is parallel to the flat surface of the ground. Planes in this direction divide the body into upper (cranial) and lower (caudal) parts. Crossing the body in this direction, planes are at right angles to the sagittal and frontal planes. The terms 'superior' and 'inferior' relate to this plane. The superior radio-ulnar joint is near to the elbow (i.e. above or towards the head), while the inferior radio-ulnar joint is adjacent to the wrist (below or towards the ground). When the limbs move in different directions, the terms superior and inferior can become confusing, e.g. if the arm is above the head. Another way of identifying structures may then be used. The terms 'proximal' and 'distal' mean nearer to the centre of the body or further away from the centre respectively. The superior radio-ulnar joint can therefore also be named the proximal joint, and likewise the inferior joint as distal.

The *axis* of movement at a joint is at right angles to the plane. Bending the elbow is a movement in the sagittal plane about an axis passing through the frontal plane at the joint. Turning the head from side to side is a movement in the horizontal plane about a vertical axis through the joint between the first and second vertebrae of the neck. It may help to understand plane and axis if you think of the plane of movement as the wheels of a car around the axle (axis) at the hub of the wheels.

Movements can be classified in terms of the three planes and axes described. Many functional activities, however, occur in diagonal planes. The leg swing in walking does not occur exactly in the sagittal plane at the hip, but in a diagonal plane between the sagittal and frontal planes, so that the foot comes to the ground near to the midline of the body. Movement at the shoulder which carries the arm forwards and slightly across the body is in a diagonal plane.

2.3 Movements at synovial joints

Most of the movements of the body occur at the synovial joints. The structure of a typical synovial joint has been described in Chapter 1.

2.3.1 Classification of synovial joints

The synovial joints are classified by the axes of movement (uniaxial, biaxial, multiaxial) and by structure as follows.

1 A **hinge joint** allows movement in one direction only, in the sagittal plane. It is a *uniaxial* joint. Examples of a hinge joint are the elbow (Fig. 2.3a) and the ankle.

2 A **pivot** is restricted to rotational movement around a vertical axis in the horizontal plane. It is a *uniaxial* joint. Examples are the atlantoaxial joint in the neck which turns the head to look sideways (Fig. 2.3b), and the joints in the forearm which allow the hand to turn so that the palm faces backwards.

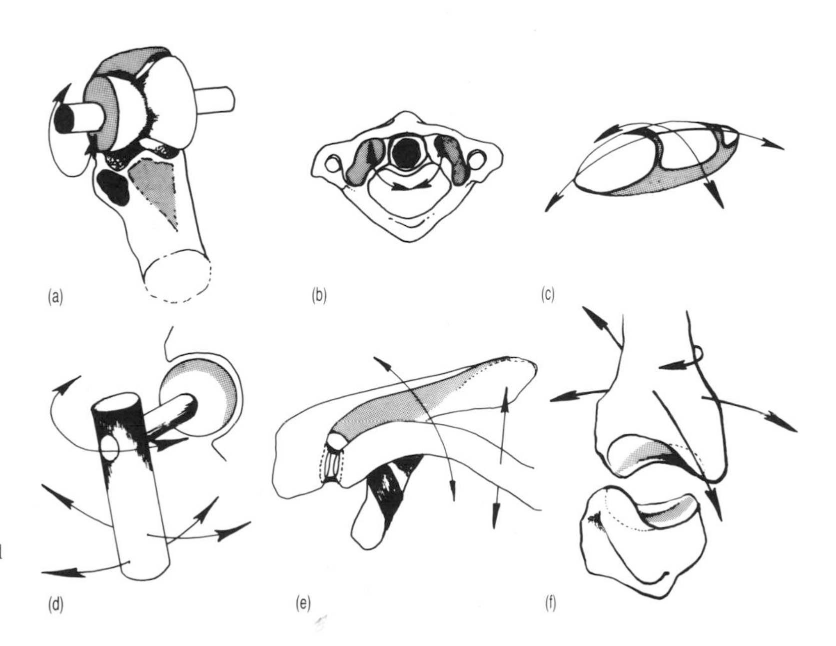

Fig. 2.3. Types of synovial joint: (a) hinge; (b) pivot; (c) ellipsoid; (d) ball and socket; (e) plane; and (f) saddle.

3 An **ellipsoid joint** has oval joint surfaces which allow movement in the sagittal and frontal planes, but no rotation. It is a *biaxial* joint. Examples are the radiocarpal (wrist) joint (Fig. 2.3c), and the joints at the base of the fingers (metacarpophalangeal joints).
4 A **ball and socket joint** allows movement in three planes (Fig. 2.3d). It is a *triaxial* or *multiaxial* joint. Examples are the shoulder and hip joints.
5 A **plane joint** has flat articular surfaces which allow limited movement around one or more axes (Fig. 2.3e). Plane joints may be arranged in series, so that the cumulative effect of the limited action at each joint gives considerable movement overall. The synovial joints between the bony arches of adjacent vertebrae are examples of plane joints, which together give the overall movements of the trunk.
6 A **saddle joint** has a surface which resembles a saddle (a concave-convexity) with a reciprocally curved surface sitting on it. The movements are in two planes with a limited range of rotation as well. The first carpometacarpal joint at the base of the thumb is a saddle joint (Fig. 2.3f).

The **range of movement** possible at each synovial joint depends on three main factors:
1 The shape of the *bony* articulating surfaces determines both the direction and extent of the movement. For example, the shallow socket of the shoulder allows a wide range.
2 The position, strength and tautness of the surrounding *ligaments* affects range. By regular stretching exercises from an early age, gymnasts and ballet dancers can stretch certain joint ligaments to achieve a greater range of movement.
3 The strength and size of *muscles* surrounding the joint affect the range. Bulging muscles around a joint halt movement when the two moving segments come into contact. For example, bending the elbow is limited by contact of the forearm with the upper arm. Other muscles may restrict movement by their position in relation to a joint. Tight hamstring muscles at the back of the thigh limit bending of the hips in touching the toes.

2.3.2 Terms for movement at synovial joints

Starting from the anatomical position, paired terms are used to distinguish the direction of movement of body segments in the three planes described (Fig. 2.4).
1 **Flexion** and **extension** are movements in the sagittal plane. Flexion movements bend the body part away from the anatomical position. Extension is movement in the opposite direction back to

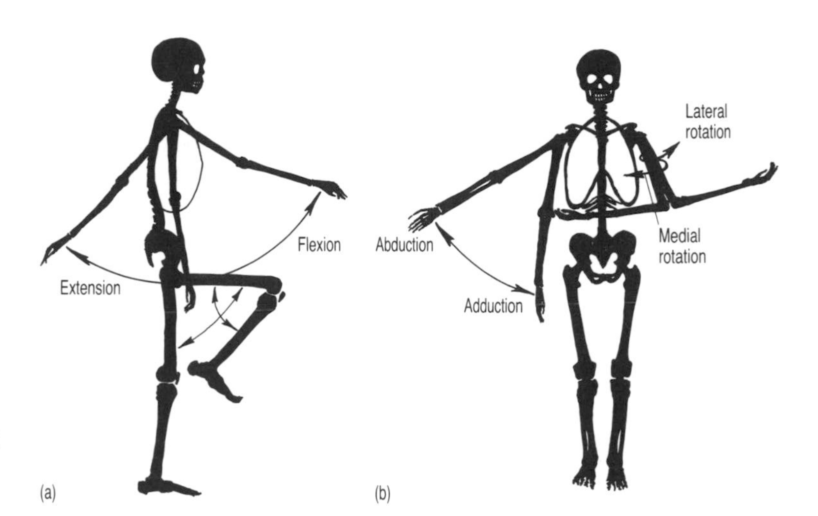

Fig. 2.4. Movements at joints: (a) flexion and extension; (b) abduction and adduction; medial & lateral rotation.

the anatomical position and beyond into a reversed position (Fig. 2.4a). In flexion, the angle between the bones is usually decreased, e.g. flexion of the elbow bends the forearm forwards and upwards towards the arm, flexion of the knee takes the leg backwards towards the thigh. Flexion movements curl the body into a ball, whilst extension stretches the body out.

2 Abduction and **adduction** are movements in the frontal plane. Abduction movements carry a body part away from the midline. Adduction is movement in the opposite direction towards the midline (Fig. 2.4b). In the hands and feet these movements are related to the central axis of the segment. The fingers move away from the middle finger, and the toes move away from the second toe.

• *DO NOT confuse a**b**duction and a**d**duction — the letter b comes first in the alphabet, and is followed by d, the return movement being adduction. The prefixes come from latin and you will recognize that 'ab' means 'away from', as in abscond — to escape; and 'ad' means 'towards', as in addition and adherent.*

When the return movement continues beyond the anatomical position, the terms 'hyperextension' and 'hyperadduction' may be used (hyper means 'more than').

3 Rotation is movement in the horizontal plane about a vertical axis. When the bone is rotating away from the midline, or towards the posterior surface, the movement is known as *lateral rotation* (or external rotation). In the reverse movement, the bone turns in towards the midline of the body, and is known as *medial rotation* (or internal rotation) (Fig. 2.4b).

Circumduction is a term used to describe a sequence of movements of flexion, abduction, extension and adduction. The bone moves round in a conical shape with the apex of the cone at the moving joint, and the base at the distal end of the bone. True circumduction does not include rotation.

The paired movement terms are also used to name the groups of muscles producing them. Muscles which bend the fingers are known as *flexors* of the fingers, whilst *extensors* straighten the fingers. The *abductors* of the hip carry the leg sideways.

- *STAND in the anatomical position. Move each of the large joints in turn, e.g. shoulder, elbow, wrist, hip, knee, ankle.*
- *RECORD the movements possible at each of the joints.*

Most body movements do not start at the anatomical position, but they are described with reference to that position. To analyse a movement, the starting position must first be defined. For example, lifting a glass from a table to the mouth starts with the shoulder in a neutral position, the elbow flexed, the wrist extended, and the fingers flexed around the glass. The changes at each joint are then described as the movement proceeds. To drink, the shoulder must be flexed and the elbow flexed further to bring the glass to the lips.

- *OBSERVE some simple everyday activities such as: standing up from sitting, climbing stairs, reaching to a high shelf, and pulling down a blind.*
- *RECORD the starting position, and list the movements made at each of the joints involved.*

2.4 Muscle attachments and group action

Muscles produce movements at joints by pulling on the bones to which they are attached. To describe a muscle, we name its attachments. One end of the muscle is usually fixed, and the attachments at the other end move. The attachment that is usually fixed or held steady is known as the *origin* of the muscle, it is usually more proximal. The moving end is called the *insertion* and is usually more distal. Some muscles can work from either end. For example, the muscle that extends the hip (gluteus maximus) pulls the thigh backwards as in climbing stairs (Fig. 2.5a). On the other hand, if the trunk is flexed forwards, this hip extensor acts in

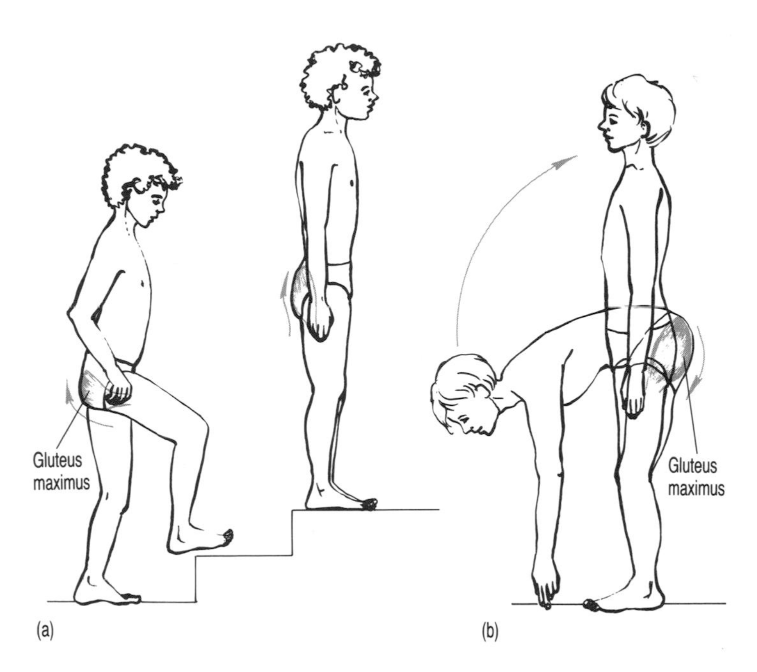

Fig. 2.5. Muscle attachments. Action of gluteus maximus in extension of the hip; (a) climb stairs, distal attachment moves; and (b) straighten up from bend forwards, proximal attachment moves.

reverse to pull the pelvis upwards and straighten the trunk on the leg (Fig. 2.5b).

Group action in muscles (Fig. 2.6)

No muscle acts alone. All the muscles arranged round a joint are involved in the movement at that joint. In the case of the elbow joint, there are four muscles crossing the front of the joint and two that lie posteriorly. The anterior group are the flexors and the posterior group are the extensors. When the elbow is actively flexed, the flexors are the *prime movers* (or *agonists*), and the extensors become the *antagonists*. The extensors are reciprocally relaxed during elbow flexion, but will act as controllers of the extent and speed of the movement. Other muscles are active to support the proximal joints, these are known as *fixators*. They are able to fix the origin of the prime movers. When biceps are active as a prime mover, the muscles attached to the trunk, scapula and humerus are active as fixators to fix the origin of the biceps. If the muscles acting as prime movers pass over more than one joint, other muscles known as *synergists* are active to prevent undesirable movements occurring at the other joints. For example, the flexors of the fingers cross the wrist and other joints in the hand. When gripping the handle of a tool or a racquet, the wrist extensors act as synergists

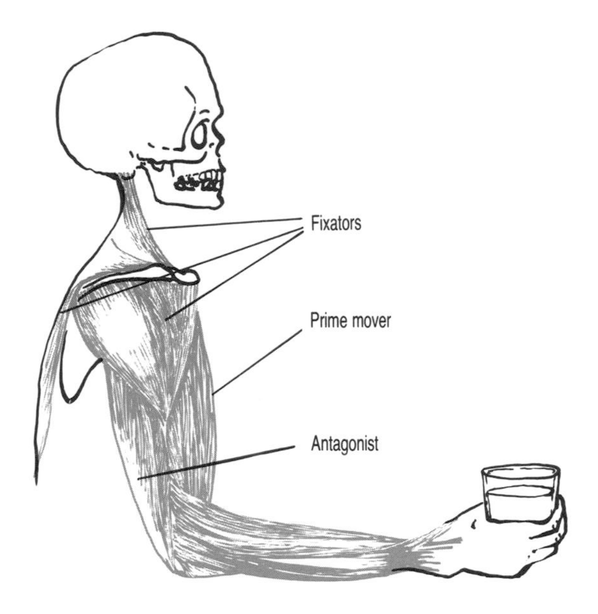

Fig. 2.6. Group action of muscles of upper limb in lifting a glass. Prime mover (biceps), antagonist (triceps), fixators (muscles attached to the scapula and humerus).

to prevent wrist flexion, and allow the finger flexors to exert maximum holding power on the handle.

2.5 Types of muscle work

Muscle action is not only used to make a body part *move*, it may also be necessary to *hold* the position of a body part, such as the forearm supporting a book in the hand. Muscles may also be needed to control the effect of an external force acting on a body part. When sitting down on to a chair, the extensors of the leg work to control the effect of gravity, which continually pulls the body down on to the seat. The term 'muscle contraction' may be a misleading one, because muscle work may involve the *shortening* of the muscle, or staying the *same length* or a controlled *lengthening* of the muscle. For this reason, muscle work is categorized into concentric, eccentric and static work.

Concentric work (now sometimes called isotonic shortening)

This applies to muscles that are shortening to produce a movement. When a saucepan is lifted off a stove, the elbow flexors are working concentrically — they shorten to lift the pan (Fig. 2.7a).

Eccentric work (now sometimes called isotonic lengthening)

An active muscle that is lengthening is doing eccentric work. The muscle activity is controlling the rate and extent of movement as

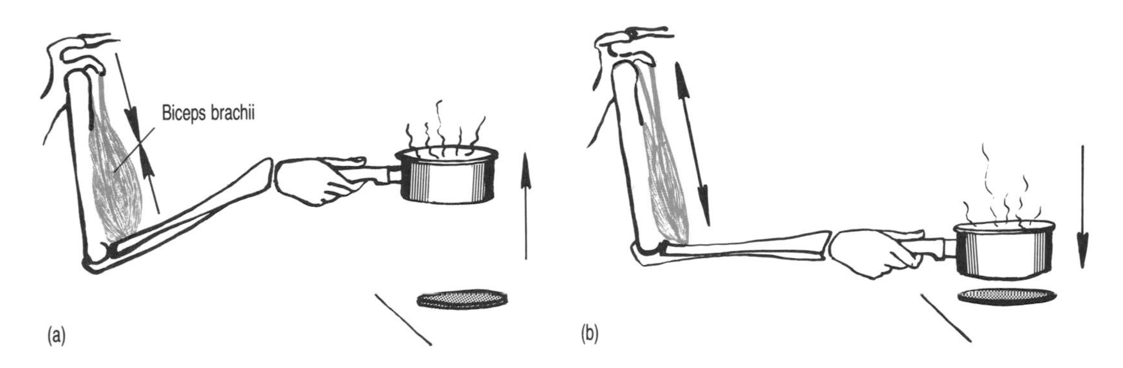

Fig. 2.7. Types of muscle work: (a) concentric, biceps flexing the elbow to lift saucepan; and (b) eccentric, biceps controlling the effect of gravity to lower saucepan.

the attachments are drawn apart by external forces, such as gravity. When a saucepan is put down on to a stove, gravity is assisting the movement, so the elbow flexors must work eccentrically to control the movement, allowing the pan to be placed carefully on the hot plate (Fig. 2.7b).

Static work (also called isometric)

The active muscles which remain the same length to hold a position are doing static or isometric work — 'isometric' means same length. If the saucepan is held still over the stove, the elbow flexors are working isometrically to prevent it from dropping down.

Static work is the most tiring form of muscle work and should not be performed for long periods without rest. Fatigue is largely due to poor blood flow and accumulation of waste products in the muscle, partly because the static state reduces the pumping action of contracting muscles on the circulation of the blood. The terms 'isometric' and 'isotonic' were first used by physiologists to distinguish two types of muscle response in isolated frog muscle. Isotonic means 'the same tension', and applies to a muscle that changes in length without a change in the tension within the muscle. In the human body, true isotonic muscle activity rarely occurs, because over the whole range of movement, changes in muscle tension occur in response to the changing effects of gravity and leverage (see Section 2.6.3 for discussion of leverage). Isometric work does occur in the body when muscles are active to hold the position of a body segment and any added load. The term is also used in exercise programmes, when the muscles are working against the resistance of springs or weights.

• *ASK a partner to lift the forearms to a horizontal position and feel the tension in the elbow flexors by palpating the muscles above the elbow.*
• *PLACE a tray in the hands with the forearms in the same position. Note the change in the tension (hardness) of the elbow flexors even though there has been no change in length of the muscles. This is because the elbow flexors are having to work more actively to support the added load of the tray.*

2.6 Biomechanical concepts

Terms used in mechanics to describe the movement of objects can be applied to the moving body. In this section, we will consider some of the concepts that help us to understand how and why mechanical principles are applied to rehabilitation and exercise programmes. The numerical calculations involved in the application of mechanical principles to the body can be found in textbooks of biomechanics.

An example of some of the mechanical terms used in a movement, such as flexion of the elbow, is as follows: The flexor muscles of the elbow produce *forces* on the bones of the forearm to bend the elbow. The amount of force depends on the *tension* developed inside the muscles at any one moment in time. The speed of the flexion movement as it progresses, will depend on the *power* of the muscle.

Forces outside the body also affect the movement. We have already referred to *gravity*, which is a constant downward force acting at the centre of mass of the body segment. Movement is the result of the *moments of force* acting around a joint, which means the *magnitude* of, and the *distance* of each force from the joint to which it is being applied (Fig. 2.8). In bending the elbow, the flexors exert

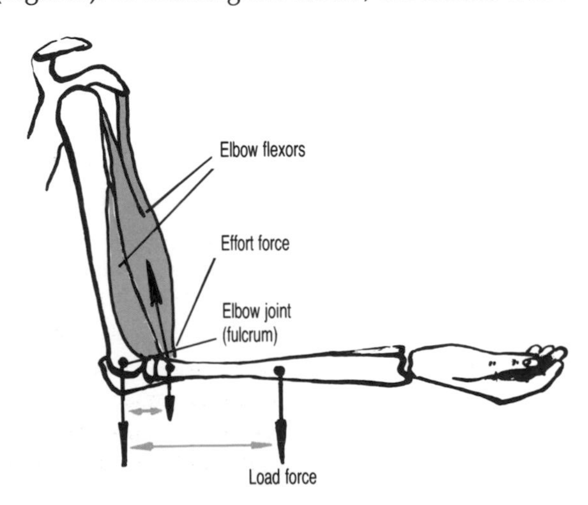

Fig. 2.8. Moment of force at the elbow. Fulcrum (elbow joint); load force (weight of the forearm and hand acting at the centre of gravity); effort force (tension in the elbow flexors acting at the point of insertion into the bones).

a moment of force which depends on the force of the muscles, and also on the distance between the insertions on the radius and ulna and the centre of the elbow joint. Gravity also exerts a moment of force in the opposite direction acting at the centre of gravity of the forearm. If the moment of force of the elbow flexors is greater than that of the force of gravity, movement occurs.

2.6.1 Gravity

The muscle work necessary to perform a particular movement depends on the position of the moving parts in relation to gravity. A controlled movement which takes place against the pull of gravity will demand different muscle work to one that is assisted by gravity. We have seen in Section 2.5 how the elbow flexors work concentrically to lift up a saucepan from a stove, and the work involved is against gravity. The flexors must work eccentrically to gently lower the load, because in this direction the forearm is assisted by gravity (Fig. 2.7). In the process of rehabilitation the upper limb may be supported by a sling, which will allow movement but will eliminate the pull of gravity. This encourages weak muscles to perform tasks that may be impossible when the full effect of gravity must be overcome. As the muscles of the upper limb become stronger, movements can be achieved without the support.

Because of the effect of gravity, the starting position for a movement determines the effort required by the active muscles.

• *FEEL the difference in the abdominal muscles when using them to (i) bend down and touch the toes; and (ii) sit up from lying.*

The same movement of the ribs towards the pelvis occurs in both (i) and (ii). In (i) the movement is assisted by gravity, and in (ii) the abdominal muscles are lifting the weight of the trunk against the force of gravity. Therefore in (ii) a greater force must be developed by the abdominal muscles.

2.6.2 Stability

An essential feature of all movement is the need to keep the body in stable equilibrium, so that we do not fall over while the body is changing position. We can all imagine the stability problems of a gymnast balancing on a beam, or a ballet dancer poised on one toe, but we take for granted the balance requirements of everyday activities. We do not have to think about balance at each step as

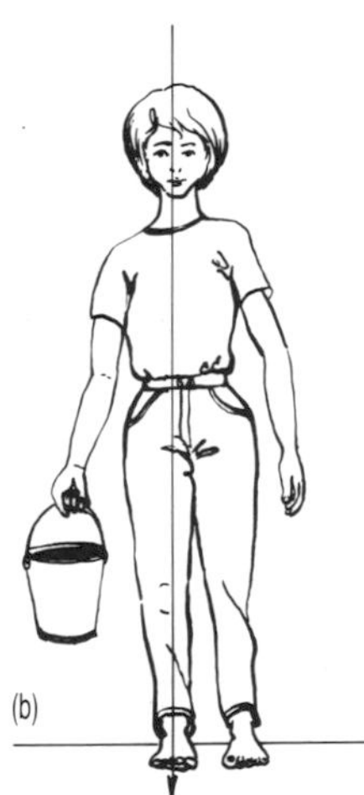

Fig. 2.9. Stability in carrying weight at the side of the body: (a) unstable, line of weight lies outside foot base; and (b) stable, body segments realigned to bring the line of weight inside the foot base.

we walk, but we may become aware of the problem when we are standing on a jolting train or bus. If the nervous system, which automatically makes the adjustments, is not functioning correctly, the balance of the body may become a constant problem and movements are difficult to perform.

An object is in stable equilibrium when its centre of gravity lies over its base of support. This means that the upright body is only stable when the line of weight from the centre of gravity falls within the foot base. If the line of weight moves outside the foot base as we move around, we will fall over. If the body was rigid like a plaster figure, the addition of a weight on one side would move the centre of gravity to that side and the figure would topple over. In the body, the postural mechanisms of the nervous system make sure this does not happen. Figure 2.9a shows how the added weight of a bucket held in the hand moves the line of weight to the right and beyond the foot base so that the body will fall to that side. Figure 2.9b shows how the body segments alter their position to move the line of weight back over the foot base and the body becomes stable again. This realignment of body segments occurs automatically.

The two main factors that contribute to stability are therefore: (i) position of centre of gravity; and (ii) size of the base of support.

Position of the centre of gravity

The centre of gravity of an object (which is the point where the force of gravity acts on it), is at the centre of mass of the object. It is relatively easy to discover the position of the centre of gravity of a symmetrical object of uniform density.

- *TAKE a piece of card of symmetrical shape — square, oblong or circular. Draw diagonals across it. The point where the diagonals meet is the centre of gravity.*
- *THREAD a string through the centre of gravity and note how the card is balanced at this point.*
- *PLACE the card flat on a table and move it towards the edge of the table. Note when the card falls off the table.*

Two facts follow from the above example.

1 The whole mass of an object acts as though it is at the point of the centre of gravity.

2 An object is only stable when its centre of gravity is supported.

The centre of gravity of the body varies slightly in different individuals. A person with heavy shoulders will have a higher centre of gravity than one with heavy hips and legs. The centre of gravity also changes position during movement. Lifting the arms raises the centre of gravity, bending the knees lowers it. The important principle to remember is that the stability of an object is greater when the centre of gravity is lower, so it follows that all efforts to help the balance of the body should be directed to positions where the centre of gravity is lowest. If we bend to pick up a baby or a box, the knees should be bent and the trunk flexed to move the centre of gravity down and over the feet. A hoist used to move a patient will be most stable if it is adjusted to the lowest position.

Base support

The upright body is least stable when the feet are parallel and close together because in this position the base support is small (Fig. 2.10a). As the feet are moved further apart the base support is increased and we are less likely to fall over (Fig. 2.10b). Walking aids such as a stick, crutches or pulpit frame all increase the size of the base support and therefore allow more swaying of the body above without falling (Fig. 2.10c). A therapist should stand with feet apart and knees bent to be in a stable position to resist the added weight of the patient she is helping to move.

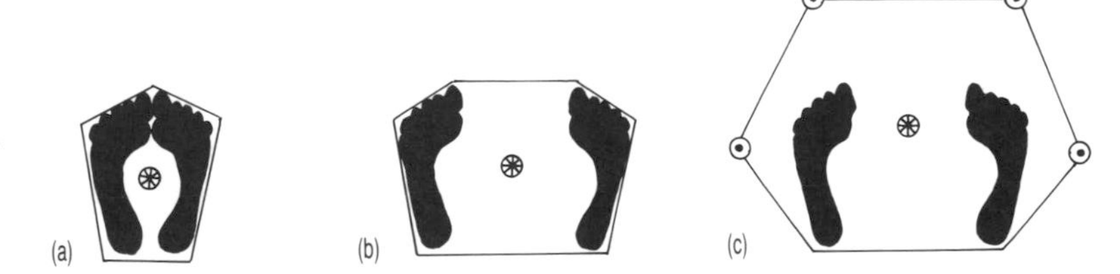

Fig. 2.10. Base support. Change in area of base support with: (a) feet together, base support small; (b) feet apart, base support larger; (c) feet with walking frame, very large base support.

2.6.3 Leverage

In the body, the bones form rigid levers and the joints form pivots or fulcra. The principle of leverage therefore applies to all movement in the body.

We have already defined 'moment of force' at the beginning of Section 2.6; it is the product of the force and its distance from the pivot. A moment of force is always trying to produce movement and a lever is only balanced when the moments of force acting around the pivot are equal and opposite.

We are all familiar with this principle when sitting on a see-saw with a small child. By putting the child at the far end on one side of the see-saw, we can balance the see-saw by sitting near to the central pivot. This shows how a large force at a short distance

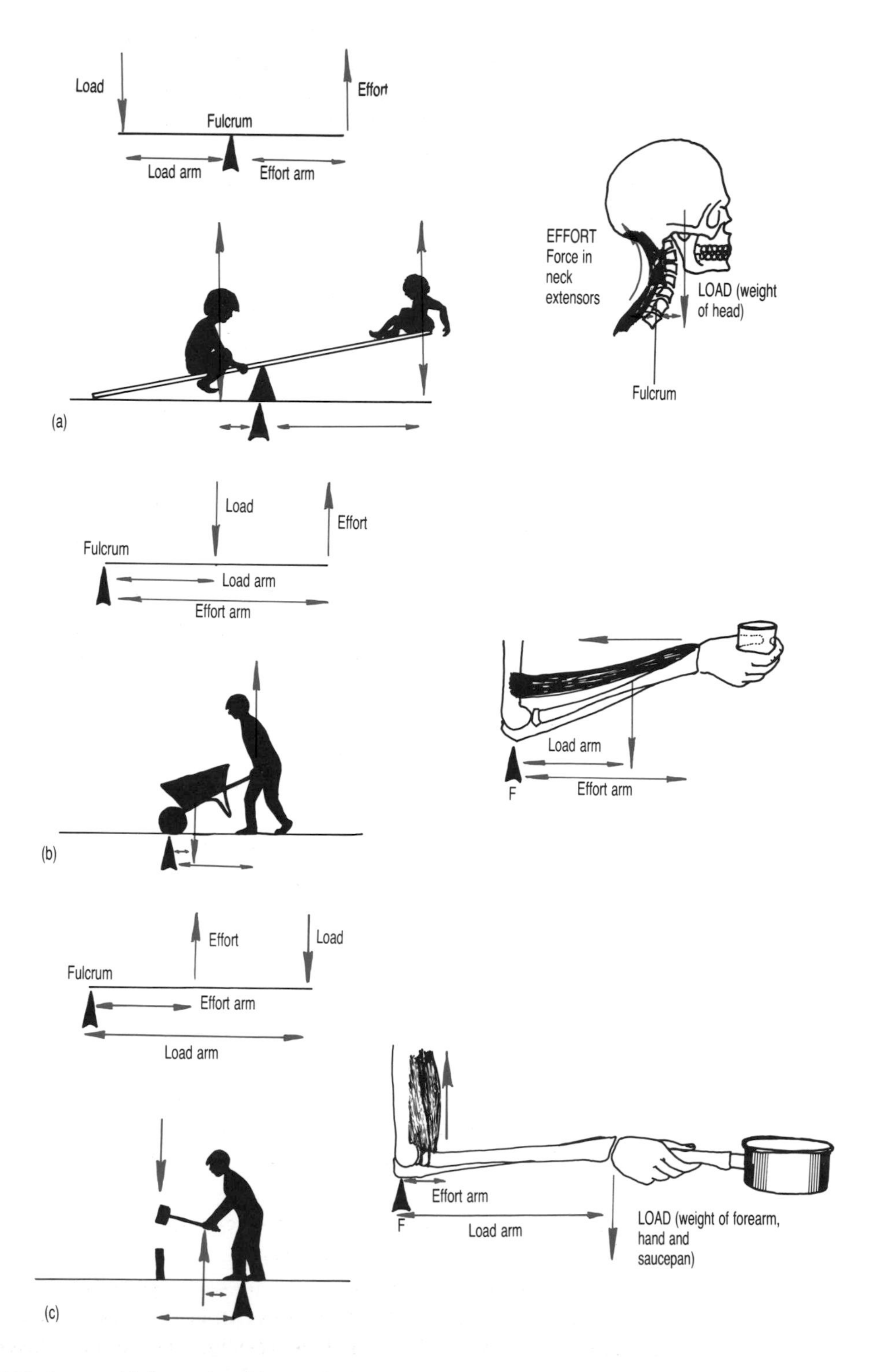

Fig. 2.11. Levers: (a) first order; (b) second order; (c) third order.

can balance a small force at a larger distance (Fig. 2.11a). Levers do not always have the pivot in the middle, the forces may both be acting on the same side of the pivot. A wheelbarrow is an example of this type of lever (Fig. 2.11b). The wheel in contact with the ground forms the pivot, the load in the barrow is near to the pivot, and the effort is applied by the hands to the handles at a greater distance from the wheel.

The muscles acting on the joints exert *effort forces* on the bony levers. The distance of the attachment of the muscles from the joint can be called the *effort arm*.

The total force of the weight of any body segment and any added weight is the *load force*. The distance of the centre of gravity of the body segment, plus any added weight from the joint can be called the *load arm*.

For movement to occur, the muscle moment (effort force times effort arm) must be greater than the gravity moment (load force times load arm). If either force, or its point of application is changed, the leverage changes.

Levers are classified into first, second and third class. Figure 2.11 shows the arrangement of effort, fulcrum and load in each of the three orders of levers, with examples of each. Most of the muscles of the body act as third order levers since the muscles are attached near to the joint they move. The advantage of this arrangement is that it gives a greater range and speed of movement which is important in throwing and swinging actions of the upper limb, as well as in walking and running actions in the lower limb. A few muscles, such as brachioradialis in the forearm, act as second class levers (Fig. 2.11b). The tension in this muscle is important to relieve the stress on the bones of the forearm when weights are held in the hand.

Figure 2.12 shows how leverage can be used to exercise muscles against gradually increasing loads. If a load is moved further away from a joint in specific stages, the force developed by the muscles at each stage must increase. Activities for weak shoulder muscles should first involve gravity assisted movement and then movements with the elbow flexed so that the load arm is short, see Fig. 2.12a. As the muscles become stronger, the shoulder can be used to move the extended upper limb (load arm longer), and eventually with a weight held in the hand (load force larger) as shown in Fig. 2.12b and c. When tools or racquets are held in the hand, they increase the length of the load arm and the magnitude of the load force (Fig. 2.12d and e). In weight training programmes, the muscles are exercised against increasing resistance placed at increasing distances from the joint which forms the centre of the

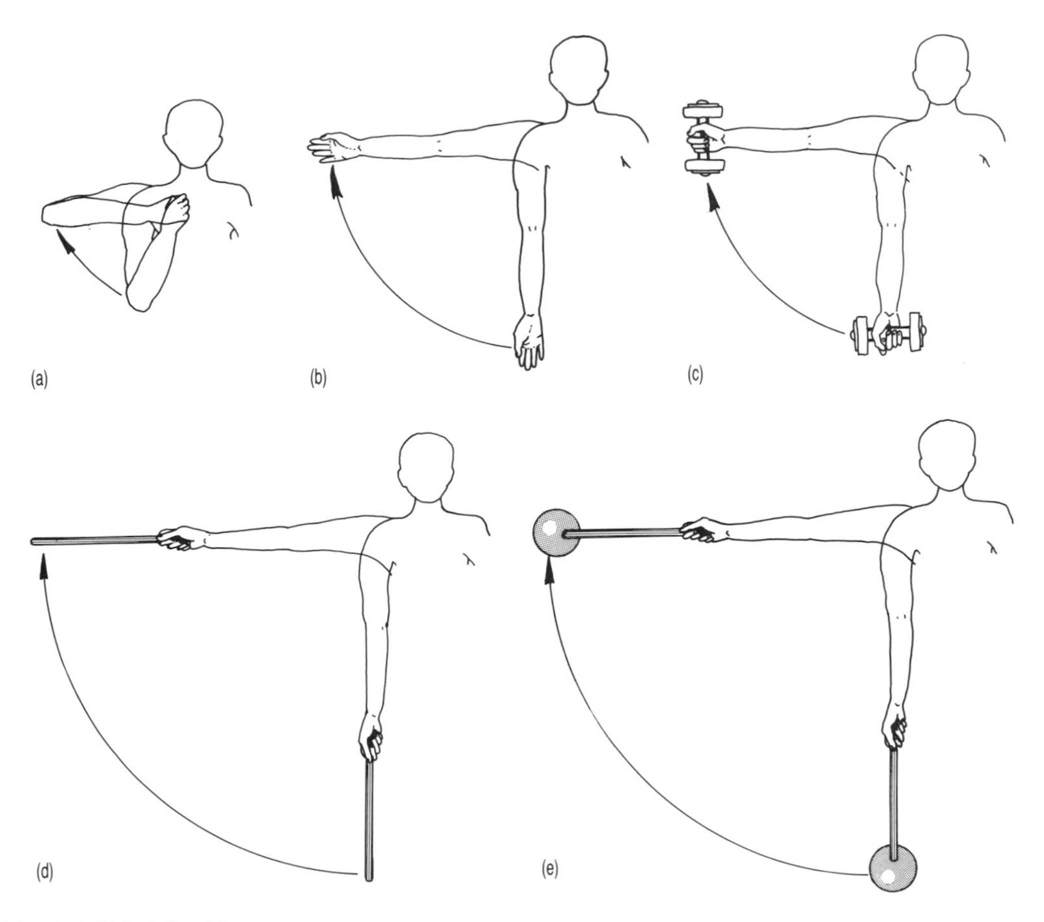

Fig. 2.12. Principle of leverage. Increase in effort required to lift the arm sideways in abduction: (a) short load arm; (b) longer load arm; (c) long load arm and added load; (d) load arm extended by tool; (e) load arm extended plus added load. (Adapted from *Simple Movement* by Laura Mitchell, published by John Murray, 1980).

movement. If we wish to increase the strength of abdominal muscles, sit ups are performed first without weights, next with a weight held in front of the chest, and finally with the weight held in the outstretched arms.

As the body moves, the centre of gravity of each body segment changes in relation to the joint at which it moves. Therefore the muscle effort needed to overcome the gravity moment of the segment will change. If we lift the arm sideways (abduction at the shoulder) the limb moves through a large arc of nearly 180°. When the limb is near the side of the body, and when it is above the head, the gravity moment is small since the centre of gravity is near to the joint. The greatest muscle effort will be required when the arm is horizontal because in this position the centre of gravity is furthest from the joint.

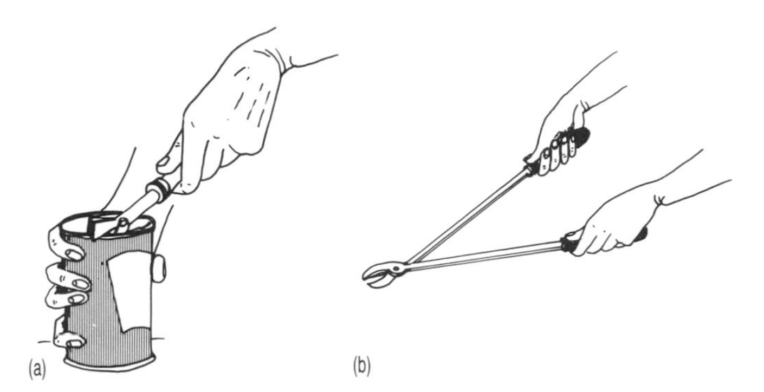

Fig. 2.13. Adaptation of tools to reduce effort: (a) tin opener with extended handle; (b) long handled shears.

• *HOLD a weight in the hand with the arm by the side. Then move the arm to horizontal. Feel the extra effort to support the same weight in this position.*

Leverage is also applied in adapting tools used in daily living for people with weak muscles or painful joints. If the lid of a jar is opened by a tool with a long handle then less effort will required than when grasping the lid itself; in the same way scissors and shears with long handles will be easier to use than those with short handles (Fig. 2.13).

It is important to understand the concept of moment of force and leverage in the choice of activities and exercises used to increase the strength of muscles in the body. In the adaptation of tools and equipment for use by people with weak muscles it is important to remember two principles.

1 Put the *load* as *near* to the pivot as possible.

2 Apply the *effort* as *far* from the pivot as possible.

At the end of this chapter, you should be able to understand the basic terminology used in the description of movement in the body, as follows:

1 Define the anatomical position, and the three major body planes of reference.

2 Name the terms for movement in the three planes at the joints of the body.

3 Classify the synovial joints based on structure. List the factors that limit movement at joints.

5 Appreciate how the attachments of muscles are described. Understand what is meant by the 'group action of muscles'.

7 Describe concentric, eccentric and static muscle work.
8 Explain how the position of the centre of gravity and the area of the base support affect body stability.
9 Understand the principle of leverage applied to:
(a) the attachments of muscles around joints; and
(b) the adaptation of equipment and tools to reduce the effort required in their use.

3 / Control Systems. The Brain and Spinal Cord

The organization of the nervous system begins centrally as folds which appear along the back of the 3 week old embryo. The folds meet to form the neural tube, and the nervous tissue destined to become the central nervous system is laid down. Folding and bending of the cranial (head) section of the tube follows to form the *brain,* whilst simpler growth changes in the remainder of the neural tube form the *spinal cord.* The adult brain presents a clear superficial appearance, and the complex arrangement of the deep brain areas is the result of the folding in the embryo. The 'first look' at the central nervous system in this chapter will be focussed on providing the names, location and overall function of the parts of the brain and spinal cord, with particular emphasis on their role in movement.

PART I/THE BRAIN

3.1 Position and relations of the main brain areas

At first glance the brain seems to only be composed of the two cerebral hemispheres; however although they are the largest feature of the brain, they conceal many other important areas. The two symmetrical hemispheres have a folded surface with their inner aspects lying close together in the midline. Underneath the posterior end of each cerebral hemisphere is the *cerebellum;* which also has two hemispheres that are joined together in the midline. Part of the *pons* is visible anterior to the cerebellum; and below the pons, is the cone shaped *medulla oblongata.* The medulla leads down into the spinal cord at the foramen magnum ('large hole') in the base of the skull. (See Fig. 3.1.)

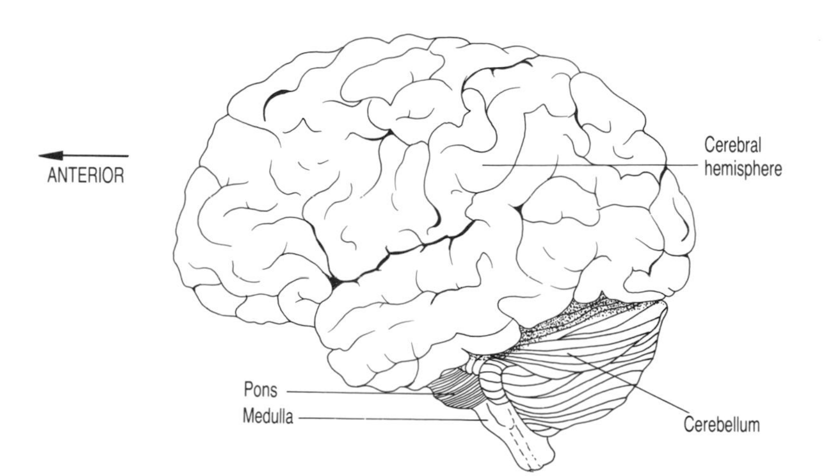

Fig. 3.1. External appearance of the brain, lateral view of left side.

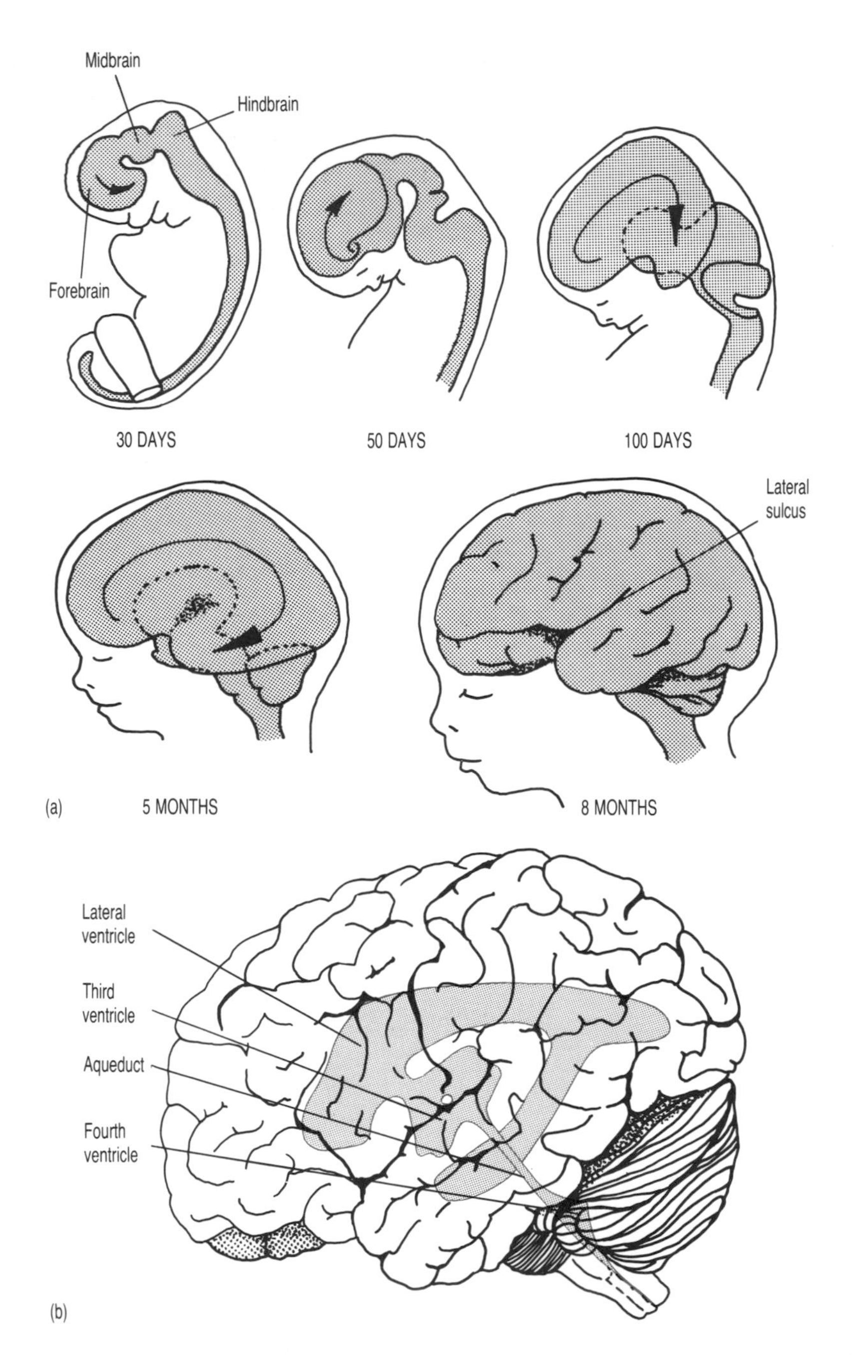

Fig. 3.2. (a) Development of the brain showing folding of the forebrain; and (b) Adult brain viewed from the left showing the position of the cavities.

A look at the development of the brain shown in Fig. 3.2a will help in the understanding of the internal form of the cerebral hemispheres. The forebrain first grows laterally and backwards. It then folds forwards on itself and takes on the appearance of a hand wearing a boxing glove with the thumb touching the palm when viewed from the side. The tube remains hollow and develops into the ventricular system containing cerebrospinal fluid. The cavity within each cerebral hemisphere follows the shape of the half clenched hand and is known as the lateral ventricle. Hidden by the extensive growth of the cerebral hemispheres, the base of the forebrain develops to form the *basal ganglia, thalamus* and *hypothalamus,* collectively known as the *diencephalon* or 'between-brain'. The structures in the diencephalon provide important links between the cerebral hemispheres and other parts of the central nervous system for both sensory and motor activity. The cavity in the centre of the diencephalon is a thin slit between the two thalami, called the third ventricle.

The **midbrain** continues in the same position during development, increasing in total size, but obscured in the external view of the brain by the lower temporal lobes of the cerebral hemispheres. In the adult, the midbrain looks like the 'waist' area with the expanded forebrain above and hindbrain below. Find the midbrain in the sagittal section of the brain (Fig. 3.3). The midbrain provides routes for pathways carrying impulses up or down to various levels of the central nervous system and is also important for analysis of information from the eyes and ears. The central cavity of the midbrain is a narrow canal called the aqueduct which leads down into the fourth ventricle, the cavity of the hindbrain. The fourth ventricle lies behind the pons and upper part of the medulla, with the cerebellum forming the roof of the cavity. Figure 3.2b shows the cavities of the brain in position in the adult brain.

The **brain stem** is the term used to describe the brain areas below the diencephalon, it includes the midbrain, pons and

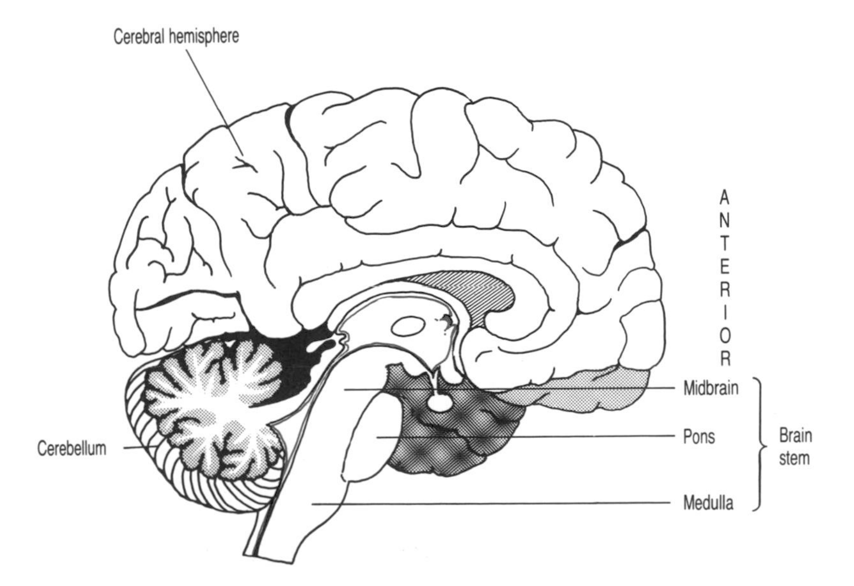

Fig. 3.3. Median sagittal section of the brain.

medulla. If the outgrowths of the cerebral hemispheres and cerebellum are removed from the brain, the complete brain stem can be seen with the diencephalon above. The functional importance of the brain stem must not be under estimated, it is concerned with all the automatic adjustments to changes in posture and movement in the body.

• *LOOK at a model of the brain, and diagrams of sections through the brain in anatomy textbooks to identify the position and relationships of the following brain areas: cerebral hemispheres, thalamus, basal ganglia, midbrain, pons, cerebellum and medulla oblongata.*

3.2 Cerebrospinal fluid

Cerebrospinal fluid is found in all the cavities of the brain, in the central canal of the spinal cord, and surrounding the brain and spinal cord, in between two of the three layers of protective connective tissue, and the meninges. These are described in Section 3.13. The main function of the fluid is to act as a shock absorber. It also carries nutrients and other essential substances to the nerve tissue. Figure 3.4 shows a sagittal section of the brain and part of the

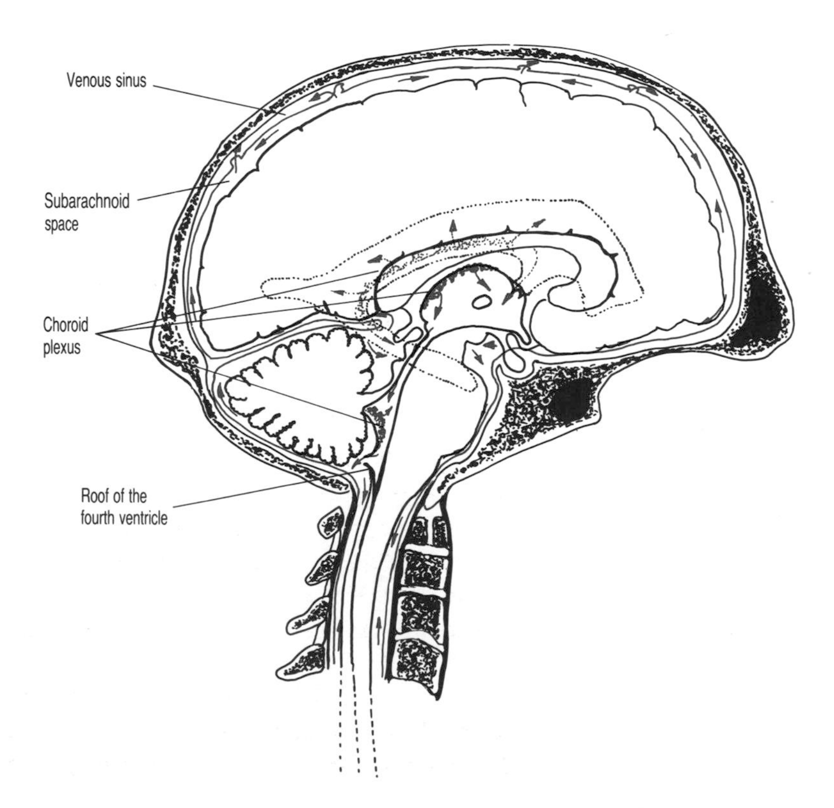

Fig. 3.4. Sagittal section of the brain to show the circulation of cerebrospinal fluid.

spinal cord to illustrate the way in which the cerebrospinal fluid circulates through the central cavities and around the outside of the central nervous system. The fluid is secreted from special patches of blood capillaries called choroid plexuses situated in each of the ventricles of the brain. The ventricles are found in the areas of greatest growth and expansion during development. Cerebrospinal fluid is formed by a process of filtration from the capillaries of each choroid plexus at the rate of 500 ml per day. Follow the arrows in Fig. 3.4 to see how the fluid flows downwards in the brain and then through openings in the roof of the fourth ventricle into the space between the coverings of the brain. The absorption of cerebrospinal fluid into the blood takes place mainly in the dural venous sinus between the two cerebral hemispheres, known as the superior sagittal sinus.

3.3 Organization of grey and white matter

The basic structure of neurones has been described in Chapter 1. Surprisingly, half the total volume of the central nervous system is not made up of neurones, but of *neuroglia* which are special support cells found in between the neurones, and of capillaries which supply the high oxygen demands of nerve tissue. The neuroglia act as transport and insulating cells, and also cooperate in the function of the neurones.

All the *cell bodies* and dendrites of the *neurones* form the core of the central nervous system known as *grey matter*. In the brain the core is not continuous, but the cells are collected together for a particular function to form many *nuclei* of grey matter. For example the thalamus is a nucleus of grey matter where sensory neurones synapse on the route to consciousness and other areas. The axons of the neurones lie in the *white matter* surrounding the nuclei in the brain stem. In the cerebral hemispheres and cerebellum an additional layer of grey matter, known as the *cortex*, lies outside the white matter. The cell bodies of the cortical neurones are laid down in organized layers. The cortical grey matter is folded to greatly increase the surface area. Each raised part seen on the surface is known as a *gyrus*. Each depression in between the gyri is called a *sulcus* (Fig. 3.5a). A very deep sulcus is sometimes called a *fissure*. In the white matter the bundles of axons lie in particular directions which are classified into four different functional categories shown in Fig. 3.5b.

1 Short association fibres connect groups of adjacent gyri. They form 'U' shaped bands which bend around the sulci.

2 Long association fibres are found lying deeper, and connect one

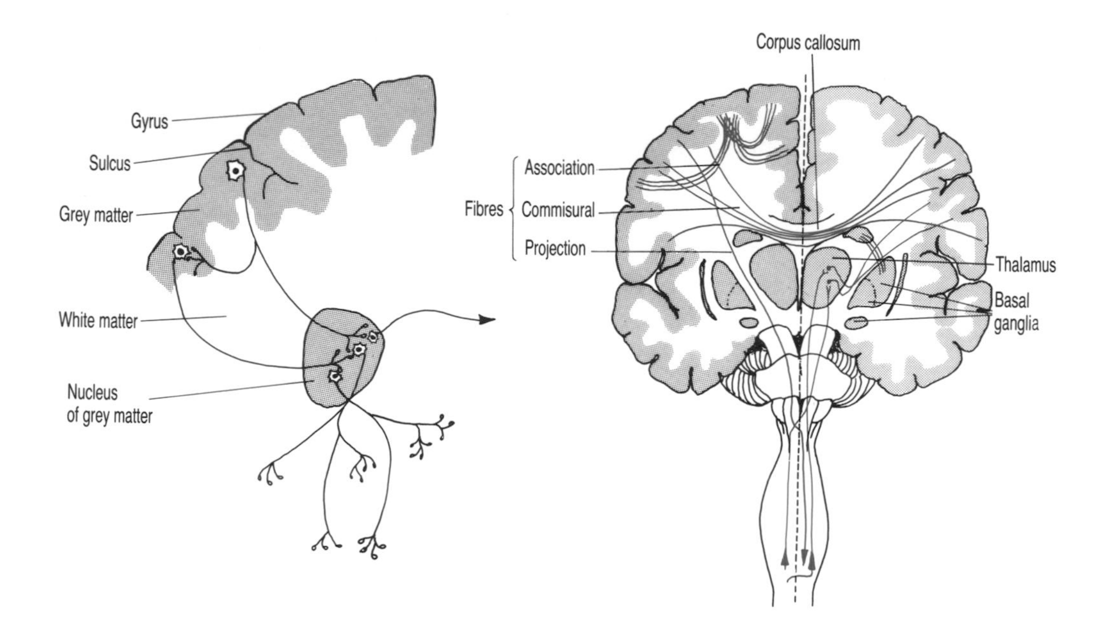

Fig. 3.5. (a) Small area of cortex of grey matter to show gyrus and sulcus, nucleus of grey matter; and (b) frontal section of the brain to show position of thalamus and basal ganglia. Projection, association and commissural fibres also shown.

part of the surface grey matter with another within the same hemisphere.

3 Commissural fibres connect cortical cells from one hemisphere with those of the opposite hemisphere. These fibres come together to form the major bridges between the right and left hemisphere. The main bridge lies above the diencephalon and is known as the *corpus callosum*. It contains an estimated one million nerve fibres. There are also smaller groups of commissural fibres located adjacent to the corpus callosum.

4 Projection fibres convey information between the surface grey matter and lower centres of the brain stem and spinal cord. Each of the projection fibres carries impulses in one direction only, either upwards or downwards.

• *LOOK at a brain model and sections of the brain to identify the following.*

1 *The cortical layer of grey matter in the cerebral hemispheres and cerebellum — it forms the outer surface like the skin of a fruit.*

2 *The nuclei of grey matter in the brain stem and at the base of the cerebral hemispheres and cerebellum.*

3 *The white matter found below the layers of cortex, and also surrounding the nuclei in the brain stem.*

It is important to build up a three dimensional picture of the shape, position and relations of the brain areas. Diagrams of sections taken through the brain at different levels can be compared with slices in various directions of a Swiss (jelly) roll or a piece of marble: each slice shows one particular colour in a different way, but the shapes can be put together to determine the three dimensional shape inside. The task of visualizing the brain in this way is not easy, but can be achieved with practice.

3.4 Cerebral hemispheres

The expansion of the cerebral hemispheres (or cerebrum) to envelop nearly all other brain areas distinguishes the primates, especially Man, from other animals. It is, therefore, not surprising that the surface of the hemispheres (cerebral cortex) has been studied extensively for over two centuries. The microscopists of the mid nineteenth century noted variations in the basic cellular architecture in different regions of the cerebral cortex, and the result of these studies was a detailed mapping into 52 numbered areas by Brodman (1909) and used clinically to this day for purposes of description (Fig. 3.6). Evidence from brain damage accumulated to suggest that different areas of the cerebral cortex have particular functions. In 1848, an American railroad worker named Phineas Gage survived an iron bar piercing right through the front of his brain. He could still move, eat and talk normally without a large area of cerebral hemisphere, but his friends commented that his personality had changed. A few years later in 1861, Broca identified a particular area in the left hemisphere concerned with speech from the postmortem examination of a patient with a severe motor

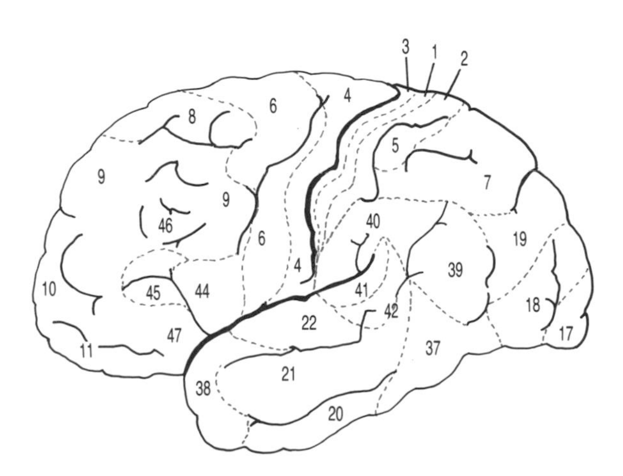

Fig. 3.6. Cerebral hemispheres with examples of Brodman areas.

speech defect. Head injuries in soldiers from the trenches in World War I, and studies of the electrical activity of the surface of the brain during surgical intervention later led to the identification of distinct motor and sensory areas. The remaining 'silent' parts of the cortex

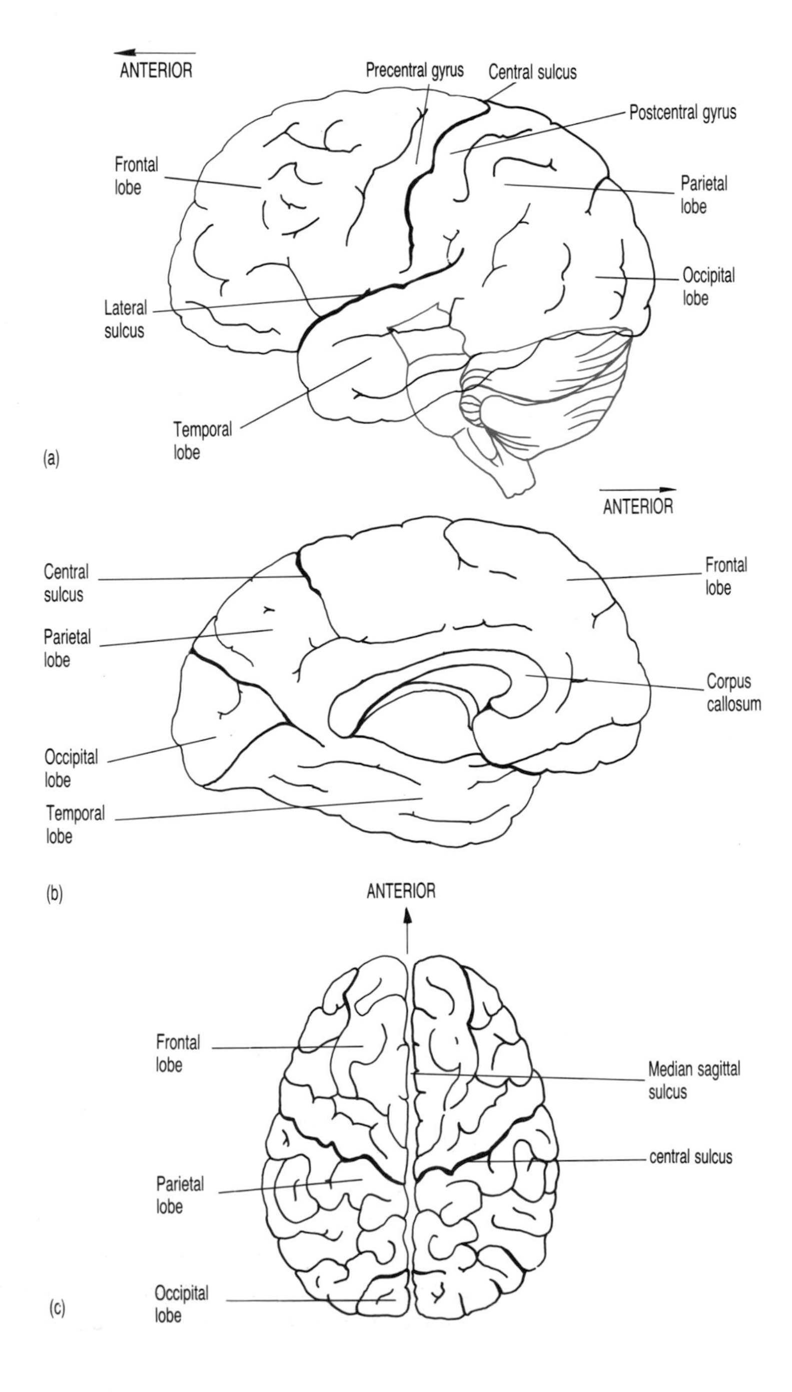

Fig. 3.7. Cerebral hemispheres: (a) lateral view; (b) medial view; and (c) view from above.

were described as association areas for interpretation and integration of the activity in the primary areas. A map of the functional areas in the cerebral cortex was developed, which is still used as a diagnostic tool in damage and disease of the central nervous system.

Each cerebral hemisphere is divided into four lobes named after the skull bones which cover them (Fig. 3.7a). In each hemisphere, the lobes are separated by two deep sulci — the central sulcus, and the lateral sulcus.

1 The **frontal lobe** lies anterior to the central sulcus.
2 The **parietal lobe** lies behind the central sulcus.
3 The **occipital lobe** is at the posterior end of the hemisphere, above the cerebellum at the base of the skull.
4 The **temporal lobe** lies below the lateral sulcus.

Each lobe continues on to the medial surface of the hemisphere, shown in Fig. 3.7b, and is separated from the opposite hemisphere by the median sagittal sulcus (Fig. 3.7c).

It is important to realize that the surface of the cerebral hemispheres extends from the level of the eyebrows in front, to the base of the skull at the back of the head, and down to the level of the ears at the side. This becomes obvious when a life size model of the brain is placed inside the cranial cavity of the skull.

The overall functions of each lobe will be described in turn, remembering that the variety of interconnections between the four lobes means that no individual lobe functions alone.

3.4.1 Frontal lobe

The frontal lobe is a large part of the cerebral hemisphere found below the frontal bone of the skull. The posterior part of the frontal lobe is a motor area involved in movement. The larger anterior end, which lies above the orbit of the eyes (supraorbital area), controls our emotional responses to changes inside and outside the body.

The posterior band of grey matter lying immediately in front of the central sulcus (precentral gyrus) is the *motor cortex,* which is concerned with the performance of movement in the opposite side of the body (Fig. 3.8). The cell bodies of the neurones in the motor cortex do not project to individual muscles, but to functional groups of muscles. Direct links to the small muscles of the hands, the feet and the face are particularly important, and damage to the motor cortex results in loss of precision movements.

There is representation of half of the body in an 'upside down' position in each motor cortex. The head is represented in the lower cortex on the lateral side, the upper limb and trunk above, and

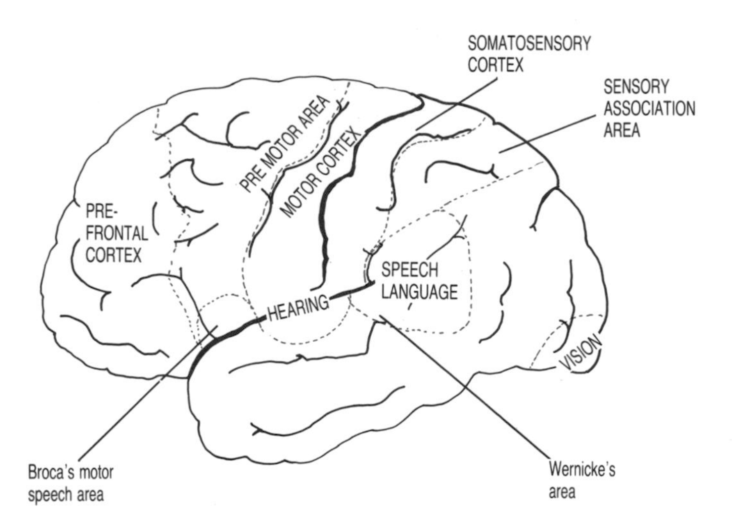

Fig. 3.8. Cerebral hemisphere. Main functional areas.

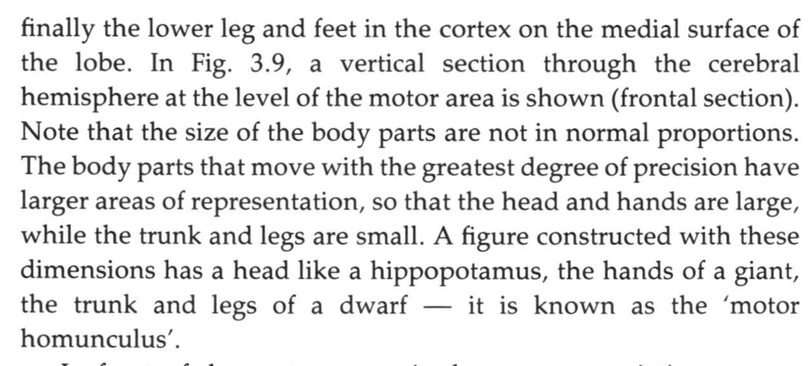

finally the lower leg and feet in the cortex on the medial surface of the lobe. In Fig. 3.9, a vertical section through the cerebral hemisphere at the level of the motor area is shown (frontal section). Note that the size of the body parts are not in normal proportions. The body parts that move with the greatest degree of precision have larger areas of representation, so that the head and hands are large, while the trunk and legs are small. A figure constructed with these dimensions has a head like a hippopotamus, the hands of a giant, the trunk and legs of a dwarf — it is known as the 'motor homunculus'.

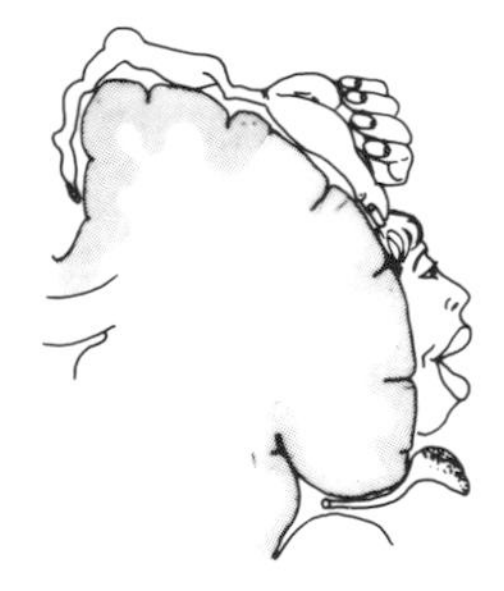

Fig. 3.9. Frontal section through precentral gyrus (motor cortex) to show representation of body parts.

In front of the motor cortex is the motor association area or **premotor cortex**, which has a large proportion of its surface extending on to the medial side of the frontal lobe. Activity in the premotor cortex results in modification of movement by coordination of groups of muscles on both sides of the body. Projection fibres from the premotor area descend directly to the spinal cord, or indirectly via the motor cortex of the same and opposite sides. The chief role of the premotor cortex is probably in the execution of learned bilateral movements such as walking.

The **motor speech area** identified by Broca lies in the lower part of the frontal lobe in the lip of the lateral sulcus (see Fig. 3.8 above). The function of this area, found in the dominant hemisphere, is in the production of fluent speech.

The **prefrontal** (or supraorbital) **cortex** occupies the large anterior area of the frontal lobe and connects with all other lobes of the cerebral hemispheres, the thalamus, the limbic system and many other brain areas. Observations of the effects of damage to the prefrontal cortex suggest that it is concerned with the expression of emotion and the modification of mood and behaviour. The prefrontal cortex seems to determine the way we react to other people and the environment from day to day, reflecting the results of learned and socially acceptable responses to external stimuli.

3.4.2 Parietal lobe

The parietal lobe lies posterior to the frontal lobe and beneath the parietal bone of the skull. The overall function of the parietal lobe is the integration of sensory input from receptors in all parts of the body, and also from the special sense organs. This gives our body awareness, and our knowledge of the environment around us.

The **somatosensory** (somaesthetic) **cortex** is the primary area which lies immediately behind the central sulcus in the postcentral gyrus (see Fig. 3.8 above). Pathways from receptors in the skin, muscles and joints of the opposite side of the body connect with the sensory cortex via the thalamus. The areas of the body are represented in an 'upside down' position in the same way as in the motor cortex. The area of cortex representing the hand is large, paticularly the palmar surface of the thumb and index finger. The lips also have a large area of representation for the complex sensory input required for speech and the mastication of food. The pathways to the sensory cortex will be described in Chapter 11.

Posterior to the somatosensory area is the **sensory association area** where integration of the primary sensations occur. An object, such as a key, held in the hand and moved about can be recognized even with the eyes closed. Information about the size, shape, weight, temperature and texture arriving at the cortex can be integrated with reference to memory, so that the exact nature of the object can be identified. This ability is called *stereognosis*, and is an important function of the parietal lobe. The integration of information from joints and muscles allows us to appreciate the position of all body segments at any moment. The sensory association area also receives input from visual and auditory areas of the cortex. Objects and sounds in the area around the opposite side of the personal space are identified and recognized. The overall input to the parietal lobe is therefore important in the interpretation of both the internal and external environment on the opposite side of the body.

The observation of patients with damage to the parietal lobe, particularly the right side, demonstrates the function of this lobe in body awareness. A patient may ignore one side of the body, for example, in dressing, only one leg may be put into trousers and only one arm into a sleeve. Objects in the opposite side of the visual space may be ignored. All this loss of body awareness affects movement, and the patient may have difficulty in finding his or her way around their home.

3.4.3 Temporal lobe

The temporal lobe, found beneath the temporal bone of the skull, is the 'hi-fi' area concerned with the reception, recording and replay of sound. This area also plays a part in long term memory.

The primary interpretation of *sound* occurs in the superior gyrus of the temporal lobe (see Fig. 3.8 above). Sound falling on the ear is transmitted by nerve impulses from the cochlea of the inner ear to the temporal lobe via relays in the brain. The primary auditory area receives these impulses, and depending on the frequency of the sound which evoked them, they are appreciated in certain areas in the superior temporal gyrus. The pathway is mainly crossed to the opposite temporal lobe, but each auditory area receives some impulses from both ears and links posteriorly with the auditory association areas which interpret the sound frequencies. In the dominant hemisphere, the extension of the auditory association area around the tip of the lateral sulcus and into the parietal lobe is known as *Wernicke's area* (see Fig. 3.8 above), which plays a role in receptive aspects of speech and language. Visual and auditory input from the written and spoken word are integrated in this area.

3.4.4 Occipital lobe

The occipital lobe lies beneath the occipital bone of the skull. All visual information transmitted from the eye is processed by the occipital lobe. Visual clues are important in keeping the balance of the body, in locating moving objects in the environment, and in the control of tools held in the hand. The primary visual cortex, known as the *striate cortex*, lies at the posterior pole of the occipital lobe and extends mainly on to the inner or medial surface, on either side of the calcarine sulcus (see Fig. 3.7b, p. 56). Impulses from the retina of both eyes arrive in each visual cortex, so that damage to this area can result in some loss of vision in both eyes.

The *prestriate cortex* is an association area which lies in front of the primary area. Processing of the visual information occurs in the prestriate cortex. Links to the parietal and temporal lobes are involved in the recognition of objects and faces, and in the understanding of the written word. Bilateral damage to the anterior end of the occipital lobe may result in the inability to recognize objects even though they can be seen clearly (visual agnosia).

A **summary** of the main functions of the lobes of the cerebral hemispheres can now be made.

1 Frontal lobe: movement, motor speech, logic, mood and emotion.

2 Parietal lobe: sensation in skin and muscles, appreciation of body image and personal space.

3 Temporal lobe: hearing, language and memory.

4 Occipital lobe: vision.

3.4.5 Right and left hemispheres

The functional asymmetry of the right and left hemispheres, first recognized by Broca in the mid nineteenth century has gained new interest from 'split brain' studies. Both hemispheres receive, process and integrate the same basic information. Only the *dominant* hemisphere (usually the left) contains the areas for speech and language, and this side is particularly concerned with analytical functions. The *non dominant* hemisphere plays a greater role in non verbal, creative activity requiring spatial appreciation (Fig. 3.10).

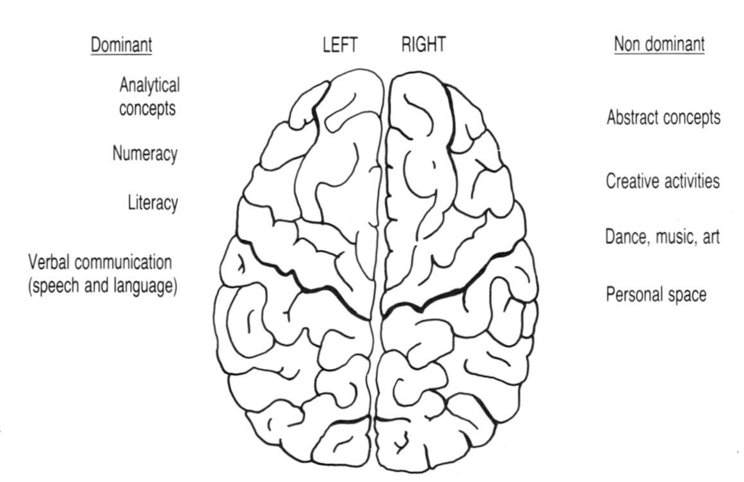

Fig. 3.10. Lateralization of function in the right and left hemispheres.

3.5 Basal ganglia

The basal ganglia (or basal nuclei) are found in the diencephalon at the base of the cerebral hemispheres and in the midbrain. Figure 3.11 has made the lateral cerebral cortex appear transparent to reveal three of the basal ganglia: the *caudate, putamen* and *globus pallidus.* (The caudate and putamen are sometimes called the *corpus striatum.* The putamen and globus pallidus are sometimes called the *lentiform nucleus.*)

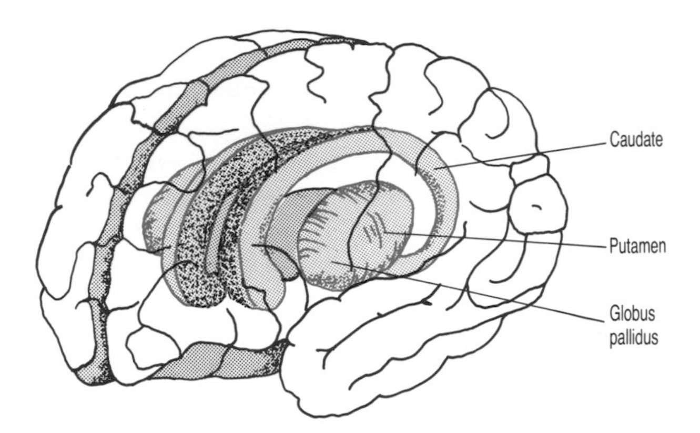

Fig. 3.11. Basal ganglia in position at base of cerebral hemisphere.

Lower down are found the *subthalamic nuclei,* and the basal nucleus in the midbrain is the *substantia nigra* (see Fig. 3.17a, p.68).

Nerve fibres linking the individual nuclei with each other form a complex interdependent system, which functions as a whole. Figure 3.12 shows some of the links between of the basal ganglia and also with the cerebral cortex and the thalamus. The main outflow is from the globus pallidus to the brain stem and the spinal cord.

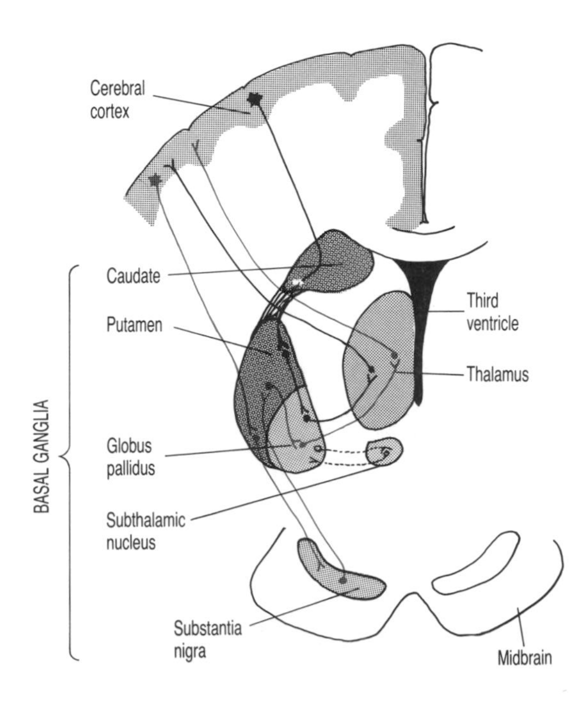

Fig. 3.12. Frontal section to show the cerebral cortex, thalamus and basal ganglia with some of the links between these areas.

Repetitive movements, such as walking, once learnt, can be performed automatically without reference to consciousness. The basal ganglia form part of the subconscious control system for such activities.

The exact functions of the basal ganglia are largely speculative. Clinical evidence suggests that the basal ganglia are essential for the planning and programming of the motor commands for movement. In diseases of the basal ganglia, movement disorders range from the rigidity and tremor characteristic of Parkinson's disease, to the uncontrollable purposeless actions when other nuclei are involved. The tremor in Parkinson's disease is present at rest and disappears with movement. There is difficulty in starting a movement, and the body may 'freeze' during the progress of a movement. For example, in walking, reaching the threshold of a door may prove an impossible barrier which cannot be crossed. The 'barrier' may be overcome if extra visual or verbal cues are given, and the movement can then proceed.

3.6 Thalamus

The thalamus lies deep in the grey matter at the base of the cerebral hemispheres, in the diencephalon, on either side of the slit-like third ventricle. Each thalamus is an oval mass of grey matter, surrounded by the basal ganglia and internal capsule (see Section 3.7). Figure 3.12 shows the thalamus on one side with the third ventricle medially and the basal ganglia laterally.

The thalamus is made up of many nuclei which project to specific areas of the cerebral cortex of the same side (ipsilateral). Sensory information arrives at the thalamus from all levels and both sides of the spinal cord and brain stem to be processed before being passed on to the sensory areas of the cerebral cortex. The projection fibres to the cortex radiate out like the spokes of an umbrella with the thalamus at its centre. Some of the incoming sensory information is **specific**, such as pressure from the skin of the hand grasping a handle, or the position of a limb from receptors in joint capsules. The thalamus processes this information, and activity is relayed to subconscious motor centres in the brain stem, as well as to consciousness in the cortex.

Non-specific sensory information arrives at the thalamus mainly from the reticular formation, which is a diffuse network of neurones found along the length of the brain stem (see Fig. 3.18, p. 70). The reticular formation acts to sift most of the sensory activity originating in the spinal cord, and alters the level of activity in the thalamus and cortex by excitation or inhibition of these areas.

The overall function of the thalamus is to act as a centre for sensation which is passed on to the cortex for further analysis. The thalamus also provides the essential background information for the motor system during movement.

3.7 Internal capsule

The internal capsule is an area of *white matter* containing the projection fibres to and from the brain stem and cerebral cortex. Figure 3.13 is a sagittal section through one cerebral hemisphere showing the corona radiata of projection fibres all converging at the

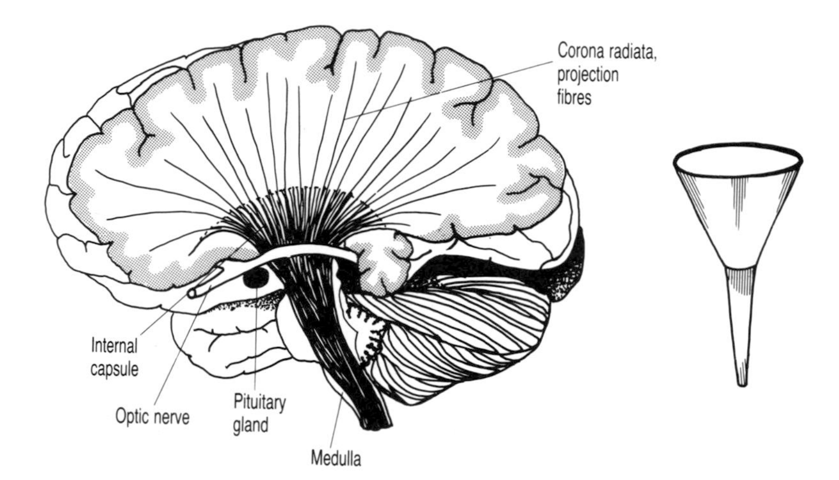

Fig. 3.13. Sagittal section to show the projection fibres converging into the internal capsule like a funnel.

base of the hemisphere to form the internal capsule. The same fibres continue in the midbrain as the cerebral peduncles (see Fig. 3.17a, p. 68). The narrow pathway of the internal capsule lies between the putamen and globus pallidus laterally, with the thalamus and caudate nucleus medially. Figure 3.14 is a horizontal section across the internal capsule to see its position in relation to these masses of grey matter in the diencephalon. Notice that in this view, the internal capsule is shaped like a boomerang. The fibres converge below to become the cerebral peduncles of the midbrain. The fibres in the internal capsule include motor and sensory nerve fibres supplying the muscles, skin and glands of the opposite side of the body (contralateral).

The result of damage to the internal capsule is extensive and may lead to loss of movement and sensation on the whole of the opposite side of the body. The term 'capsule' is a misleading one. The internal capsule is like the flattened stem of a funnel through

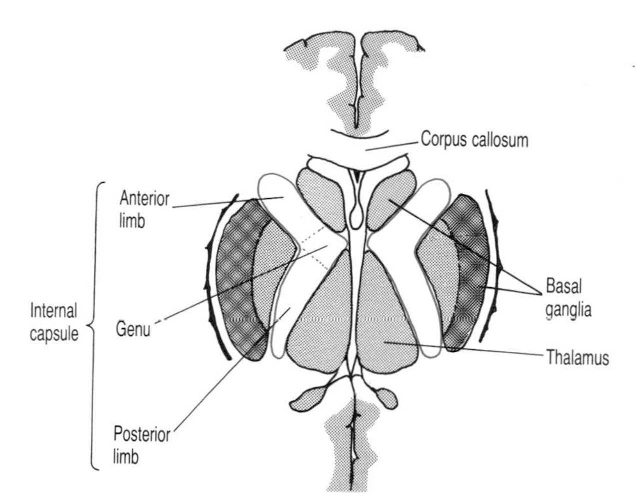

Fig. 3.14. Horizontal section at the level of the internal capsule.

which all the fibres to and from the cortex pass. If the stem is interrupted, the funnel becomes useless.

3.8 Hypothalamus and limbic system

The **hypothalamus** is smaller than the thalamus and lies beneath it in the floor of the third ventricle (Fig. 3.15a). Like the basement of a house with thermostats and stopcocks, the hypothalamus contains groups of neurones for the control of body temperature and body water. The output from the hypothalamus is to the autonomic division of the peripheral nervous system which controls: (i) the diameter of blood vessels; (ii) the secretion of sweat glands; and (iii) the release of hormones from the pituitary gland.

The hypothalamus is the highest control area for all the mechanisms that maintain homeostasis in the body. This area can be referred to as the 'visceral brain'.

The **limbic system** as a whole is a complex series of interconnected structures lying in the forebrain and midbrain linked by a large cable of white matter known as the fornix (Fig. 3.15b). The limbic forebrain includes an area of cerebral cortex (cingulate gyrus) lying medially above the corpus callosum, and the hippocampus lying buried in the temporal lobe. The loss of memory for events occuring after damage to the temporal lobe indicate that the hippocampal part of the limbic system has a function in the retention of memory. The limbic system, by its connections with the prefrontal cortex and the hypothalamus, is sometimes called the 'emotional brain'. Feelings of pleasure and anger produce physiological responses via activity in the hypothalamus to the autonomic nervous system (see Chapter 4). Links between the limbic system

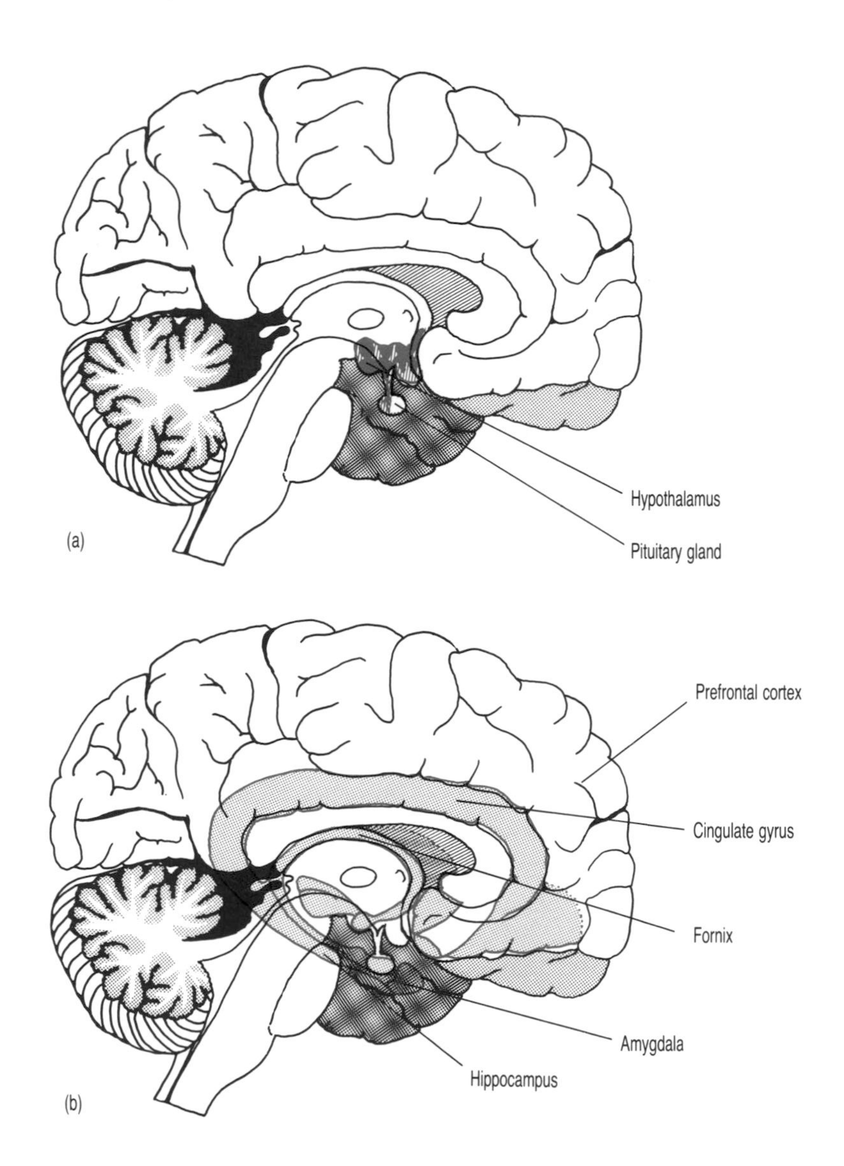

Fig. 3.15. Medial view of left side of brain to show the position of: (a) hypothalamus; and (b) limbic system.

and the motor system, particularly the basal ganglia, have been identified and suggest a function in motivation for movement.

3.9 Brain stem

When the cerebral hemispheres and the cerebellum are removed from the brain, the whole of the *brain stem* is revealed (Fig. 3.16). From above downwards, the brain stem consists of: *midbrain, pons* and *medulla oblongata,* with the *reticular formation* continuous through all three areas. The white matter of the brain stem contains bundles of fibres that form direct routes between the cerebral cortex and the spinal cord. Some of these routes also branch to link with nuclei of grey matter in the brain stem and also to link with the cerebellum. This can be compared to a system of roads with direct motorways and branch routes to other areas.

Functionally, the brain stem acts as a unit in regulating the balance of the body during movement.

3.9.1 Midbrain

The midbrain is the upper part of the brain stem joining the diencephalon above to the pons below.

The posterior (dorsal) surface of the midbrain is known as the roof or *tectum*, which contains four rounded elevations of grey matter, *the colliculi* seen in Fig. 3.16. Movements of the head and eyes in response to visual input from the eyes (the two superior *colliculi*), and sound from the ear (the two inferior colliculi), are integrated in the tectum. Below the inferior colliculi, stalks of white matter link the brain stem with the cerebellum on either side, the superior cerebellar peduncles.

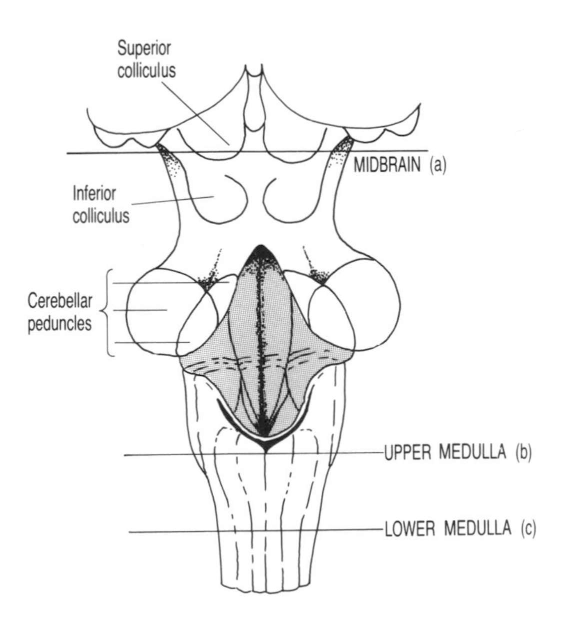

Fig. 3.16. Posterior view of the brain stem. Sections (a), (b) and (c) are shown in fig. 3.17.

The **red nucleus** is an important area of grey matter in the mid brain where pathways from the cerebellum relay before linking to the cerebral hemispheres above and the spinal cord below.

The **substantia nigra**, one of the basal ganglia described in Section 3.5, is a prominent area of grey matter seen in the section of midbrain in figure 3.17a. The remainder of the anterior and lateral part of the midbrain is known as the cerebral peduncles, which

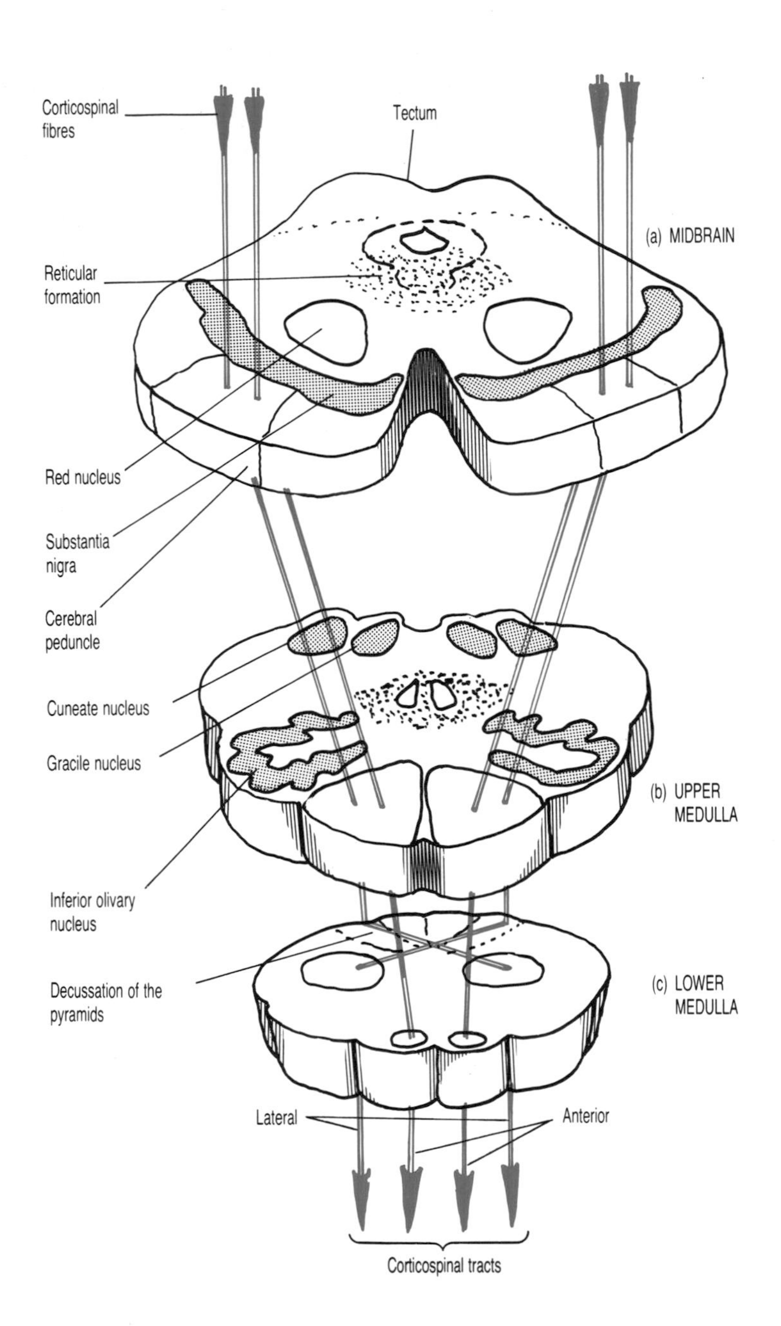

Fig. 3.17. Transverse sections through: (a) midbrain; (b) upper medulla; and (c) lower medulla.

contain ascending and descending fibres linking the cerebral cortex with the pons and spinal cord below.

3.9.2 Pons

The pons can be easily identified on the anterior side of the brain stem where a bulge is formed by the transverse fibres linking the two halves of the cerebellum. These fibres converge posteriorly to form a pair of stalks of white matter entering the cerebellum, the middle cerebellar peduncles.

The anterior pons contains the small diffuse *pontine nuclei* which form relay stations between the cerebral cortex and the cerebellum. The *vestibular nuclei* (of the eighth cranial nerve) are found in the pons and continue down into the medulla below (see Fig. 4.12, p. 94).

3.9.3 Medulla oblongata

The medulla is the cone shaped lower end of the brain stem that leads down into the spinal cord.

The white matter of the medulla contains ascending and descending pathways between the cerebral cortex and the spinal cord. Some of these routes cross to the opposite side in the medulla. The descending fibres of the corticospinal tracts cross on the anterior surface of the medulla forming a raised triangular area, the decussation of the *pyramids* (Fig. 3.17b and c).

In the upper medulla, lateral to the pyramids there is a smooth oval elevation, known as the *olive*. The major part of the olive is the inferior olivary nuceus, which relays input from the motor cortex and from the muscles into the cerebellum.

Posteriorly, the medulla contains the posterior column, an ascending sensory pathway from the spinal cord. The posterior column ends in the *gracile* and *cuneate nuclei* (Fig. 3.17b), and then crosses to the opposite side in the brain stem to reach the thalamus.

The lower stalks of the cerebellum leave the medulla posteriorly forming the *inferior cerebellar peduncles*.

• *LOOK at the transverse sections through the upper and lower medulla shown in Fig. 3.17b and c to identify the position of the gracile and cuneate nuclei, the olive and the decussation of the pyramids.*

3.9.4 Reticular formation

The reticular formation is a diffuse network of neurones in the core of the brain stem extending from the midbrain to the medulla. Some groups of neurones are collected together in nuclei, but in general the reticular formation, unlike other brain areas, consists of scattered cell bodies with the fibres lying in between. The network receives branches from all the ascending pathways through the brain stem except the medial lemniscus pathway (see Chapter 11).

Descending tracts from the reticular formation form part of the motor system and affect the activity of the lower motor neurones which supply muscles producing movement.

Neurones of the reticular formation in the midbrain project to all areas of the cerebral cortex and form the ascending reticular activating system (ARAS) shown in Fig. 3.18. The activity in these neurones affect our level of arousal and attention. The ARAS controls the 'body clock' which alternates the cycles of sleeping and waking.

The reticular formation forms the link system with the 'visceral brain' (hypothalamus and limbic system). In the lower pons and medulla it contains the 'vital centres' which control the heart from the cardiac centre, the blood pressure from the vasomotor centre and breathing from the respiratory centres. These vital centres respond to changes in blood composition and the activity in sensory nerves from receptors in blood vessels and the lungs. Continuous or intermittent activity in the centres results in stimulation of the heart muscle, the blood vessel walls, and muscles involved in breathing such as the diaphragm.

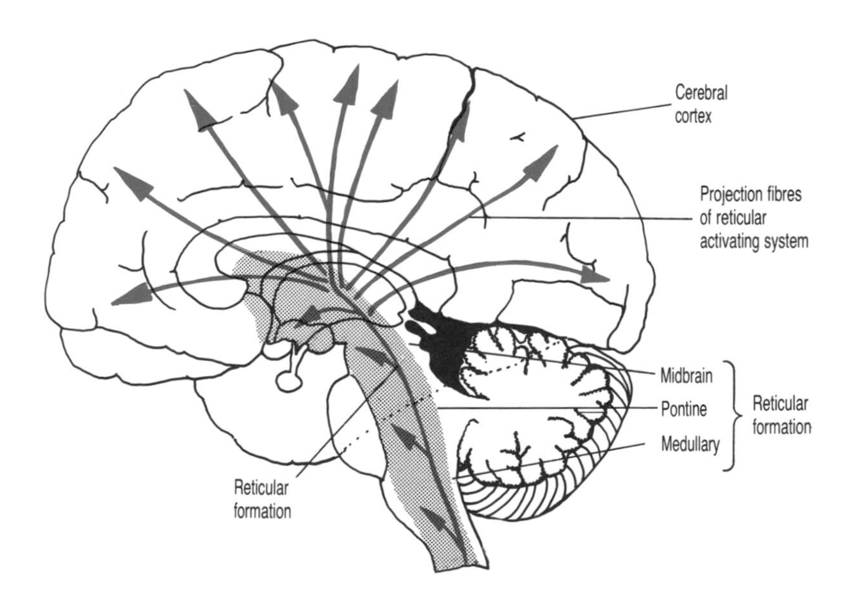

Fig. 3.18. Medial view of right side of brain to show the reticular formation.

3.10 Cerebellum

The cerebellum is formed from a posterior outgrowth of the brain stem in development and lies below the occipital lobe of the cerebral hemispheres in the posterior cranial fossa of the skull. The cerebellum has two halves which are connected by a central area known as the *vermis*. On the inferior surface of the vermis is the central *nodule* which extends on either side into the *flocculus*.

The outer layer of grey matter of the cerebellum is folded into uniform narrow gyri. The inner white matter forms a tree shape with the folded grey matter as the leaves. This was called the arboretum vitae (the tree of life) by the early neuroanatomists. The cerebellum also has a number of deep nuclei, the largest being the *dentate nucleus.* Three pairs of stalks of white matter, known as *peduncles,* connect the cerebellum to the brain stem as follows: (i) the superior peduncles with the midbrain; (ii) the middle peduncles with the pons; and (iii) the inferior peduncles with the medulla. Figure 3.19 shows diagrammatically the cerebellum and the three cerebellar peduncles linking it to the brain stem. Information on the position of the head from the vestibule of the ear enters the flocculonodular lobule, which is concerned with equilibrium and balance. The position of all body segments, at rest and during movement is relayed from the proprioceptors in muscles and joints to the anterior lobe of the cerebellum. Output from the motor cortex concerning intended movement reaches the cerebellum via nuclei in the pons and via the inferior olivary nucleus in the medulla.

The cerebellum cannot initiate movement, but it coordinates and regulates muscle activity on the same side of the body. There is no direct link between the cerebellum and the spinal cord. The output to the muscles relays in the brain stem nuclei (vestibular nucleus and red nucleus) and then into descending pathways to the

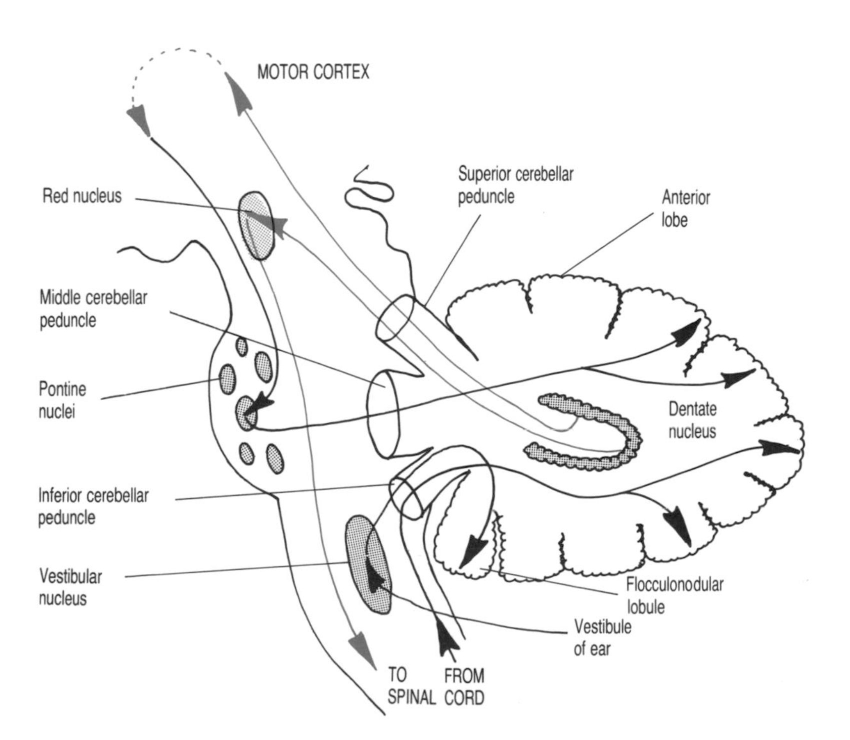

Fig. 3.19. Cerebellum with main pathways in and out (diagrammatic).

spinal cord (see Chapter 12). The output from the cerebellum makes appropriate adjustments in muscle activity during the progress of a movement from the start to the finish, and maintains the balance of the body during the movement.

Damage to the cerebellum usually results in marked *ataxia* (incoordination and unsteadiness) on the *same* side of the body. There is difficulty in walking on a straight line, or making accurate movements with the upper limb, such as placing the finger on the nose.

3.11 Summary of brain areas

1 Motor cortex: Performance of voluntary movement.

2 Premotor cortex: Bilateral movement.

3 Thalamus and **Sensory cortex**: Integration of background sensation.

4 Basal ganglia: Strategic planning, selection of motor commands for movement.

5 Hypothalamus and **limbic system**: Behavioural and visceral aspects of movement.

6 Cerebellum: Coordination, start and stop movements, balance and equilibrium.

7 Midbrain: Coordination of sight and sound in relation to movement.

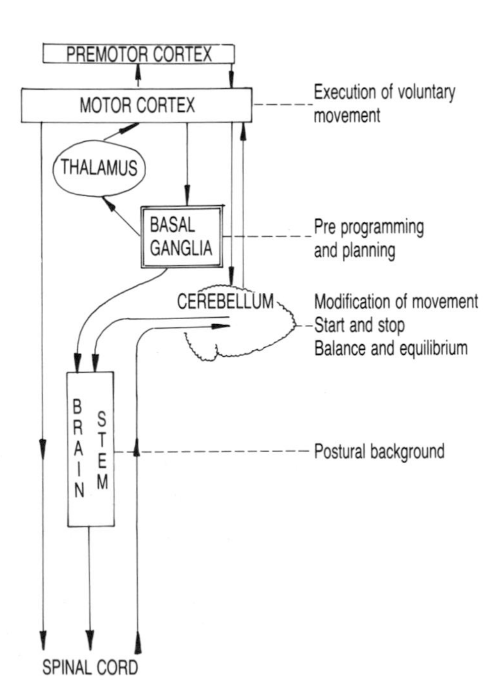

Fig. 3.20. Summary of brain areas.

8 Pons: Relay for information on head position, and for motor commands from the cortex to the cerebellum.
9 Medulla: Cardiovascular and respiratory control.
10 Reticular formation: Arousal and attention level during movement.
11 Brain stem (midbrain, pons and medulla): Postural background.

Figure 3.20 shows a summary of the main connections between the brain areas concerned in movement.

PART II/THE SPINAL CORD

The spinal cord appears to be a simple structure by comparison with the brain, but its role in the function of the central nervous system is nevertheless very important. Basic movement patterns of the limbs and trunk are integrated in the spinal cord. Most of the body's sensory information is received by the spinal cord and is passed on to higher levels in the brain.

13.12 The position and segmentation of the spinal cord

The embryonic neural tube grows in diameter and length with the bony vertebrae developing round it. The internal cavity of the tube remains as a small central spinal canal containing cerebrospinal fluid. A pair of spinal nerves grow out from the developing spinal cord between adjacent vertebrae. The segment of the cord that gives rise to each pair of spinal nerves is named in relation to the corresponding vertebra, for example the segment lying under the first thoracic vertebra is known as T1. There are 31 segments in the spinal cord, named as follows: eight cervical (C1–C8), twelve thoracic (T1–T12), five lumbar (L1–L5), five sacral (S1–S5) and one coccygeal.

There are eight cervical segments: The first pair of cervical nerves lie between the skull and the first cervical vertebra, and C1–C7 all emerge above the corresponding vertebra.

The eighth pair of cervical nerves emerge between the seventh cervical and first thoracic vertebrae, so that all the nerves below C7 emerge below the corresponding vertebra.

The vertebral column grows in length more rapidly than the spinal cord, so that in the adult the lower end of the spinal cord lies at the level of the disc between the first and second lumbar vertebrae. The lower end tapers to a point and is attached by a strand of connective tissue (filum terminale) to the lower end of the sacrum and to the coccyx.

• *LOOK at an articulated skeleton, or the individual vertebrae loosely strung together. Put a piece of plastic tubing 45 cm long into the vertebral canal and note where the lower end lies. The tubing should be thicker towards the upper and lower ends to truly represent the spinal cord. Note that the vertebral canal is larger in the cervical and upper lumbar regions to accomodate the cervical and lumbar enlargements of the spinal cord.*

• *IDENTIFY the intervertebral foramina between adjacent vertebrae where the spinal nerves emerge. Starting at the skull, see how the spinal segment and pair of spinal nerves C8 appear.*

• *LOOK at Fig. 3.21, a sagittal section through the spinal cord and vertebral column with the spinal nerves emerging from the cord. The cervical and lumbar enlargements accomodate the large number of neurones that supply the upper and lower limbs respectively.*

• *LOOK at Fig. 3.22 to see a transverse section of the spinal cord lying in position surrounded by the bony vertebra. The right and left sides of the spinal cord are symmetrical and are separated by two longitudinal sulci, one anteriorly and one posteriorly.*

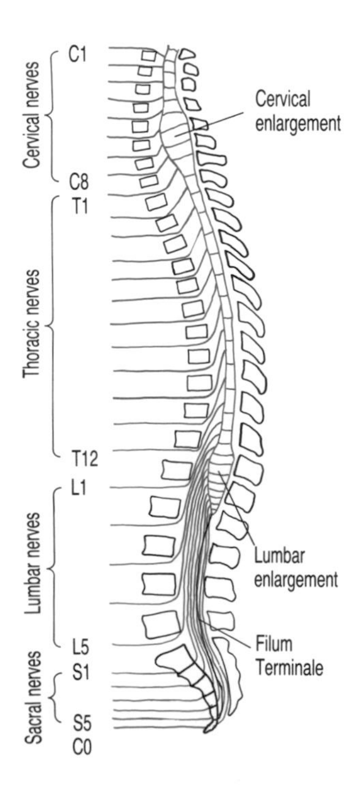

Fig. 3.21. Spinal cord and spinal nerves in position in relation to the vertebral column.

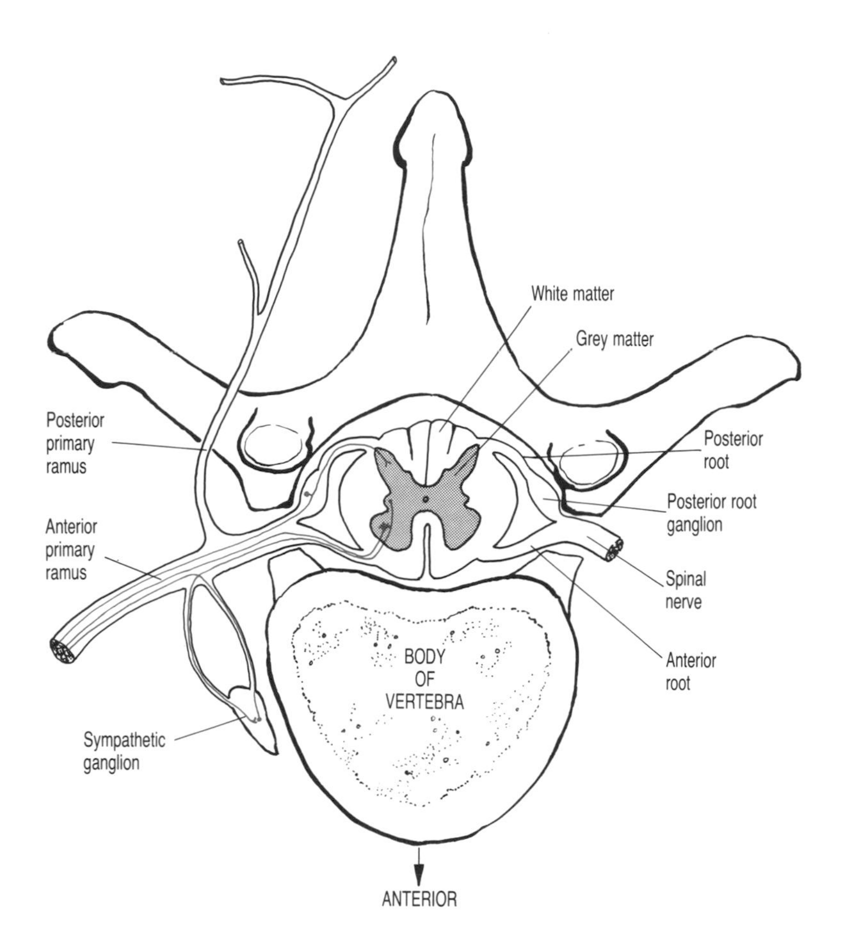

Fig. 3.22. TS spinal cord surrounded by vertebra.

3.13 Meninges

The spinal cord is protected externally by three membranes of connective tissue which are also continuous over the surface of the brain. The three layers from superficial to deep are the dura mater, arachnoid mater and pia mater (Fig. 3.23).

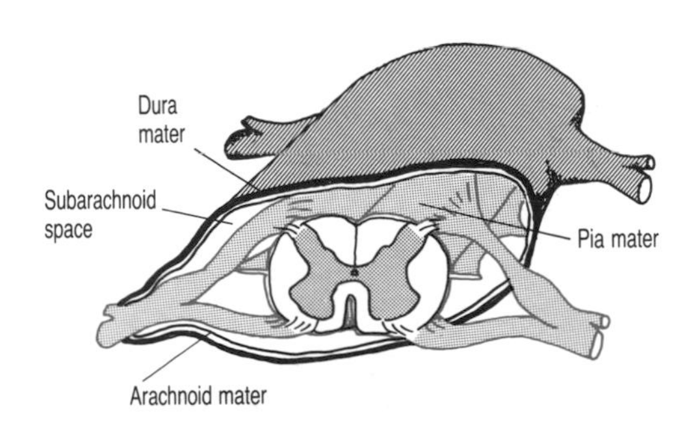

Fig. 3.23. Meninges surrounding the spinal cord.

The **dura mater** is a thick layer densely packed with collagen fibres and some elastin which lines the cranial vault of the skull and the vertebral canal of the spine as far down as the level of the second sacral vertebra.

An **epidural space** lies between the dura mater and the periosteum and ligaments of the vertebral column. Anaesthetics injected into the epidural space of one spinal segment may spread upwards or downwards to affect the spinal nerves emerging from adjacent segments.

The **arachnoid mater** is a thin membrane lying in close contact with the dura mater, separated by a thin film of fluid. Deep to the arachnoid mater is the subarachnoid space containing cerebrospinal fluid. The arachnoid mater ends at the level of the second sacral vertebra. This means that between the third lumbar vertebra (where the spinal cord ends), and the second sacral vertebra, cerebrospinal fluid can be extracted for examination without risk of damaging the spinal cord. This procedure, a 'lumbar puncture', is usually done by inserting a blunt needle between the laminae of the third and fourth lumbar vertebrae (Fig. 3.24).

The **pia mater** is a loose membrane of connective tissue which covers the whole surface of the brain and spinal cord, and dips down into all the sulci. There is a rich network of blood vessels associated with the pia mater providing a major part of the blood supply to the brain and spinal cord.

The meninges protect the spinal cord and brain from infection, and the cerebrospinal fluid acts as a shock absorber.

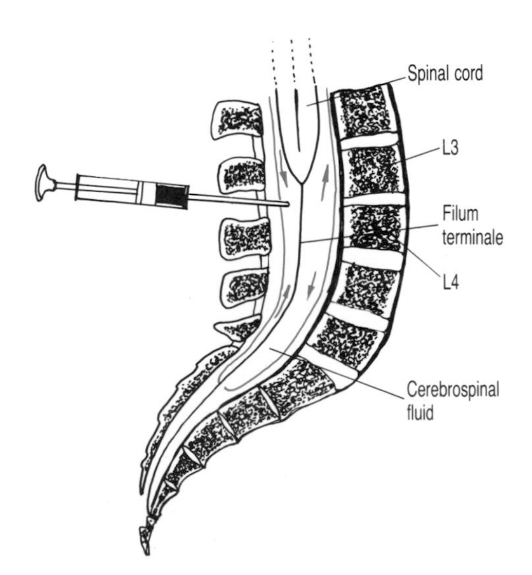

Fig. 3.24. Lower end of spinal cord showing the position for a lumbar puncture.

3.14 Organization of grey and white matter

The internal structure of the spinal cord is organized into an H shaped central core of grey matter with anterior and posterior horns, surrounded by white matter. Transverse sections of the spinal cord at different levels can be seen in Fig. 3.25. The anterior horn is large in the cervical and lumbar regions where the lower motor neurones supplying limb muscles are found. The grey matter in all the thoracic segments, the lumbar segments 1 and 2, and sacral segments 2, 3 and 4, has a lateral horn where the cell bodies of neurones are found, which form part of the autonomic nervous system supplying organs, glands and blood vessels. The white matter containing groups of nerve fibres carries impulses up (*ascending tracts*) or down (*descending tracts*) the spinal cord (Fig. 3.26b).

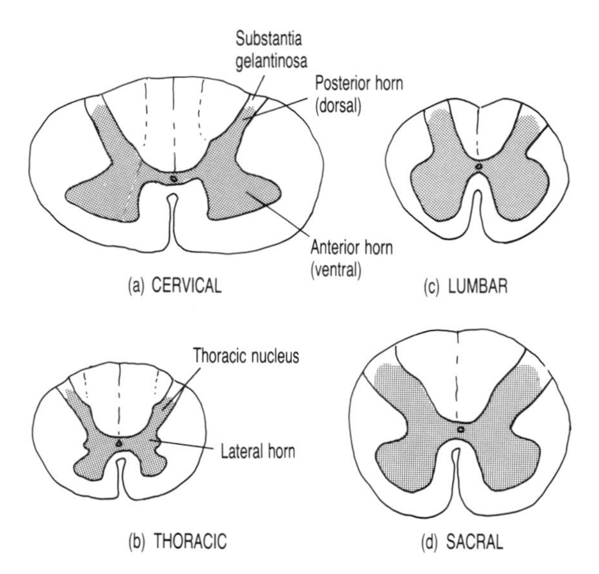

Fig. 3.25. Transverse sections of spinal cord; (a) cervical; (b) thoracic; (c) lumbar; and (d) sacral.

The white matter of the spinal cord is largest at the upper end and smallest at the lower end. Fibres leave the descending tracts at each segment to enter the grey matter. Sensory neurones in spinal nerves link (either directly or after synapsing in the posterior horn), into the ascending tracts of the white matter at every level. Figure 3.26a shows how the posterior column of white matter increases in size as it receives fibres from successive spinal nerves.

3.14.1 Grey matter

The nerve cells which form the grey matter can be divided into the following.

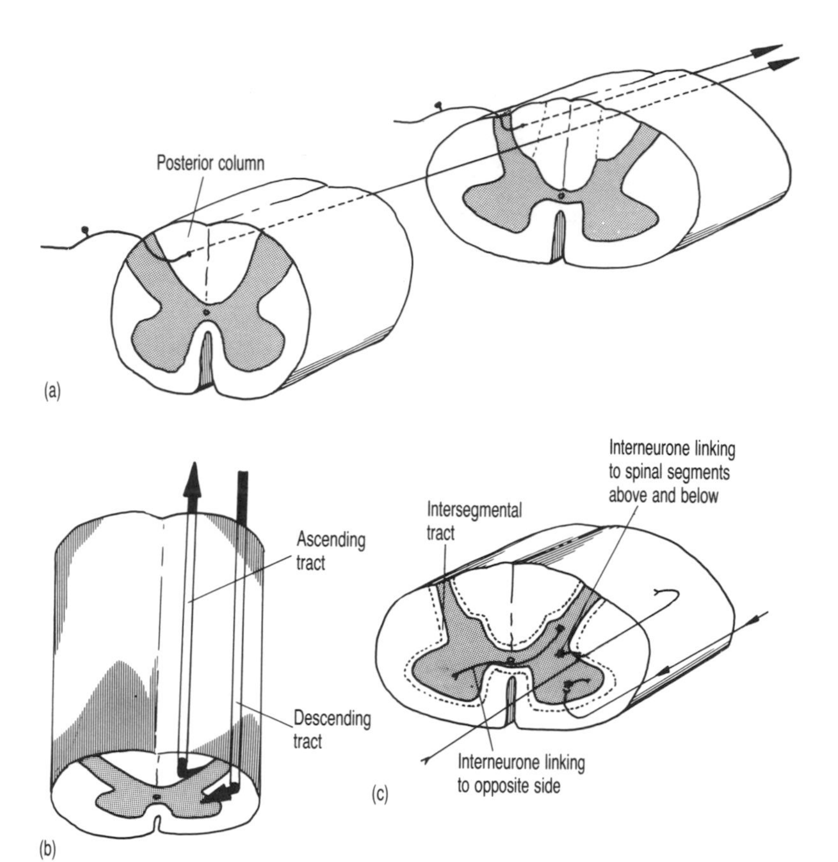

Fig. 3.26. Transverse sections of the spinal cord to show:
(a) fibres entering the white matter to form ascending tract in spinal cord;
(b) ascending and descending tract in spinal cord; and
(c) interneurones.

1 Motor cells whose axons are distributed to all parts of the body including the muscles. The motor nerve cells are largely found in the *anterior* (ventral) horn and include the alpha and gamma motor neurones described in Chapter 1.
2 Tract cells which receive impulses from the *sensory* neurones entering the cord at all levels from the skin, muscles and joints. The tract cells are found in the *posterior* (dorsal) horn, and their axons enter the ascending tracts of the white matter.
3 Interneurones lie in the intermediate area between the posterior and anterior horns. The interneurones receive impulses from sensory neurones entering the posterior horn, and from descending tracts in the white matter from the brain. In bilateral activities, interneurones which link across the cord are important. All the segments of the spinal cord are connected by interneurones whose fibres lie in the intersegmental tract (see Section 3.14.2), so that activity can spread to other spinal levels above and below.

Figure 3.26c shows the position of the different types of interneurones in the spinal cord.

The motor neurones of the *anterior horn* are organized into pools of neurones (see Fig. 1.14, p. 23) which supply particular groups of muscles acting on one joint. There is some evidence that the neurones supplying more distal muscles in the limbs lie lateral and posterior to those supplying proximal muscles. The mapping of neurone pools in the human spinal cord is still uncertain.

The *posterior horn* has been divided into areas with particular functions. The *substantia gelatinosa* is the most distal region where incoming fibres from the skin relay before entering ascending pathways are found. More centrally in the posterior horn, cell bodies are collected in the *thoracic nucleus* (Clarke's column). The axons of cells in the thoracic nucleus carry sensory information about the length of muscles and the position of joints in ascending tracts to the cerebellum. The thoracic nucleus is present below T1 in the spinal cord and is absent in the cervical cord.

3.14.2 White matter

The white matter is divided for description into three columns or funiculi; posterior (dorsal), lateral and anterior (ventral). The ascending and descending tracts form the white matter. Each tract links two particular areas of the central nervous system and is usually named after these two areas. For example, the spinothalamic tract is an ascending pathway that links the spinal cord with the thalamus; the corticospinal tract is a descending route from the cerebral cortex to the spinal cord. Figure 3.27 shows the position of

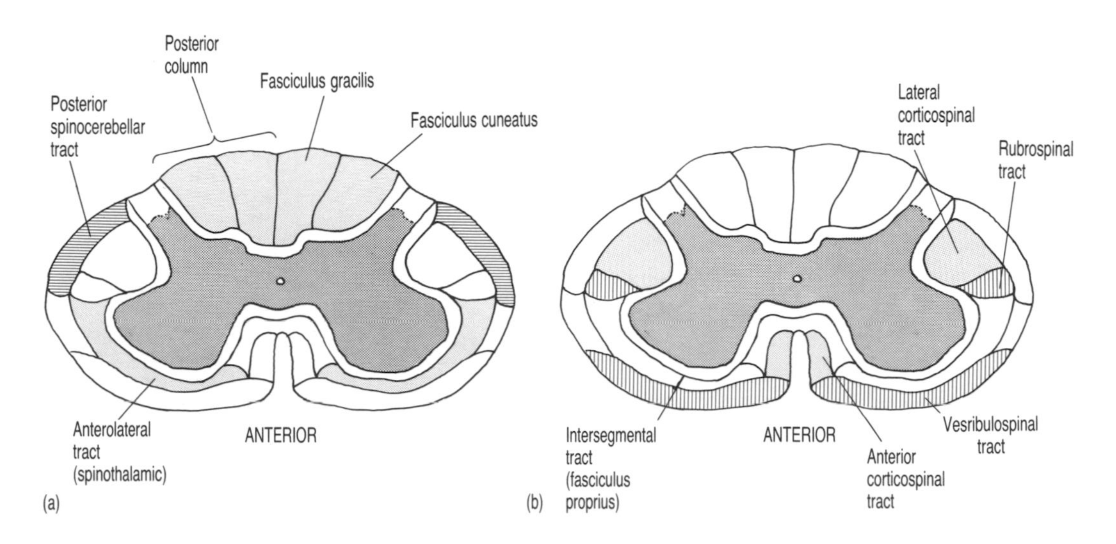

Fig. 3.27. TS spinal cord. Position of main tracts in white matter: (a) ascending tracts; and (b) descending tracts.

the main tracts, with the ascending tracts in Fig. 3.27a and the descending tracts in Fig. 3.27b. It must be remembered that all the tracts are present on both sides of the spinal cord. Detail of the function of these tracts will be discussed in Chapters 11 and 12. At this stage the general way in which the white matter is organized should be appreciated. Each tract is rather like a cable of wires, but evidence indicates that there is some overlap between the individual tracts.

A narrow band of white matter surrounds the whole of the central core of grey matter. The fibres of this band, known as the *fasciculus proprius* or *intersegmental tract*, connect different segments of the spinal cord (Fig. 3.26c). The fibres vary in length, some pass from one segment to another and others pass nearly the whole length of the cord, branching up, down and across the cord.

3.14.3 Spinal reflex movements

A spinal reflex action is one that involves the stimulation of receptors, impulses passing into the spinal cord in sensory neurones, and relay to motor neurones to produce a response. The stretch reflex, described in Chapter 1 is an example of a spinal reflex. In other spinal reflexes, when several groups of muscles are active in the response, there is spread of impulses to more than one

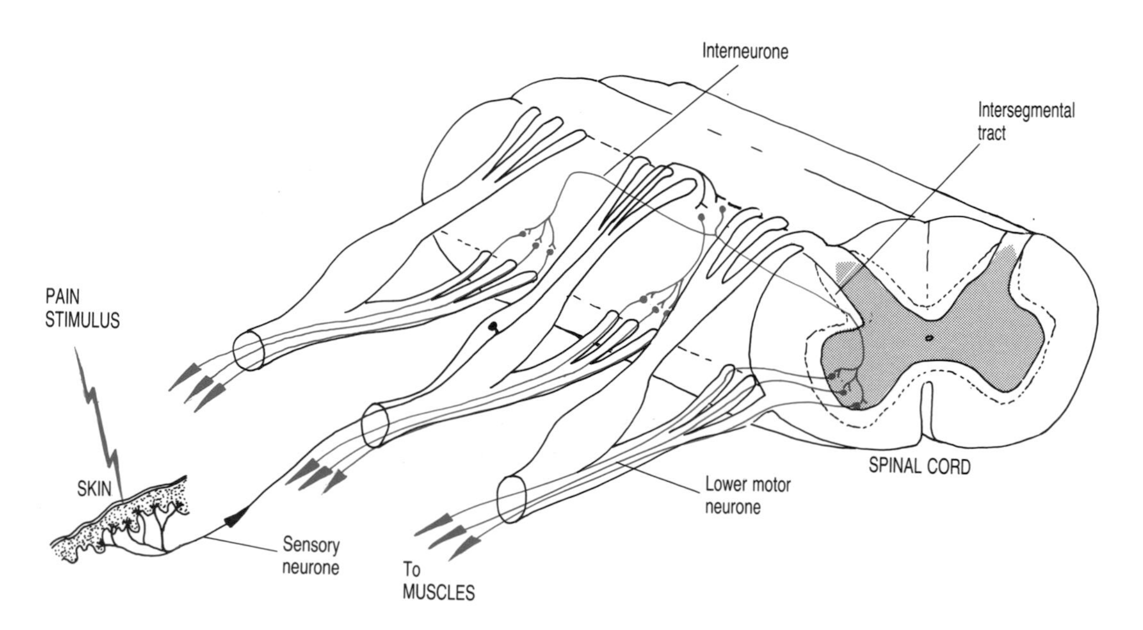

Fig. 3.28. Flexor withdrawal reflex. Arrangement of neurones to show spread of activity to three spinal segments.

spinal segment. The intersegmental tract of the spinal cord is then involved. An example is the *flexor withdrawal reflex*, which withdraws a limb away from a painful stimulus. In this reflex, the whole of the damaged limb is flexed, and changes in the muscles of the opposite limb prevent the body from falling over. Impulses pass via interneurones in the intersegmental tract to stimulate movement in the other limbs (Fig. 3.28). This flexor withdrawal reflex is experienced when we tread on something sharp that stimulates one foot. The leg on that side is flexed and at the same time the opposite leg extends to help us balance. If the stimulus is strong, the arms may be raised as well.

The spinal cord alone is concerned with basic movement responses to external stimuli. These movements can be seen in the new born baby when the influence from higher levels of the nervous system has not yet been developed. As the control of activity by higher centres in the nervous system develops, these basic movements are modified, and more complex movements become possible. Nevertheless, the spinal cord remains the centre for the final pathway to the muscles and for receiving the sensory information which forms the background to all movement.

At the end of this chapter, you should be able to:

1 Outline the development of the central nervous system.

2 Describe the organization of neurones in the brain and spinal cord into grey and white matter.

3 Describe in outline the meninges and the circulation of cerebrospinal fluid.

4 Identify in position the main brain areas and spinal cord as seen in sections of the central nervous system made at various levels.

5 Describe the overall structure and discuss the main functions of the following areas of the central nervous system:

(a) cerebral hemispheres,
(b) basal ganglia,
(c) thalamus,
(d) hypothalamus and limbic system,
(e) brain stem — midbrain, pons, medulla and reticular formation.
(f) cerebellum,
(g) spinal cord.

4 / Link Systems. Peripheral Nervous System

The peripheral nervous system provides the link between the central nervous system and all the parts of the body. Links are required to activate muscles for movement, and to monitor ongoing changes in the muscles as movement proceeds. Links from the skin and the sense organs give information about changes in the environment around the body. Links to blood vessels, glands and organs regulate the internal environment to meet the metabolic demands of the muscles. The peripheral nervous system provides all these links.

The peripheral nerves, containing sensory and motor nerve fibres, are arranged in a bilateral system of paired nerves leaving the central nervous system. Incoming signals are conducted to the brain and spinal cord, and after processing, the responses are carried out by outgoing signals in the peripheral nerves.

The axons found in the peripheral nerves can be divided into two functional categories. The *somatic* component consists of all the sensory and motor axons associated with activity in the muscles, the joints and the skin. The *visceral* component is all the axons carrying impulses to the glands, organs and blood vessels. The visceral nerve fibres are part of the autonomic nervous system.

The importance of the peripheral nervous system during movement is to provide the structural framework for the following:

1 The monitoring of changes in the external environment.

2 The monitoring of changes in length and tension in the muscles.

3 The activating of muscle fibres.

4 The regulation of the internal environment to maintain the oxygen demands of the muscles.

Damage to the peripheral nervous system nerves at any point, from their origin in the central nervous system to their terminations inside the muscles, will result in loss of muscle function. Trophic changes, such as flushing and dryness of the skin, will also occur if the visceral fibres are damaged.

4.1 The position and location of cranial and spinal nerves

The paired nerves of the peripheral nervous system leaving the central nervous system are divided into the *cranial nerves* leaving the brain; and the *spinal nerves* leaving the spinal cord.

The **cranial nerves** consist of the twelve pairs of nerves whose cell bodies are located in the brain. Seen most clearly in a ventral view of the brain, the pairs of cranial nerves appear to be formed at irregular intervals. The final position of each pair is a result of the changes in development of the head and neck, and the folding of the embryonic neural tube which forms the brain (Fig. 4.1).

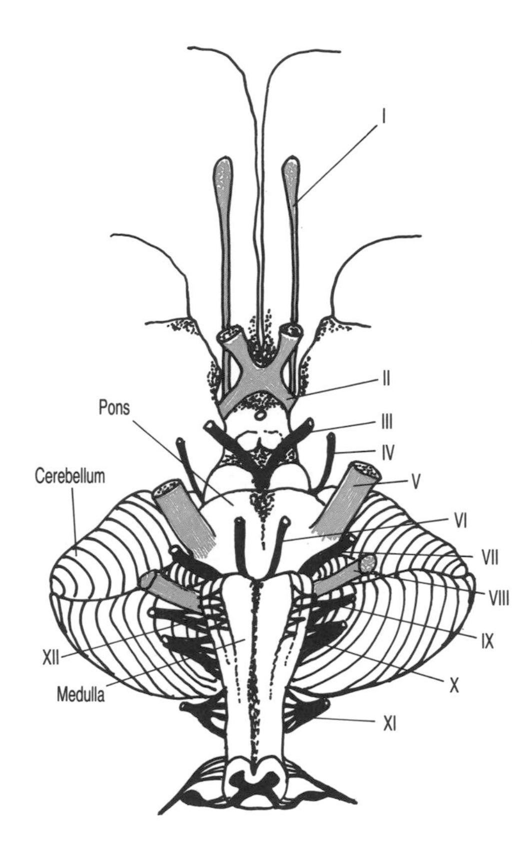

Fig. 4.1. Ventral view of the brain showing the origins of the twelve pairs of cranial nerves.

The **spinal nerves** consist of the 31 pairs of nerves leaving the spinal cord. Each pair of spinal nerves emerges from the spinal canal of the vertebral column between adjacent vertebrae at the intervertebral foramina. The lower end of the spinal cord in adults lies at the level of the disc between the first and second lumbar vertebrae (see Chapter 3, Section 3.12). The lower spinal nerves therefore lie in the spinal canal below this level before emerging at their corresponding level. This sheath of lumbar and sacral nerves is known as the *cauda equina.*

• *LOOK at an articulated skeleton and return to Fig 3.21 and 3.22 (pp. 74–75) to revise the emergence of the 31 pairs of spinal nerves from the vertebral column.*

4.2 Spinal nerves

Each spinal nerve begins at the spinal cord by two roots: the *anterior* (ventral) *root,* and the *posterior* (dorsal) *root.* Each root consists of a series of rootlets which eventually join. Illustrations of pathways in the nervous system represent the roots as single trunks for clarity.

The **anterior roots** consist of axons that grow out from multipolar nerve cells in the spinal cord, the lower motor neurones. Axons from the anterior horn cells of the spinal cord are large diameter alpha fibres which activate skeletal muscle fibres, and smaller gamma fibres innervating the intrafusal fibres of the muscle spindles lying within the skeletal muscles. The autonomic motor fibres have their cell bodies in the lateral horn of certain segments of the spinal cord (see Section 4.5).

The **posterior roots** develop in a different way. A ridge of cells on each side of the neural tube in the embryo forms a pair of ganglia (cells) for each segment of the spinal cord. Fibres grow centrally from each ganglion into the spinal cord, and also laterally to join the fibres of the anterior root. The neurones of the posterior root are all sensory, carrying information from the receptors in the skin, muscles and joints. The cell bodies lie in the posterior root ganglion, (isolated from the hundreds of synaptic connections possible for cell bodies of neurones in the grey matter of the spinal cord). Axons of the sensory neurones enter the spinal cord, branch to segments of the cord above or below, or turn into the posterior white matter to reach the brain stem before synapsing.

Remember the following:

1 Anterior roots are motor.
2 Posterior roots are sensory.
3 Injury to anterior roots leads to loss of movement.
4 Injury to posterior roots leads to loss of sensation.

The spinal nerve is the common nerve trunk formed by the anterior and posterior roots joining distal to the posterior root ganglion.

4.2.1 Divisions of the spinal nerve

The complete spinal nerve is only a few millimetres long. It then divides, giving off a branch posteriorly known as the posterior (dorsal) primary ramus. This first branch passes close to the articular processes of the vertebra. The posterior primary rami of all the spinal nerves together supply the deep muscles of the back and the skin covering them (Fig. 4.2).

Each spinal nerve continues as the anterior (ventral) primary ramus. These branches of the spinal nerves supply all the skin and muscles other than those supplied by the posterior primary rami.

Fibres of the autonomic system in the anterior primary ramus connect with ganglia which lie on the sides of the bodies of the vertebrae, by grey and white rami. Figure 3.22 shows these connections (p. 75). The white rami carry preganglionic myelinated fibres

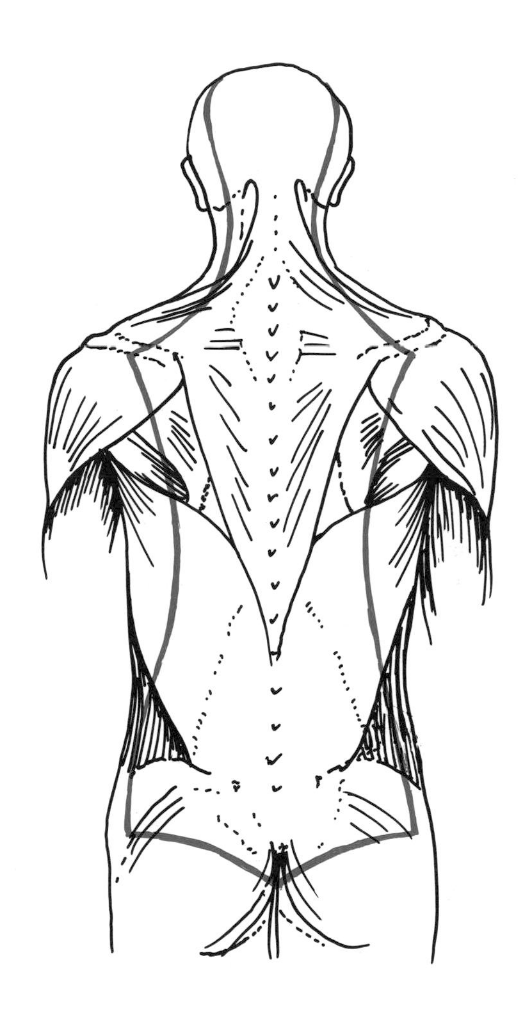

Fig. 4.2. Posterior view of the head and trunk. Area of skin supplied by posterior primary rami of the spinal nerves.

from the spinal nerve to the ganglion. Grey rami carry non-myelinated post ganglionic fibres from cell bodies in the ganglion to the spinal nerve. The autonomic fibres will be considered again in Section 4.5.

The content and distribution of the anterior primary ramus from this point is called the spinal nerve. Each spinal nerve contains all the somatic and visceral fibres which supply the corresponding body segment. The thoracic spinal nerves are like the basic plan described. The other spinal nerves show considerable mixing, branching and joining before passing as peripheral nerves to their destination. This regrouping of fibres occurs in a *plexus,* for example the brachial plexus is shown in Fig. 7.1 (p. 156). Fibres from one spinal nerve may eventually lie alongside other fibres from a different spinal nerve in one peripheral nerve. Figure 4.3 shows how two spinal nerves (C5 and 6) contribute fibres to the peripheral nerve supplying the biceps muscle in the arm.

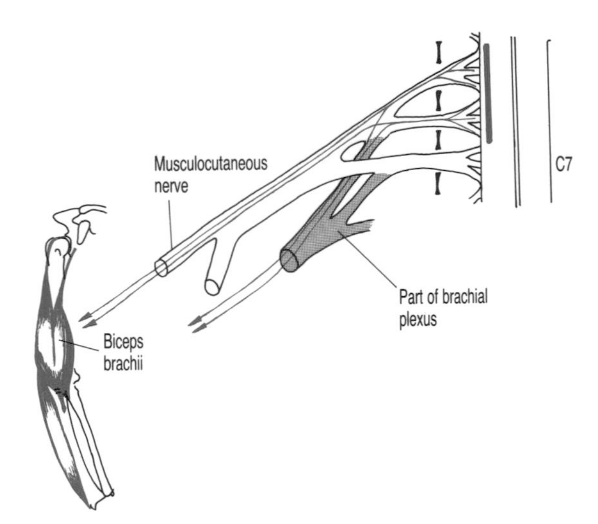

Fig. 4.3. Formation of a peripheral nerve (musculocutaneous) from two spinal segments (C5 and C6).

There are four major plexi formed by the anterior primary rami of the spinal nerves.

1 C1–C4 is the cervical plexus to the muscles of the neck.
2 C5–T1 is the brachial plexus to the muscles of upper limb.
3 L1–L4 is the lumbar plexus to the muscles of the thigh.
4 L4–S4 is the sacral plexus to the muscles of the leg and foot.

Figure 4.4 shows the plexi formed by the spinal nerves.The lumbar and sacral plexus can be considered together as the lumbosacral plexus supplying the whole of the lower limb.

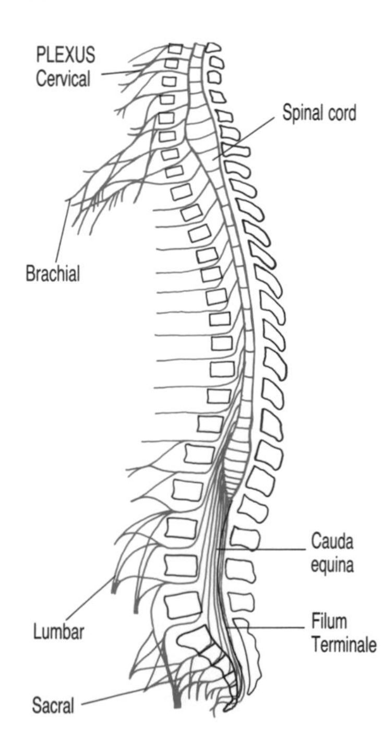

Fig. 4.4. Spinal cord in relation to the vertebral column. Spinal nerves forming the cervical, brachial, lumbar and sacral plexi.

4.2.2 Dermatomes and myotomes

A **dermatome** is an area of skin supplied by all the sensory nerve fibres of one spinal nerve. An example of a dermatome is a band of skin around the trunk innervated by the sensory fibres of the second pair of thoracic nerves. A map of the dermatomes of all the spinal nerves is seen in Fig. 4.5. In the trunk, the dermatomes form a series of bands, one for each spinal nerve from T1 to L1 in order.

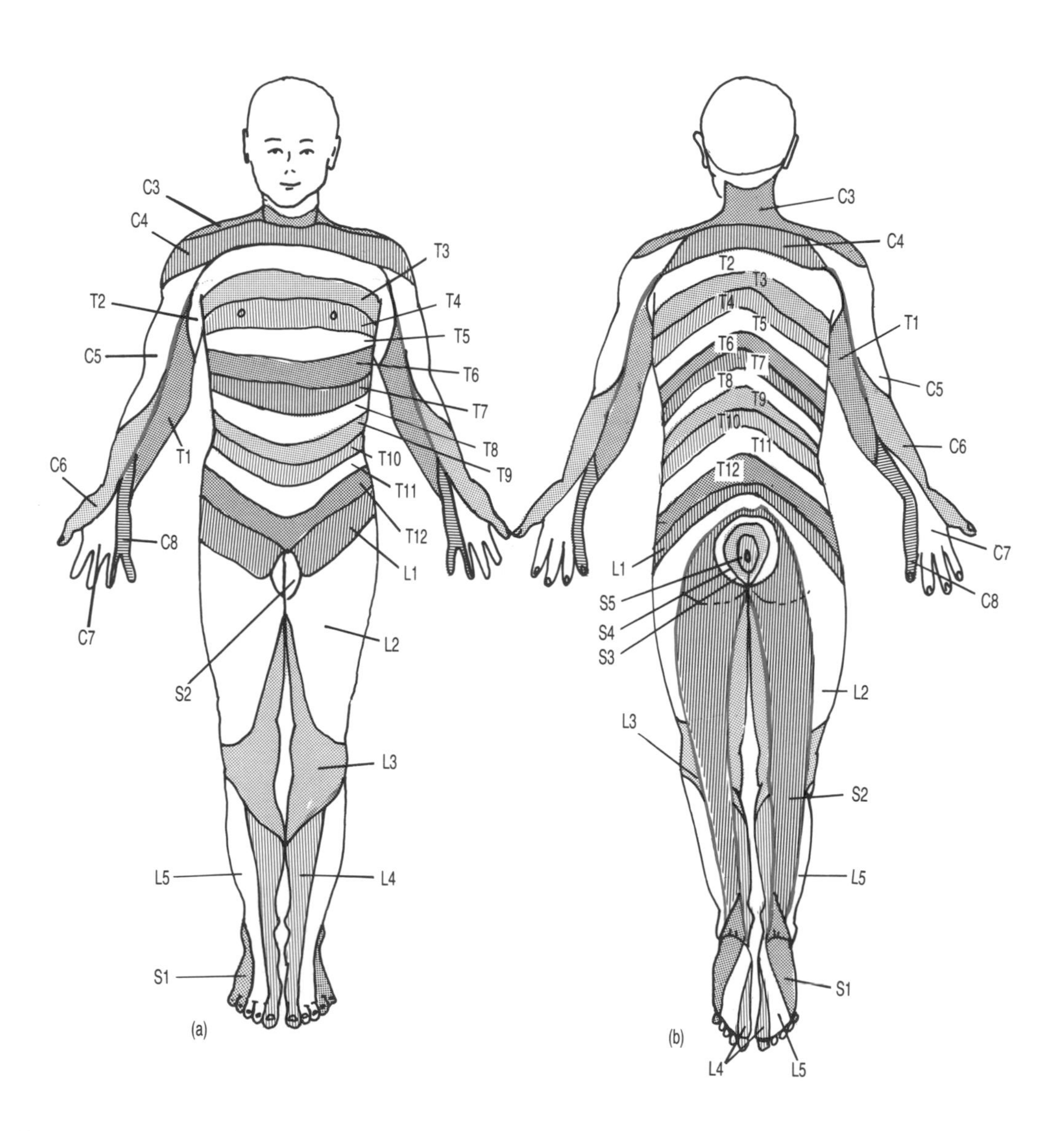

Fig. 4.5. Dermatomes. The cutaneous distribution of the spinal nerves in (a) anterior, and (b) posterior view.

There is some overlap, and each dermatome may receive fibres from three or four spinal nerves. In the limbs, the arrangement of dermatomes is more complicated. Each limb develops from a bud, which grows out in the embryo, and some dermatomes are carried to the ends of the limb. C7 and C8 are carried in this way to the hand, while L5, S1 and S2 reach the skin of the foot. From the diagram, you can see that damage to the posterior roots of spinal nerves in the upper part of the neck (C5, C6) will give loss of sensation around the shoulder, while severance of lower roots (C7, C8) will affect sensation in the hand.

A **myotome** is all the muscles supplied by one spinal segment and its pair of spinal nerves. For example, fibres from the first thoracic nerve (T1) are distributed to a long finger flexor in the forearm and some of the intrinsic muscles of the hand. Each individual muscle, however, receives fibres from two or three spinal nerves (Fig. 4.3), so that injury to one spinal segment may only have a limited effect on one particular muscle.

4.3 Peripheral nerves. Composition and distribution

Branches of spinal nerves are distributed to all parts of the body as peripheral nerves. The structure of a peripheral nerve is shown in Fig. 4.6. Note that half the total bulk of a nerve is connective tissue, and each nerve has its own blood supply, which branches along the length of the nerve in both directions.

Note: peripheral nerves are 'mixed', i.e. contain motor, sensory and autonomic fibres.

In their course along a limb, peripheral nerves branch to enter muscles, joints, walls of blood vessels and connective tissue. Some

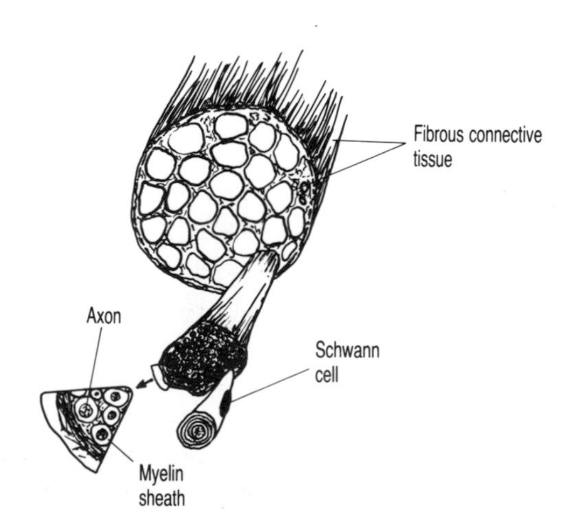

Fig. 4.6. Transverse section of a peripheral nerve showing the connective tissue and axons.

of the peripheral nerves pierce the deep fascia of connective tissue around the body deep to the skin. These nerves are *cutaneous nerves* which supply all the structures in the skin.

4.3.1 Muscular branches

The nerves that enter muscles contain motor nerve fibres which supply the muscle fibres; without these motor fibres, the muscles cannot function, and no reflex or voluntary movement is possible if the lower motor neurones to a muscle are damaged. The nerves which enter muscles also contain sensory nerve fibres from the proprioceptors in the muscle; the proprioceptive fibres provide the cental nervous system with information on the length and tension of the muscles, and the angulation of the joints.

4.3.2 Cutaneous branches

The cutaneous nerves contain sensory fibres from the various receptors in the skin. One nerve often connects with other cutaneous nerves in the same area, so that severing one cutaneous nerve may reduce sensation in the area but does not abolish it.

There are also motor fibres in cutaneous nerves which are part of the autonomic system supplying blood vessels, sweat glands and the small muscles at the base of hair follicles. Damage to the vasomotor fibres leads to flushing and dryness of the skin. One example of a cutaneous nerve is the superficial terminal branch of the radial nerve. This nerve pierces the deep fascia above the wrist, and branches to supply an area of skin on the back of the hand. Damage to this nerve usually results in a very small area of sensory loss due to overlap from the other cutaneous nerves in the hand.

4.3.3 Injury to peripheral nerves

The changes in movement and sensation that occur when a nerve is damaged may be different in every case. If the nerve is stretched or crushed, but no axons are actually severed, there may still be some conduction of nerve impulses, but it will be poor due to swelling or haemorrhage (Fig. 4.7a): some loss of movement and muscle tone occurs, but the sensation of touch and pain remains: recovery may begin after a few days. If the axons are severed, there may be complete loss of movement and sensation, as well as a flushing of the skin (as a result of the loss of the vasomotor fibres to blood vessels): recovery will depend upon the extent of involvement of the sheaths round the axon, Schwann cell sheath and endoneurium

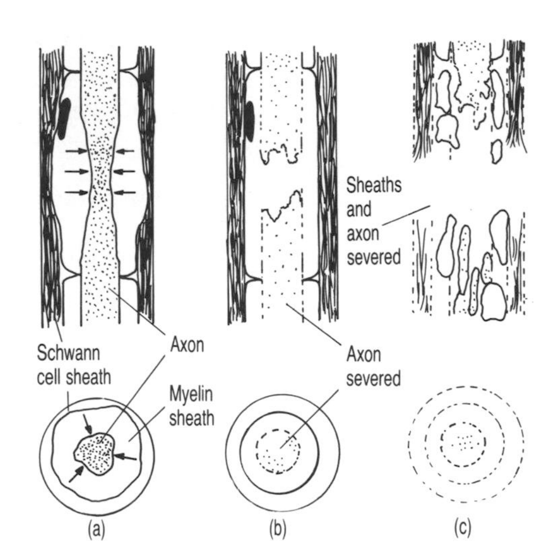

Fig. 4.7. Injury to peripheral nerve: (a) axon and Schwann cell sheath intact, swelling of the myelin sheath; (b) axon severed, sheaths intact; and (c) axon and sheaths severed.

(Fig. 4.7b and c). Axons growing into intact sheaths will complete regeneration in a few weeks. If new tubules must be formed by the Schwann cells, recovery may take months.

4.4 Cranial nerves

In general, the components of a cranial nerve are the same as those of spinal nerves, i.e. they contain sensory, motor and autonomic fibres. There are some differences, however, from the plan of the spinal nerves already described: not all cranial nerves are 'mixed', some contain sensory fibres only, for example the optic nerve from the retina of the eye.

All the fibres of one cranial nerve emerge together from the brain either as a single bundle, or as a row of filaments which join together at a short distance from the brain stem. (Note that in each mixed spinal nerve, motor and sensory fibres were separated into two distinct roots leaving the spinal cord.)

Each cranial nerve has one or more nuclei of grey matter in the brain stem, where motor fibres originate and sensory fibres terminate. The sensory fibres of some cranial nerves, particularly from the sense organs, synapse in other brain areas before relaying in the nucleus of the specific cranial nerve. The nuclei of cranial nerves are not found in an orderly sequence in the brain stem. Migration of these nuclei occurs in development from primary segments to a final position in the mature brain.

The **functions** of the cranial nerves include the following.

1 Conveying information from the special senses — the eyes, ears, organs of balance, nose and tongue — to the brain for integration

and eventual interpretation in consciousness.
2 Providing the pathways for brain stem reflexes essential for the orientation of the head, movements of the eyes, and other reflexes such as sneezing and coughing.
3 Carrying activity in autonomic fibres which control the size of the pupil of the eye, the muscle of the heart, and the activity of the digestive organs (motility and secretion).

The cranial nerves particularly concerned with movement and posture will now be discussed in outline.

4.4.1 Movement and sensation of the face

The facial VII and trigeminal V nerves cooperate in movements of the face and the mouth for the expression of mood and emotion (Fig. 4.8), the production of speech, and in mastication.

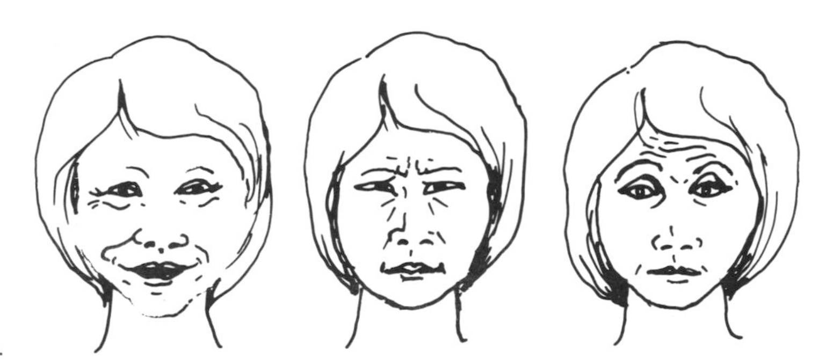

Fig. 4.8. Facial expressions resulting from activity in various muscles of the face.

The **facial nerve** contains motor fibres to the muscles of facial expression. Descending corticobulbar (nuclear) fibres from the opposite motor cortex of the brain, synapse in the motor nucleus of the nerve in the pons. The facial nerve emerges from the pons and leaves the skull through a foramen in the temporal bone close to the middle ear. The nerve then passes through the parotid salivary gland just in front of the ear and divides into five branches, arranged like the digits of a goose's foot (Fig. 4.9). Between them the branches supply all the muscles of the scalp and face, except the muscles of mastication, which receive motor fibres of the *trigeminal* nerve. Movements of the lips and tongue, essential for speech, are made by coordination of activity in the facial nerve with the *hypoglossal nerve* to the muscles of the tongue.

Figure 4.10 shows the positions of the main muscles of the face. Combining in different ways, they produce all the movements involved in facial expression, in speech and in mastication of food.

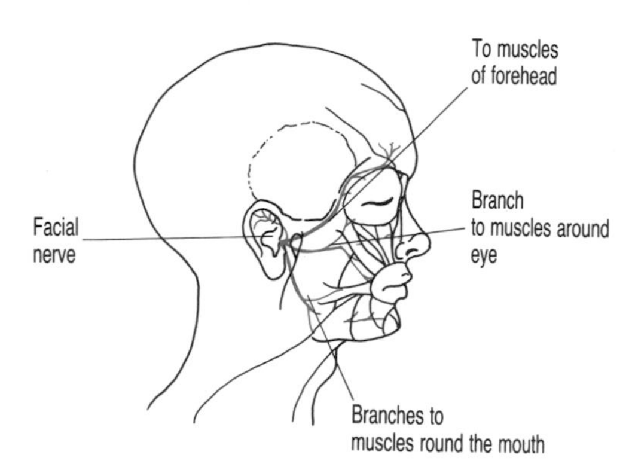

Fig. 4.9. Lateral view of the head to show the facial nerve and branches to the muscles of the face.

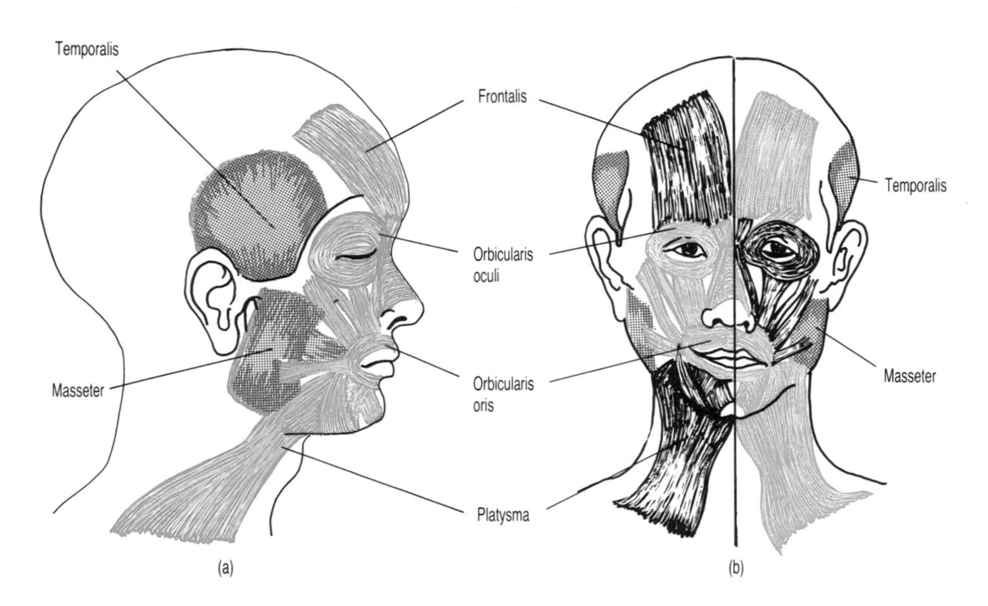

Fig. 4.10. Muscles of the face; (a) viewed from the side; and (b) anterior view (left side of face shows a facial palsy).

The **trigeminal nerve** is important for sensation in facial skin. The three divisions of this nerve supply particular areas (Fig. 4.11). The ophthalmic branch enters the orbit and then branches to the skin of the forehead and the front of the scalp. The maxillary branch passes through the floor of the orbit and then turns downwards to the skin over the cheek and to the teeth of the upper jaw. The mandibular branch supplies the skin over the side of the head and the lower jaw. Loss of sensation in the face leads to difficulty in activities such as shaving and putting on make up. Motor branches of the nerve supply the temporalis and masseter muscles used in mastication of food (Fig. 4.10a).

A summary of all the twelve pairs of cranial nerves is given in Appendix A3.1.

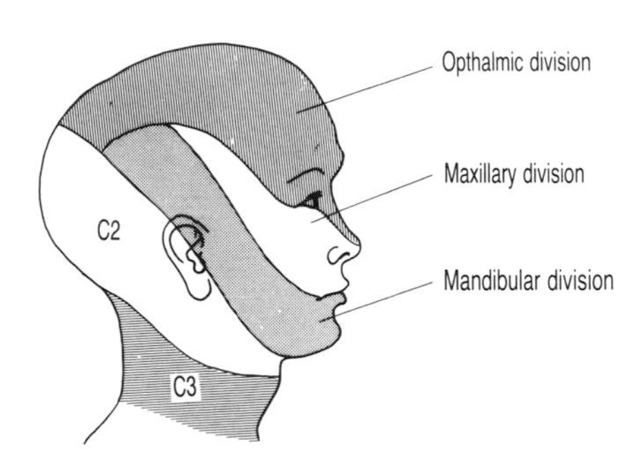

Fig. 4.11. Side view of the face to show the distribution of the three divisions of the trigeminal nerve.

4.4.2 Movement of the head and eyes

The **vestibular nerve**, part of the eighth cranial nerve, plays an important part in the automatic adjustment of the head in relation to the body during movement. The receptors in the utricle, saccule and semicircular canals of the inner ear send impulses along the vestibular nerve to the vestibular nucleus in the pons and medulla. (More detail of the receptors is given in Chapter 11.) Changes in the position of the head in relation to gravity are signalled via the brain stem to the spinal cord. Appropriate changes in postural muscle activity are then made to keep the body in equilibrium (Fig. 4.12).

Movement of the head is also coordinated with the activity of the muscles at the back of the eye, so that an object can remain in focus on the retina of the eyes. As the head turns to one side, the eyes are turned in the opposite direction to keep a constant image on the retina. Turning the head alters the balance of sensory impulses from the semicircular canals along the vestibular nerve. Impulses pass from the vestibular nucleus via the brain stem to the nuclei of the three cranial nerves that supply the muscles of the eye (Fig. 4.12). This is known as the *vestibulo-ocular reflex* which keeps an object in view as the head turns during movement. If the head continues to turn, the eyes will move rapidly in the same direction of head movement to focus on a new fixed point. This combination of slow eye movement in the opposite direction followed by a rapid movement in the same direction is known as 'nystagmus'. The same eye movements occur when the body remains still and the field of view is moving, for example looking out of a window while sitting in a moving train.

The movements described are examples of reflex actions that involve sensory and motor neurones lying in cranial nerves, so they are known as *brain stem reflexes*.

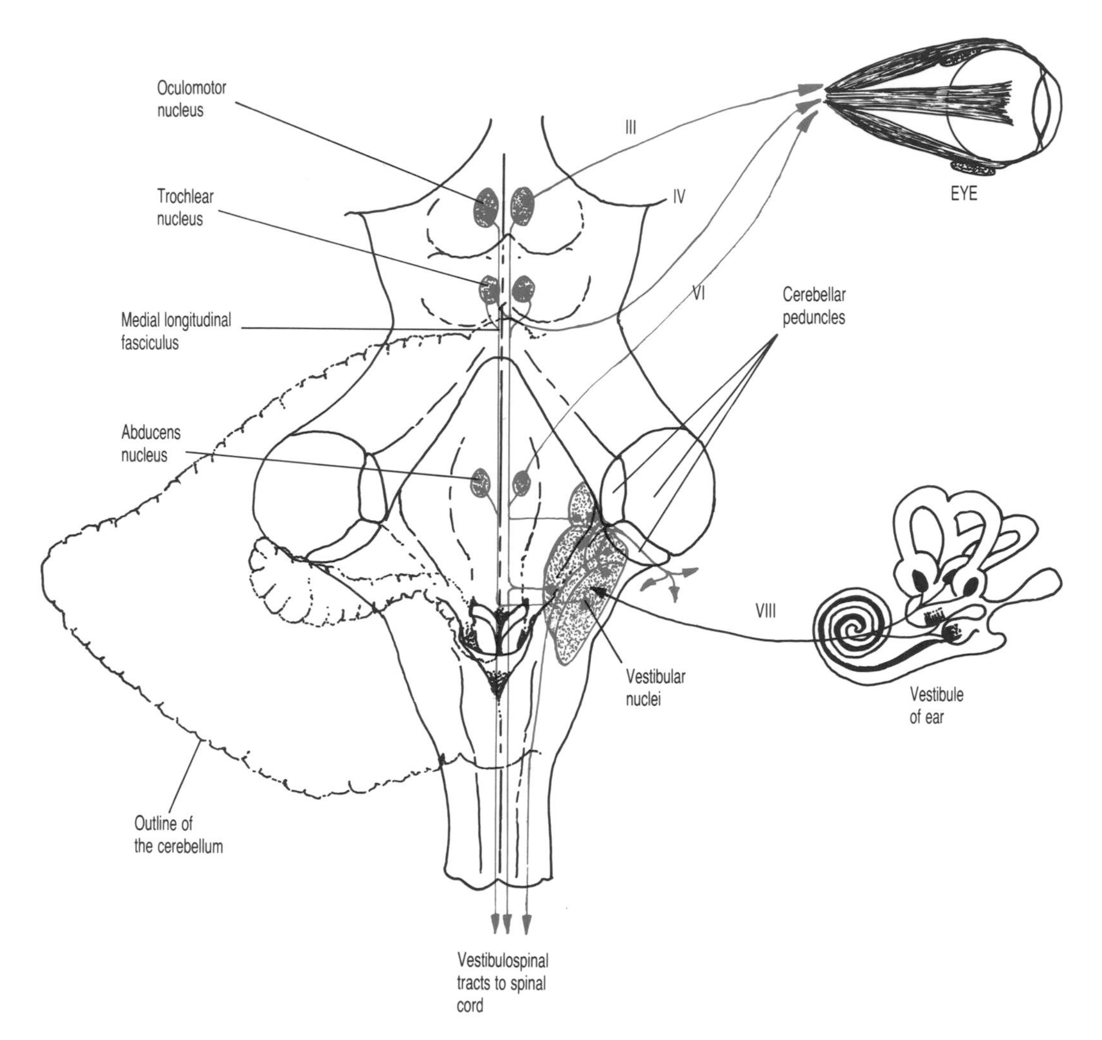

Fig. 4.12. Posterior view of the brain stem showing the components of the vestibulo-ocular reflex.

4.5 Autonomic nervous system

The overall function of the autonomic nervous system is to influence the activity of internal organs and glands, and to regulate the blood circulation to keep the internal environment of the body constant. Conduction of impulses in the autonomic fibres is slower than in the somatic component of the peripheral nervous system, since the axons are of a smaller diameter. Also there are two neurones between the central nervous system and the effector organ, so there is delay at the synapse between them. The junction between the two neurones is located in an autonomic ganglion. The preganglionic neurones originate in the brain stem or spinal cord and their fibres lie in cranial or spinal nerves. Many of the autonomic ganglia are linked together, so that activity in the post ganglionic neurones can be compared and spread to several effector organs and glands.

Another difference between somatic and autonomic parts of the peripheral nervous system is the chemical transmitter substances secreted at the synapses. In the pathways to skeletal muscles, the neurotransmitter substance is acetylcholine. While all the preganglionic fibres of the autonomic system are also cholinergic, the post ganglionic fibres of the sympathetic division secrete noradrenalin (norepinephrine).

The autonomic nervous system is divided into the following two divisions:

1 Sympathetic which mobilizes the body to release energy for activity.

2 Parasympathetic which conserves energy and maintains the homeostatic balance of the body at rest.

Most organs and glands receive fibres from both divisions, which cooperate to control the level of activity at any one moment. For example, the heart rate is the result of the parasympathetic division slowing the heart via the vagus nerve, and the sympathetic nerves increasing the rate.

Sympathetic nervous system

Nerve fibres of the sympathetic nervous system originate in the lateral horn of the grey matter of all the thoracic segements and the first two lumbar segments of the spinal cord. These preganglionic fibres lie in the spinal nerves T1–L2 and synapse in the chain of ganglia lying on the bodies of the vertebrae (Fig. 4.13). The postganglionic fibres link to all spinal levels via the sympathetic chain, so that stimulation of the sympathetic nervous system can have a widespread effect in all regions of the body.

Stimulation of the sympathetic system prepares the body for action by the following effects.

1 Stimulation of cardiac muscle to increase the heart rate and the force of contraction, and constriction of blood vessels to raise blood pressure.

2 Relaxation of the smooth muscle of the walls of the bronchioles of the lungs so that a greater volume of air can be ventilated by the lungs.

(All these effects increase the oxygen supply to the active muscles carried by the blood circulating at a higher volume per minute.)

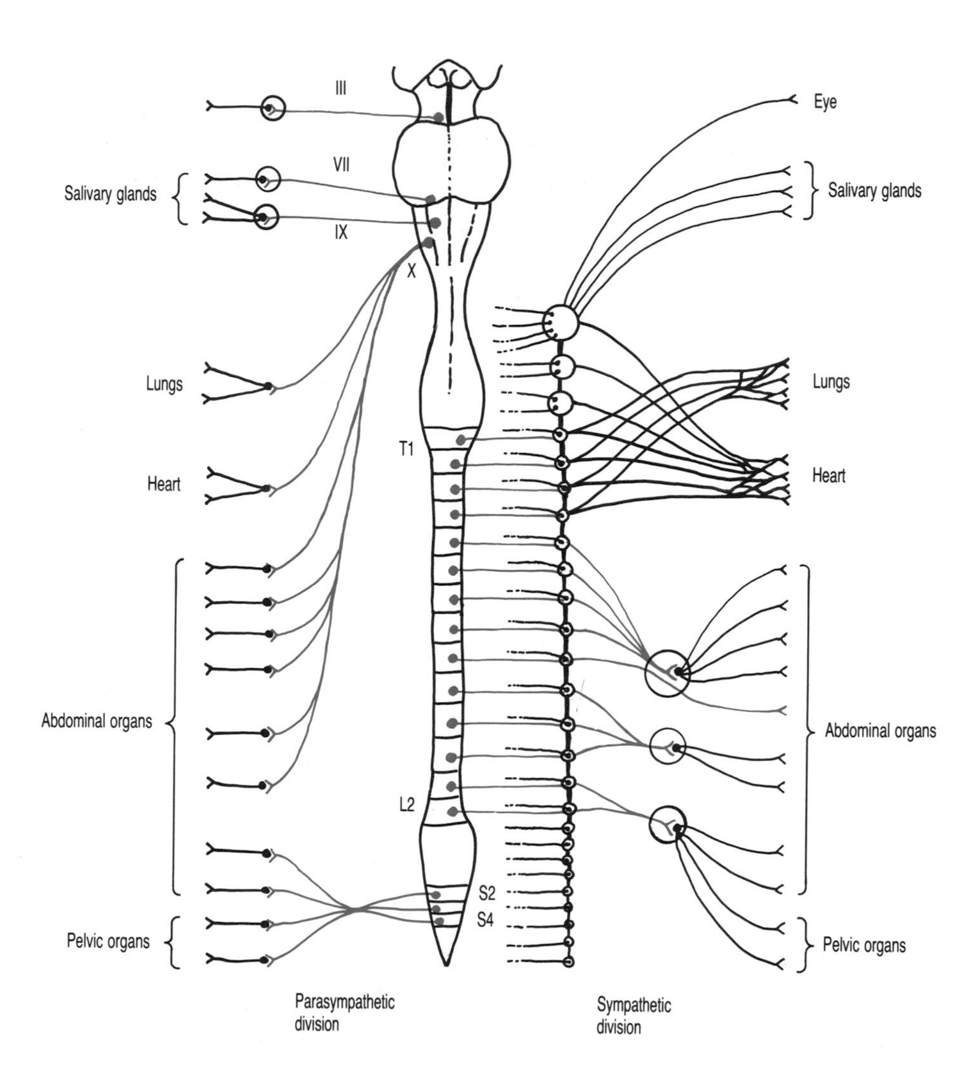

Fig. 4.13. Spinal cord and brain stem. General plan of the autonomic nervous system. Sympathetic division on the right, and parasympathetic division on the left of the diagram.

3 Liver glycogen is mobilized to raise the glucose level of the blood and supply extra nutrients to the contracting muscles.
4 Dilation of the pupil of the eye to allow more light to enter.
5 Stimulation of sweat glands in the skin to lose the extra heat generated from the muscles, and keep the body temperature constant.

The sympathetic system is controlled by the hypothalamus which responds to changes in the external environment and also to emotional changes such as fear and anxiety.

Parasympathetic nervous system

The parasympathetic system acts in localized regions of the body, unlike the mass reaction of the sympathetic. The preganglionic fibres originate from two widely separated regions of the central nervous system.

1 Four of the cranial nerves;
Occulomotor nerve III — to the sphincter muscle of the iris and ciliary muscle controlling the convexity of the lens.
Facial nerve VII — to the lacrimal glands of the orbit and two pairs of salivary glands.
Glossopharyngeal nerve IX — to parotid salivary glands.
Vagus nerve X — to the heart, respiratory bronchioles & all the digestive tract as far as the transverse colon.
2 Spinal segments S2, S3, and S4. The pelvic splanchnic nerves leave the spinal cord to supply the descending colon and rectum, the bladder and reproductive organs (Fig. 4.13).

The preganglionic fibres of the parasympathetic are long and the ganglia are found near to the structure supplied. The postganglionic fibres are short and multibranching.

The effects of stimulation of the parasympathetic division are designed to conserve and restore the energy sources of the body. The parasympathetic system conserves energy by the following methods.
1 Decreasing heart rate and force of contraction, so lowering the blood pressure.
2 Constricting smooth muscle of the respiratory bronchioles.
3 Constricting the pupil of the eye in response to bright light.

In addition, the parasympathetic system increases the activity of the digestive system, and is active during emptying of the bladder and rectum. (Recently it has been found that neurones of the digestive tract receive postganglionic fibres from the sympathetic system as well as preganglionic fibres of the parasympathetic, so that a separate 'enteric' system has been described).

A summary of the cranial nerves, and spinal segments supplying the upper and lower limb muscles is in Appendix 3.

At the end of this chapter you should be able to:
1 Describe the position, arrangement and overall function of the cranial and spinal nerves.

2 Describe the components of a typical spinal nerve and its major divisions.
3 Define dermatome, myotome, peripheral nerve, and plexus.
4 Summarize briefly the effects of injury to a peripheral nerve.
5 Explain what is meant by a 'brain stem reflex' and give an example.
6 Outline the nerve supply to the skin and muscles of the face.
7 Summarize briefly the structural and functional differences between the somatic and autonomic parts of the peripheral nervous system.
8 Outline the functions of the sympathetic and parasympathetic divisions of the autonomic nervous system.

Further Reading

Angevine JB & Cotman CW (1981) *Principles of Neuroanatomy*. Oxford University Press, New York.

Brown TS & Wallace PM (1980) *Physiological Psychology*. Academic Press, London.

Galley PM & Forster AL (1982) *Human Movement*. Churchill Livingstone, Edinburgh.

Gowitzke BA & Milner M (1980) *Understanding the Scientific Basis of Human Movement*. Williams & Wilkins, London.

Hay JG & Reid JG (1982) *The Anatomical and Mechanical Bases of Human Motion*. Prentice-Hall, Hemel Hempstead.

Holmes RL, Sharp JA & Brown M (1969) *The Human Nervous System, a Developmental Approach*. J & A Churchill Ltd, London.

Lamb JF, Ingram CG, Johnston IA & Pitman RM (1980) *Essentials of Physiology*. Blackwell Scientific Publications, Oxford.

Loewy AG & Siekevitz P (1971) *Cell Structure and Function*. Holt, Rinehart & Winston, New York.

MacKinnon P & Morris J (1986) *Oxford Textbook of Functional Anatomy*. Vol. 1. Oxford Medical Publications, Oxford.

McClintic JR (1985) *Basic Anatomy and Physiology*. John Wiley & Sons, Chichester.

Mitchell L & Dale B (1980) *Simple Movement. The How and Why of Exercise*. John Murray, London.

Netter FH (1972) *The Ciba Collection of Medical Illustrations. Vol. 1, Nervous system*. Ciba, New Jersey.

Scientific American (1979) *The Brain*. WH Freeman & Co., New York.

Wirhed R (1984) *Athletic Ability and the Anatomy of Motion*. Wolfe Medical, London.

Section 2 The Anatomy of Movement

ORGANIZATION AND COOPERATION IN DAILY LIVING

5 / Positioning Movements. The Shoulder and Elbow

PART I/MOVEMENTS OF THE SHOULDER

5.1 Functional movements of the shoulder

(a)

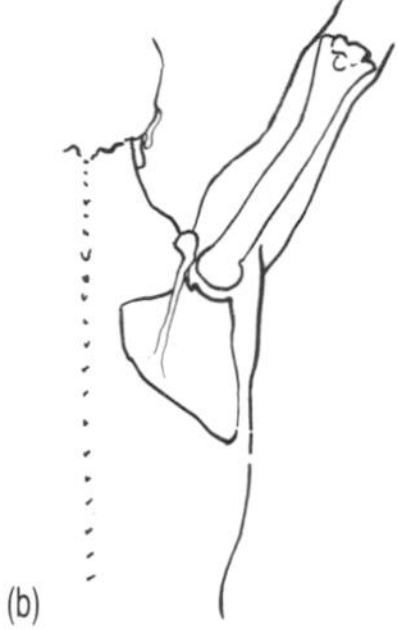

Fig. 5.1. Posterior view of right scapula and humerus: (a) anatomical position; and (b) abduction of the humerus with lateral rotation of the scapula.

The shoulder forms a foundation from which the whole of the upper limb can move. Acting like the cab of a crane, the shoulder allows the hand to be placed in all directions around the body, in the same way as the jib of a crane places its load. The wide area that can be covered in positioning the hand is due largely to movement at the glenohumeral joint between the head of the humerus and glenoid cavity of the scapula (see Appendix 2). Range is further increased by movement of the scapula on the chest wall in the same direction as the humerus, which allows the humerus to move further. Figure 5.1 shows the combined movement of the humerus and scapula in abduction of the arm in order to reach the vertical position.

The muscles arranged around the shoulder not only perform the wide range of movement, but also anchor the arm to the trunk, supporting the weight of the upper limb as it moves. When the hand performs precision movements, the shoulder muscles hold the arm steady.

All the muscles involved in movements of the shoulder region are attached to the humerus, scapula and clavicle. The proximal attachments of the larger muscles extend on to the sternum, ribs and vertebral column.

The pectoral girdle

The bony clavicle and the scapula form the pectoral girdle which provides the link between the upper limb and the trunk.

Two articulations are involved in this link: the *sternoclavicular joint* between the medial end of the clavicle and the clavicular notch on the manubrium of the sternum; and the *acromioclavicular joint* between the lateral end of the clavicle and the acromium process of the scapula (see Appendix 2). The sternoclavicular joint is stabilized by the strong costoclavicular ligament, and by an intra-articular disc attached to the clavicle above and to the first costal cartilage below.

• *PALPATE the sternoclavicular joint while the arm moves at the shoulder in all directions. Note that movement at the sternoclavicular joint occurs each time the humerus moves.*
• *PLACE your hand on the scapula of a partner. Feel the movement of the scapula in the same direction as the humerus, while your partner moves the arm to place the hand all around the body.*

The muscles that move the shoulder girdle and the glenohumeral joint can be divided into three.

1 The muscles stabilizing the glenohumeral joint.

2 The muscles acting on the glenohumeral joint.

3 The muscles moving the pectoral girdle.

Remember that the muscles in each section do not act in isolation, but combine in various ways, grouping and regrouping as the movement proceeds. The division into three sections is for the purpose of description only.

5.2 Muscles stabilizing the glenohumeral joint

In the glenohumeral joint, the shallow glenoid cavity of the scapula and the hemispherical head of the humerus, joined by a thin loose capsule, permit a wide range of movement, but present a poor prospect for stability. The only strong ligament linking the bones is the coracohumeral ligament extending from the lateral border of the coracoid process to the greater tuberosity of the humerus.

The most effective provision of support for the joint is from the four muscles surrounding it and blending closely with the capsule. These muscles are the *supraspinatus, infraspinatus, teres minor* and *subscapularis,* which act like guy ropes holding the humerus in contact with the scapula, and are known as the 'rotator cuff' muscles (Fig. 5.2). The lesser tuberosity of the humerus receives the subscapularis tendon, covering the joint anteriorly. The other three muscles are inserted into the greater tuberosity, with supra-

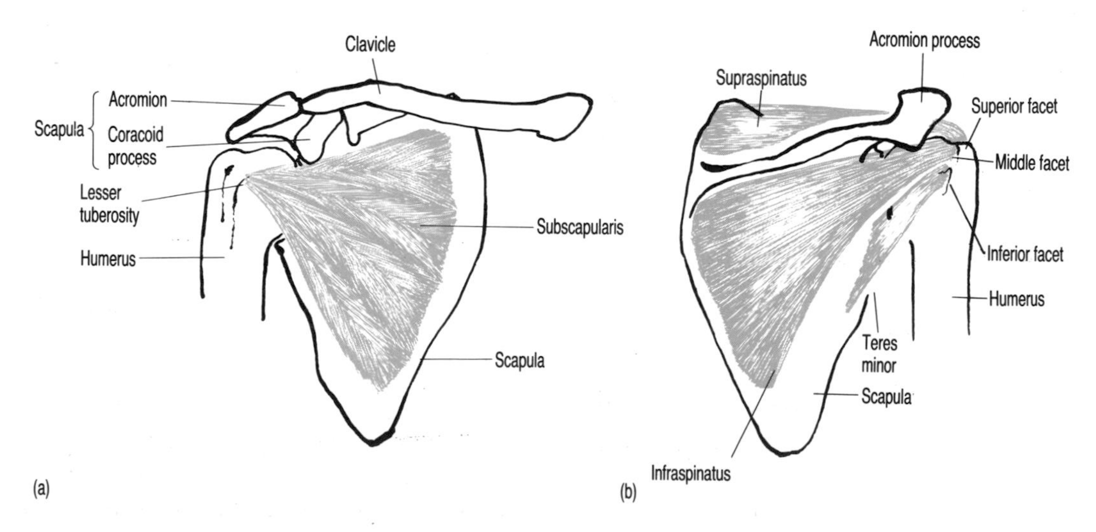

Fig. 5.2. Right scapula and humerus to show the 'rotator cuff' muscles: (a) anterior view; and (b) posterior view.

spinatus superiorly, then infraspinatus and teres minor below and posteriorly. The absence of any additional support inferiorly means that dislocation is usually downwards and forwards, under its own weight as the arm hangs by the side, or during abduction movement. General weakness of shoulder muscles, for example in a stroke patient, commonly leads to subluxation of the shoulder joint.

The 'rotator cuff' muscles have weak action as prime movers since their insertions are close to the joint, but they function as stabilizers in all movements of the shoulder joint. The supraspinatus initiates abduction of the shoulder before the deltoid can exert its pull on the lateral shaft of the humerus. The other three muscles act as rotators of the humerus, the subscapularis medially, infraspinatus and teres minor laterally.

5.3 Muscles acting on the glenohumeral joint

Three large muscles, surrounding the glenohumeral joint, move the joint through its wide range. Their attachments cover a wide area of the pectoral girdle and trunk, and converge to insert on to the humerus.

The three muscles are the deltoid; the pectoralis major; and the latissimus dorsi. The teres major and coracobrachialis are two less important muscles that will be considered together.

Deltoid

The deltoid muscle gives the rounded shape to the shoulder and has the overall shape of an inverted triangle. The margins of the muscle can be clearly seen in athletes and swimmers. Lack of use after injury may lead to wasting which gives the shoulder a 'squared appearance'.

- *LIFT a saucepan or book down from a high shelf and feel the continuous activity in the deltoid as the arm is raised and then lowered. If the deltoid was relaxed as the arm came down, the movement would be rapid and uncontrolled, and you would probably drop the book on the floor.*
- *PALPATE the origin of the deltoid in a partner with the arm relaxed by the side. Start anteriorly at the lateral end of the clavicle to feel the anterior fibres. Next cross the acromium process of the scapula where the middle fibres arise. Continue along the spine of the scapula to find the posterior fibres. All the fibres converge to insert on the lateral shaft of the humerus about half way down (Fig. 5.3).*

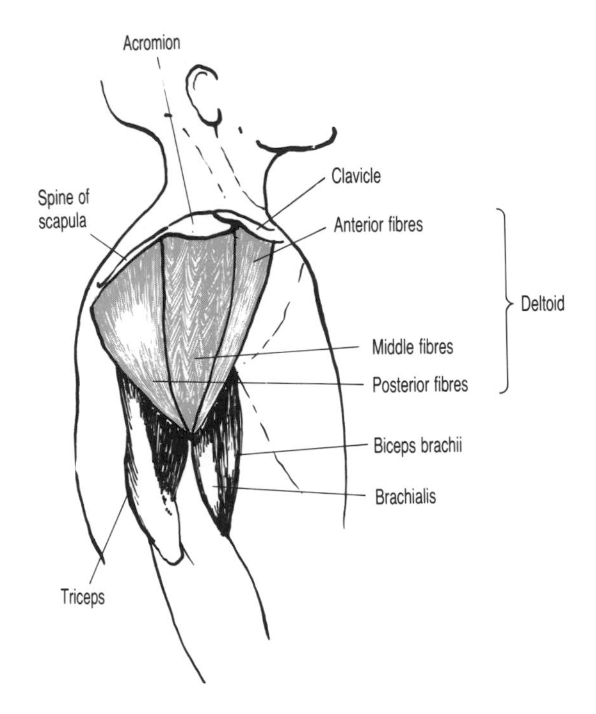

Fig. 5.3. Side view of the right shoulder to show the position of the deltoid muscle.

• *INSPECT the skeleton to find the deltoid tuberosity formed by the pull of the deltoid on the humerus.*

The deltoid is a powerful abductor of the arm lifting the arm sideways and up above the head. It is also active when the arm is lowered back down to the side, working eccentrically to control the effect of gravity. All movements reaching forwards (Fig. 5.4a), and above the head (Fig. 5.4b), involve the action of the deltoid muscle.

The anterior fibres flex and medially rotate the glenohumeral joint, while the posterior fibres extend and laterally rotate it. Both sets can work together to prevent forwards and backwards movement during abduction of the arm by the strong middle fibres. Part or all of the deltoid is used in most movements of the humerus on the scapula. The muscle also acts as a support sling for the shoulder, especially when the upper limb is carrying heavy loads such as a suitcase or shopping bag.

Pectoralis major

The pectoralis major is a large triangular muscle whose base lies vertically along the midline of the thorax, and the apex is attached to the humerus (Fig. 5.5). The lower border of the triangle can be

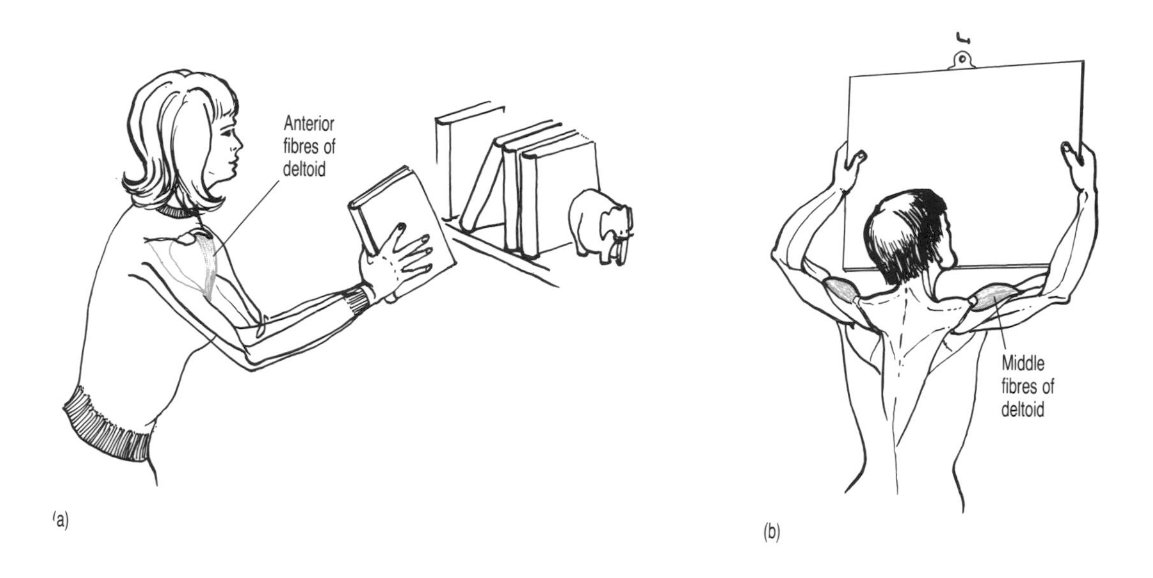

Fig. 5.4. Functions of the deltoid: (a) reaching forwards; and (b) reaching above the head.

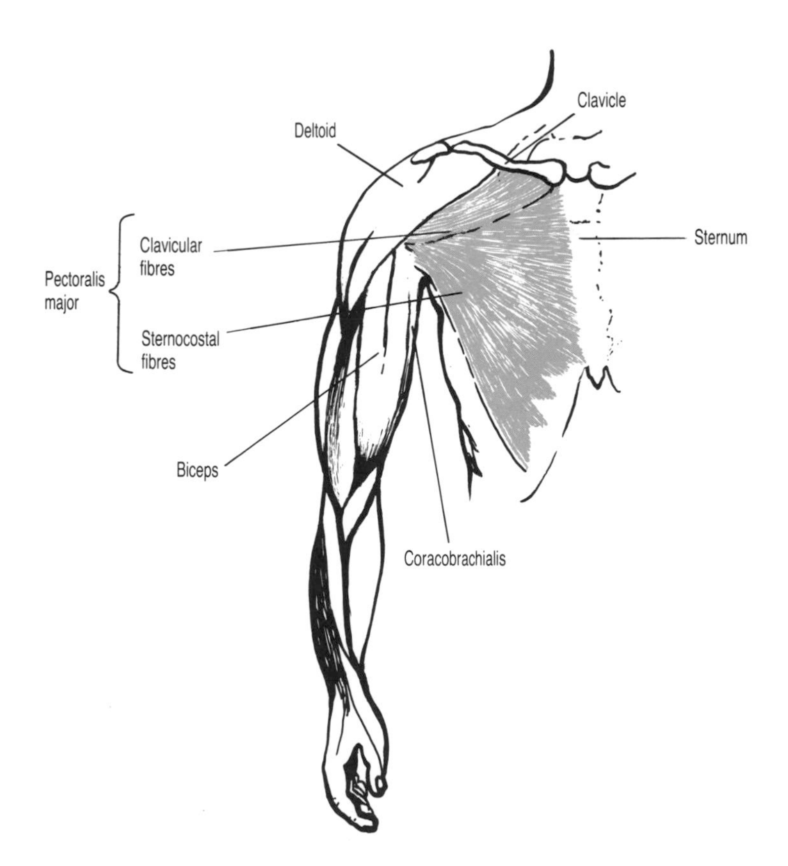

Fig. 5.5. Right thorax and upper limb showing the position of pectoralis major.

felt in the anterior wall of the axilla. The main bulk of the muscle is difficult to observe in women as the breast covers some of its surface.

• *PRESS the hands together in front of the body to put the muscle into action. The muscle can now be palpated in the axilla by a partner.*

The uppermost fibres of the pectoralis major arise from the clavicle medial to the anterior fibres of the deltoid. The remainder of the base of the triangle is formed by fibres arising from the anterior surface of the sternum and the costal cartilages of the first six ribs. All the fibres converge to the insertion on the anterior humerus in the groove between the two tubercles, (intertubercular sulcus or bicipital groove), with the clavicular fibres lying superficial to the sternocostal fibres. The clavicular fibres act with anterior fibres of the deltoid to flex the shoulder to a right angle. The lower costal fibres act with the posterior deltoid to pull the arm downwards in extension. Pulling down a window roller blind is an extension movement against resistance. Acting as a whole, the pectoralis major is an adductor and medial rotator of the shoulder drawing the arm across to place the hand on the opposite side of the body,

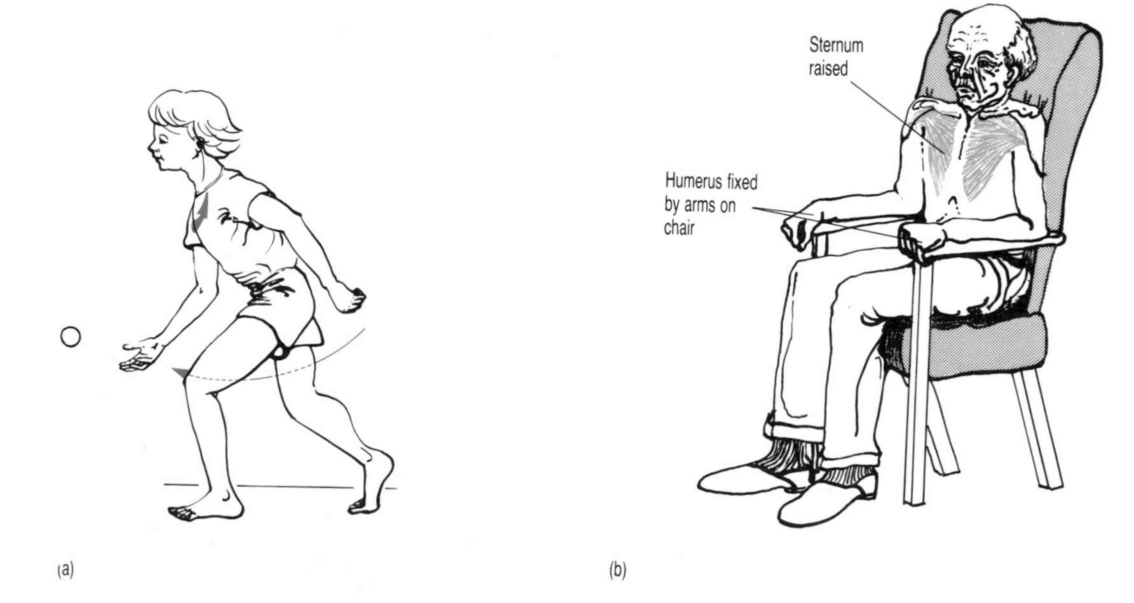

Fig. 5.6. Functions of pectoralis major: (a) throwing a ball — flexion and adduction; and (b) assisting breathing with humerus fixed — the sternum and ribs are lifted.

as in lifting a saucepan to one side, or moving a pile of books. The pectoralis major is used to pull the arm forwards in throwing a ball (Fig. 5.6a), javelin or discus. In tennis and squash, the pectoralis major draws the racquet forwards to hit the ball in a forehand drive. Another function of the pectoralis major is to assist in deep breathing. When the humerus is fixed, the muscle pulls the sternum upwards and outwards to enlarge the thorax and draw more air into the lungs. Figure 5.6b shows the position of the arms used to assist breathing whilst sitting in a chair. (Two of the shoulder girdle muscles, the serratus anterior and pectoralis minor, act with the pectoralis major in this position to increase the lung ventilation.)

Latissimus dorsi

A shoulder muscle arising from a large origin in the lower back and thorax, wraps round the trunk and converges towards the shoulder forming the posterior wall of the axilla. (Note: the pectoralis major forms the anterior wall.) The medial attachment of the latissimus dorsi is by an aponeurosis from the spines of the lower six thoracic, all the lumbar, and the upper sacral vertebrae. Some fibres also arise directly from the posterior half of the iliac crest. The uppermost fibres cross the inferior angle of the scapula holding it down. From the wall of the axilla, the tendon passes underneath the glenohumeral joint to end on the anterior end of the humerus in the floor of the intertubercular sulcus (Fig. 5.7).

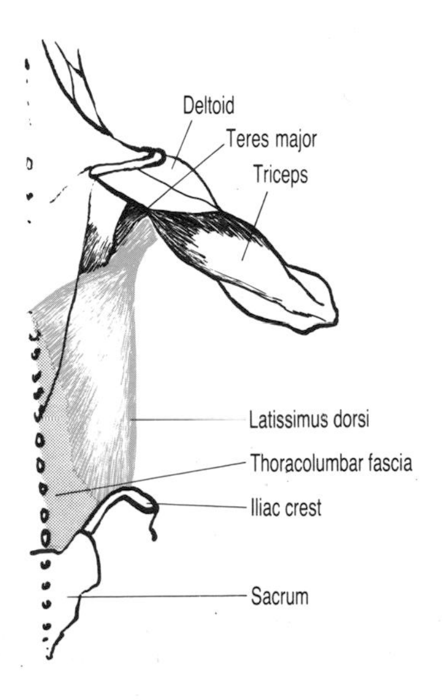

Fig. 5.7. Posterior view of right trunk and upper limb showing the position of latissimus dorsi.

(a)

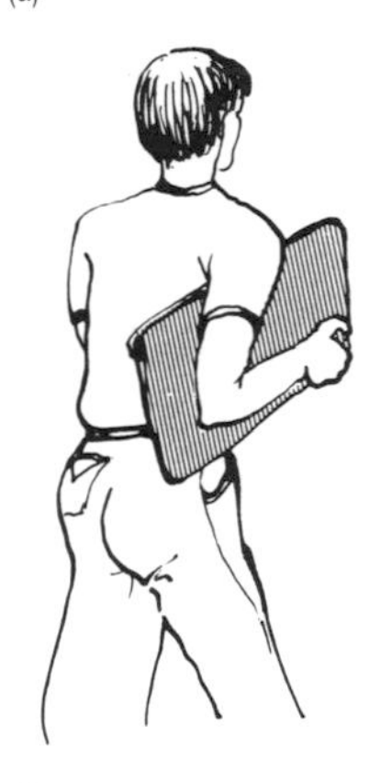

(b)

(c)

Fig. 5.8. Functions of latissimus dorsi: (a) tieing an apron — extension and medial rotation; (b) holding a document case against the side — adduction; and (c) rock climbing — adduction with humerus fixed.

• *HOLD the arm up, palpate the posterior wall of the axilla, and work out how the tendon reaches the anterior aspect of the arm on the humerus.*

The actions of the latissimus dorsi are extension, adduction and medial rotation of the shoulder joint. When the hand is above the head, the latissimus dorsi (working with the lower fibres of the pectoralis major) pulls the arm downwards and backwards against resistance, as in pulling down a blind. Continuation of this movement together with medial rotation takes the hand behind the body, as in tieing an apron (Fig. 5.8a). Working statically, the latissimus dorsi adducts the arm against the body to hold objects such as a bag or file (Fig. 5.8b). In climbing, the hand is placed above the head, and the muscle works strongly to pull the trunk up towards the arm, and lift the body upwards (Fig. 5.8c).

The latissimus dorsi is an important muscle for anyone with loss of function in the lower limb resulting from weak muscles or stiff joints. If the body cannot be raised from sitting by extension of the legs, the hands can be placed on the seat or arms of the chair, and the body lifted off the seat using the adduction action of the latissimus dorsi to hitch on the pelvis. Wheelchair patients rely heavily on this muscle to transfer them from the chair to a bed or toilet seat. In crutch walking, the latissimus dorsi helps to support the weight of the body on the hands. The muscle can also be trained to lift one side of the pelvis, so that the leg clears the ground in the swing phase in walking, 'hip hitching', the method used to teach paraplegic patients in long leg calipers to walk.

Teres major and coracobrachialis

These are two strap like muscles with a weaker individual action on the glenohumeral joint.

The teres major is attached to the lower lateral border of the scapula and lies in the posterior wall of the axilla. The insertion is with the tendon of latissimus dorsi on the anterior of the humerus. The two muscles act together on the glenohumeral joint.

The coracobrachialis takes origin from the coracoid process of the scapula and inserts into the rough area on the medial shaft of the humerus. The action of the coracobrachialis is flexion of the shoulder from the hyperextended position, i.e. humerus behind the trunk. There is evidence that the muscle functions to swing the arm forwards in walking and running. This muscle also adducts the arm on to the trunk when holding items such as a newspaper or purse under the arm.

5.4 Muscles moving the pectoral girdle

The scapula is a triangular blade of flat bone lying on the posterior wall of the rib cage. Muscles cross the anterior and posterior surfaces of the scapula, and are attached to its borders and processes. The muscles covering the anterior surface are sandwiched between the scapula and the ribs, and are loosely separated by connective tissue and fat, which allows the scapula to move freely on the chest wall. As the scapula is able to follow the direction of movement at the glenohumeral joint, it contributes to the wide range of movement of the upper limb on the trunk.

It is important to understand clearly the position of the scapula in relation to the vertebral column, ribs and walls of the axilla, in order to appreciate the direction of pull of the muscles which turn the scapula in various directions.

Firstly the terms used to describe the movements of the scapula on the chest wall will be described.

1 **Elevation:** the scapula moves upwards as in 'shrugging the shoulders'.
2 **Depression:** the scapula moves down to its resting position.
3 **Protraction:** the scapula moves forwards round the chest wall.
4 **Retraction:** the scapula moves backwards towards the spine.
5 **Lateral rotation:** the inferior angle of the scapula moves laterally and the glenoid fossa points upwards.
6 **Medial rotation:** the inferior angle of the scapula moves medially and the glenoid cavity returns to rest.

There are six muscles attached to the triangular scapula that combine to produce these movements. The muscles are the *trapezius, levator scapulae, rhomboid major* and *minor, serratus anterior* and *pectoralis minor.* By pulling together in different combinations, the muscles can elevate, depress, protract, retract, and rotate the scapula on the chest wall.

Trapezius (Fig. 5.9)

The two sides of the trapezius form a kite shaped area of muscle, the most superficial muscle of the back.

Each muscle is a triangle, with its base in the midline from the base of the skull down to the 12th thoracic spine.

The upper fibres originate from the occipital bone of the skull and the ligamentum nuchae, which covers the cervical spines in the neck. The fibres pass downwards and forwards across the neck to the lateral end of the clavicle and continue on to the acromion of the scapula. Acting as a suspension for the pectoral girdle from the skull and neck vertebrae, contraction of these upper fibres lifts the shoulders in *elevation.* In addition, they give support to the shoulders when carrying heavy loads. Awareness of the static work of the trapezius is felt when carrying heavy luggage or shopping.

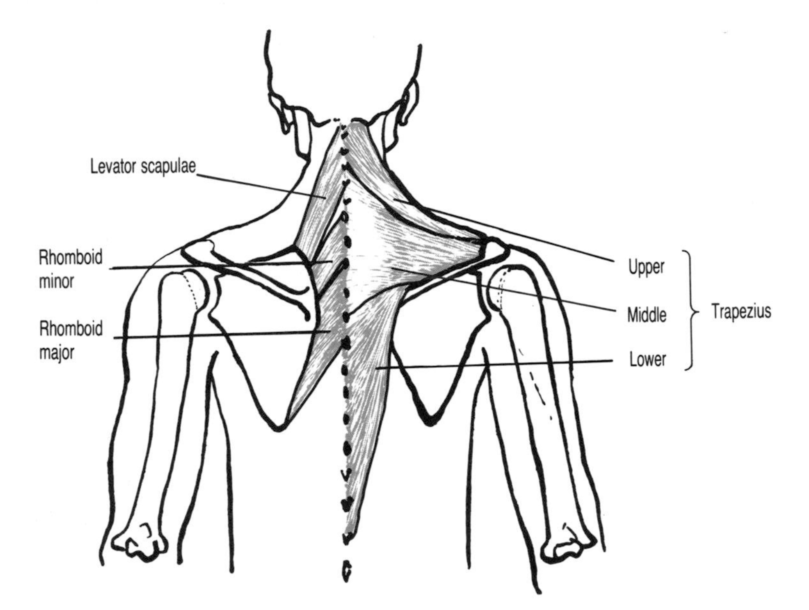

Fig. 5.9. Posterior view of the neck, thorax and arm to show the position of trapezius, levator capulae and the rhomboids.

Fig. 5.10. Retraction of the scapula to reach behind the head.

The middle fibres pass horizontally from the upper thoracic spines to the length of the spine of the scapula. Contraction of these fibres pulls the scapula towards the spine, and the scapula *retracts,* to reach behind the head. Activities involving this movement include reaching behind the head to comb the hair (Fig. 5.10) and to grasp a car seat-belt.

The lower fibres pass upwards from the lower thoracic vertebrae into a tendon that inserts into the base (medial end) of the spine of the scapula. Acting alone, these fibres will *depress* the shoulder when it has been raised. More important is the action of the lower fibres with the upper fibres to rotate the scapula, turning the glenoid fossa upwards.

Levator scapulae

The transverse processes of the first four cervical vertebrae provide the attachments for the levator scapulae, and the fibres descend to the vertebral border of the scapula above the spine (Fig. 5.9). The levator scapulae lies deep to the upper fibres of trapezius and works with them to *elevate* the scapula.

Rhomboid major and minor

These two muscles form a continuous layer deep to the middle fibres of the trapezius, originating on the spines of the upper thoracic vertebrae and inserting into the medial border of the scapula (Fig. 5.9). The rhomboids can be considered as one muscle which pulls the scapula backwards in *retraction.*

Serratus anterior

This has a saw toothed origin from the upper eight or nine ribs, clearly apparent in male swimmers and boxers with powerful shoulder muscles. From this wide origin, the fibres wrap round the thorax and underneath the scapula to be inserted in the vertebral border of the scapula (Fig. 5.11).

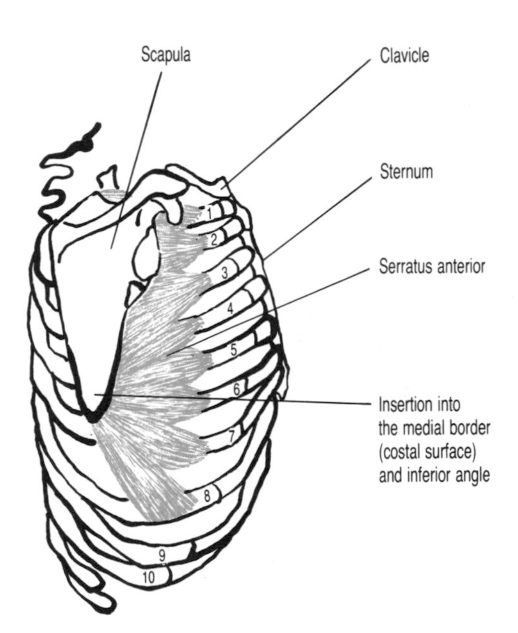

Fig. 5.11. Side view of the right thorax to show the position of serratus anterior.

The action of the whole muscle pulls the scapula forwards around the chest in *protraction.* This movement increases the forward reach of the upper limb, adds to the force of a punching action and pushes the arm forwards against a resistance, such as a door (Fig. 5.12).

The lower fibres of the serratus anterior converge on to the inferior angle of the scapula, and their action will rotate the scapula laterally to turn the glenoid fossa upwards to allow full abduction of the humerus. In lateral rotation, the serratus anterior works with the upper and lower fibres of the trapezius.

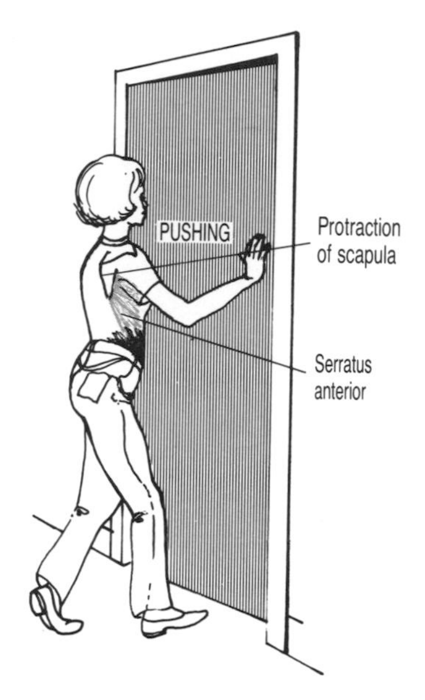

Fig. 5.12. Functional use of serratus anterior — pushing a door (protraction).

• *LOOK at an articulated skeleton to appreciate the exact position of the serratus anterior. Lying deep to the scapula it separates the subscapularis from the chest wall.*

Figure 5.13 shows serratus anterior and the rhomboids seen in a transverse section across the thorax. Identify the vertebral border of

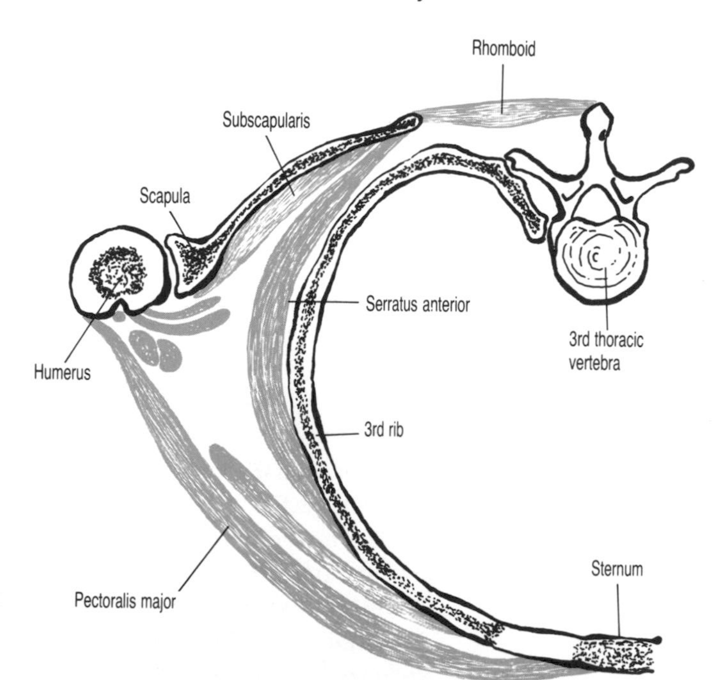

Fig. 5.13. Transverse section through the upper thorax at the level of the third rib to show the direction of pull of the serratus anterior and the rhomboids.

the scapula and note how the serratus anterior and the rhomboids pull on the scapula in opposite directions to protract and retract the scapula respectively.

Pectoralis minor

This is a small muscle lying in the anterior wall of the axilla deep to the pectoralis major, but with no action on the shoulder joint. The fibres of the pectoralis minor are attached to the coracoid process of the scapula and descend to the 3rd, 4th and 5th ribs (Fig. 5.14). By pulling on the coracoid process, the pectoralis minor can depress and medially rotate the scapula.

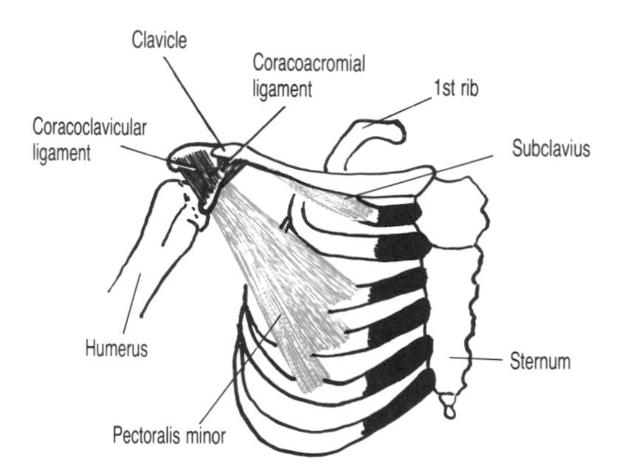

Fig. 5.14. Anterior view of right upper thorax to show the position of the pectoralis minor and subclavius.

- *LIFT the arm of a partner through the full range of abduction to reach above the head, then full adduction back to the side. Palpate the scapula during this action: lateral rotation can be felt as the arm is raised, then medial rotation as the arm is lowered.*
- *MOVE the arm to the horizontal and then round a wide circle forwards and backwards. Palpate the scapula during this action and feel the movement of protraction as the arm swings across the front of the body, and retraction as it swings behind the body.*

All the muscles attached to the clavicle and scapula combine in different ways to produce the movements of the pectoral girdle. The clavicle, spine and acromium of the scapula can be considered as two sides of a triangle, completed by a line across the root of the neck (Fig. 5.15a). This triangle moves in elevation, depression, protraction and retraction, with the sternoclavicular joint acting as the pivot. The scapula itself is a triangle, which moves in the same directions as the upper triangle when pulled simultaneously at two of its angles. When three angles of the scapula are moved by muscle

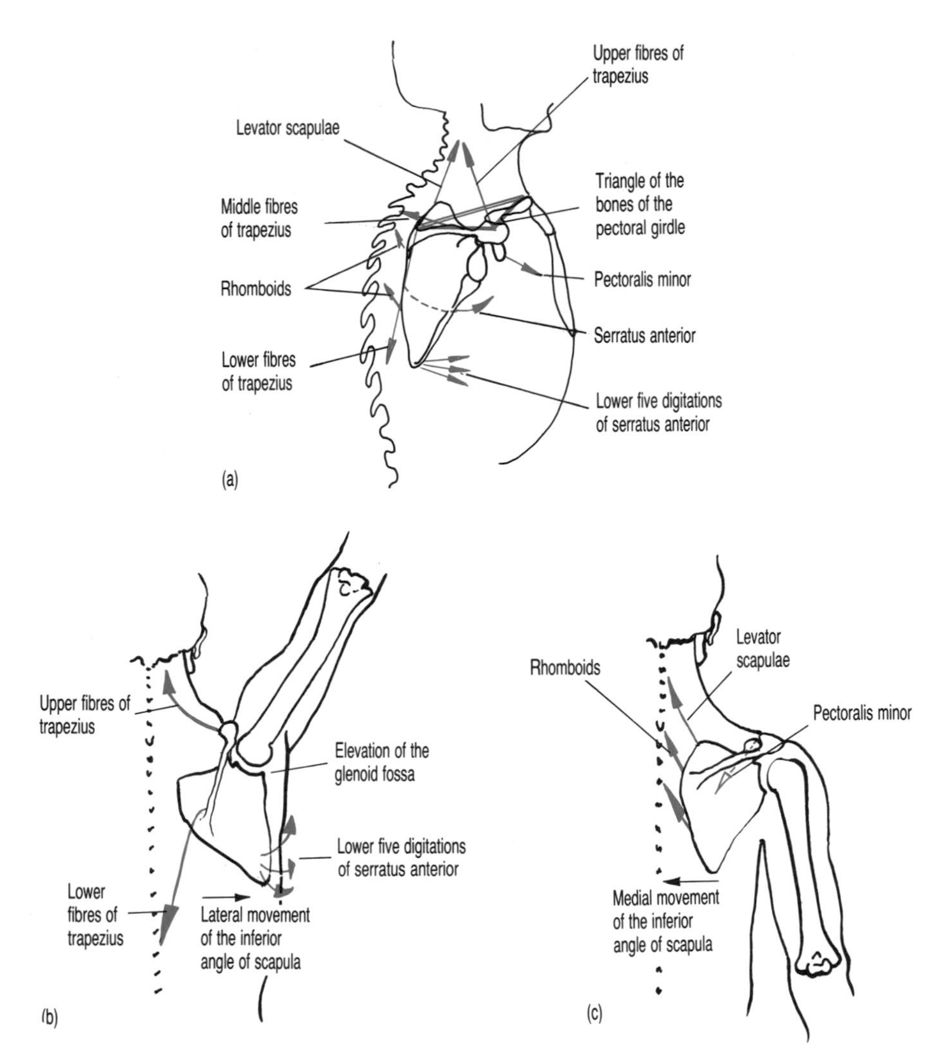

Fig. 5.15. Direction of pull of the muscles of the shoulder girdle: (a) clavicular scapular and scapular triangles of the pectoral girdle; (b) lateral rotation of the scapula; and (c) medial rotation of the scapula.

action, the scapula rotates, either medially or laterally (Fig. 15.5b and c). The axis of rotation lies just inferior to the spine of the scapula, midway along its length.

• *OBSERVE the following functional activities, then record the directions of movement of the scapula and name the muscles involved.*

1 *Reach up to a high shelf.*
2 *Push open a door.*
3 *Reach behind to grasp the seat-belt in a car.*
4 *Turn over a page of a newspaper on a table.*
5 *Pull open a drawer.*

5.5 Summary of the muscles involved in shoulder movement

Glenohumeral movement	Muscles producing movement	Scapular movement	Muscles of scapula	Function
Flexion	Deltoid — anterior fibres Pectoralis major — clavicular fibres (Coracobrachialis)	Protraction	Serratus anterior (Pectoralis minor)	Reach and push forward
Extension	Deltoid — posterior fibres Pectoralis major — sternocostal Latissimus dorsi and teres major	Retraction	Rhomboids Trapezius — middle fibres	Pull down and back
Abduction	Deltoid Supraspinatus	Lateral (upward) rotation	Trapezius — upper and lower fibres Serratus anterior	Reach sideways and upwards
Adduction	Pectoralis major Latissimus dorsi Teres major and coracobrachialis	Medial rotation	Rhomboids Levator scapulae	Pull down to side
Medial rotation	Deltoid — anterior fibres Pectoralis major Latissimus dorsi and teres major Subscapularis			Turn hand with forearm inwards
Lateral rotation	Deltoid — posterior fibres Infraspinatus and teres minor			Turn hand with forearm outwards

PART II/MOVEMENTS OF THE ELBOW

5.6 Functional movements of the elbow

Movement at the elbow brings the hands towards the head and the body. The elbow flexors are the 'hand to mouth' muscles. Try splinting the elbow in extension and find out how much we depend on elbow *flexion* for daily activities, such as washing, dressing, eating and drinking. The opposite *extension* movement of the elbow enables the hand to push against resistance, for example a lawn mower, a swing door or a pram. With loss of function of the lower limb, a person relies on elbow extensors to lift the body weight on the hands to rise from a chair or walk with crutches. (Note: the latissimus dorsi and lower fibres of the trapezius are also involved.)

The elbow is a synovial hinge joint moving through *flexion* and *extension* in the sagittal plane only. The head of the radius and trochlear notch of the ulna articulate with the lower end of the humerus. Collateral ligaments strengthen the capsule and stabilize the joint (see Appendix 2).

• *WATCH the elbow in action during the following activities.*
1 *Using a saw or a hammer.*
2 *Operating a keyboard, e.g. typewriter, microcomputer.*
3 *Lifting a box or tray from below.*

In all these activities, the forearm as a whole moves like a hinge at the elbow, i.e. about a single axis. The difference in **1**, **2**, and **3** is the position of the hand which is the result of rotation of the lower end of the radius around the ulna carrying the hand with it. The rotation movement of the forearm occurs at the joints between the radius and ulna (see radio-ulnar joints in Appendix 2).

Pronation is when the hand turns medially to face backwards from the anatomical position, or downwards when the hand is in front of the body.

Supination is the return movement when the hand turns to face forwards in the anatomical position, or upwards when the hand is in front of the body.

Pronation and supination will be discussed in more detail in Chapter 6, but it is important at this stage to distinguish between forearm and elbow movement, and to note that they often occur together as the hand is used.

• *LOOK again at the elbow activities* **1**, **2**, *and* **3** *looking this time at the position of the forearm. In* **2** *the forearm is in pronation, and in* **3** *it is in supination. In* **1** *the forearm is in the mid position between pronation and supination, called mid prone.*

5.7 Muscles moving the elbow joint

The muscles that move the elbow lie mainly in the arm above the elbow. They are found in anterior and posterior compartments which are separated on the lateral and medial sides by thick sheets of fibrous tissue, known as intermuscular septa. The muscles in these compartments are the *biceps brachii* and *brachialis* (anterior) (flexors); the *triceps brachii* and *anconeus* (posterior) (extensors). Two other muscles that are found in the forearm and assist in elbow flexion are *brachioradialis* and *pronator teres*.

5.7.1 Flexors of the elbow

Biceps brachii

The biceps muscle is the bulge in the arm we use to demonstrate our muscle strength. The muscle is easy to see in the relaxed state in those who have done some weight training. The biceps has no attachment to the humerus, and the origin of biceps is by two tendons from the scapula. The long head is a tendon attached to the superior part of the glenoid cavity within the shoulder joint, and emerges from the capsule to lie in the intertubercular sulcus (bicipital groove) of the humerus. The short head is a tendon from the coracoid process of the scapula, closely connected to the tendon of coracobrachialis. The tendons of the two heads join to form one muscle belly in the lower part of the arm, and the muscle inserts into the tuberosity on the medial side of the radial shaft just below the elbow. The tendon of insertion can be felt when the forearm

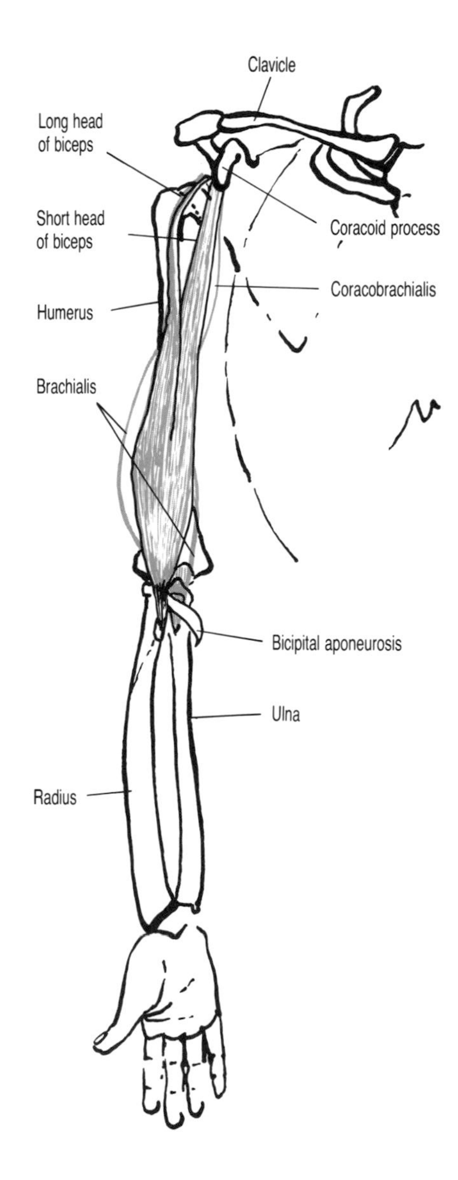

Fig. 5.16. Anterior view of the right upper limb to show the position of the biceps brachii, brachialis and coracobrachialis.

rests on a table and the muscle is relaxed. A flat band of fibrous tissue, known as the bicipital aponeurosis, extends medially from the tendon and blends with the fascia covering the medial side of the forearm (Fig. 5.16).

The flexor action of the biceps is obvious; contraction draws the radius towards the humerus. The muscle is most effective when the forearm is in supination. Working eccentrically, the biceps controls the lowering of the forearm and the hand holding a tool or utensil (see Chapter 2).

The biceps also acts as a powerful supinator, turning the forearm and hand to exert force on (for example) a door handle or screwdriver. Pulling the cork from a bottle with a corkscrew uses both actions of the biceps, supination followed by flexion. Note: the biceps is the 'party muscle'!

• *LOOK at an articulated skeleton and turn the lower end of the radius and the hand into full pronation. Notice how the radial tuberosity has moved posteriorly. The pull of the biceps tendon will now rotate the proximal end of the radius back to the anatomical position, performing an unwinding action of the forearm.*

Brachialis

The brachialis muscle lies deep to the biceps in the lower half of the arm. If the relaxed biceps is lifted and moved from side to side, the brachialis can be located below. The brachialis fibres arise from the anterior shaft of the humerus below the level of insertion of the deltoid. Passing over the anterior side of the elbow joint, the fibres insert by a broad tendon into the ulnar tuberosity below the coronoid process of the ulna (Fig. 5.16).

The Brachialis can flex the elbow efficiently in all positions of the forearm and hand. The ulna does not move in pronation and supination, so the direction of pull of the brachialis tendon always produces flexion. When the elbow flexors increase in size in response to weight training, the brachialis contributes most to the arm bulge.

5.7.2 Extensors of the elbow

Triceps brachii

The posterior compartment contains the three heads of triceps. The long head is a broad tendon attached to the inferior part of the glenoid cavity, outside the capsule, but blended with it. (Note: the long head of biceps lies inside the joint.) The two other heads of triceps arise from the shaft of humerus. The lateral head takes origin from an oblique line below the greater tuberosity on the posterior shaft. The medial head is deep and attached to the lower posterior shaft of the humerus, corresponding to the origin of the brachialis anteriorly.

The long and lateral heads join to form one layer, which unites with the deep medial head, and all three end as a broad tendon inserted into the olecranon of the ulna (Fig. 5.17).

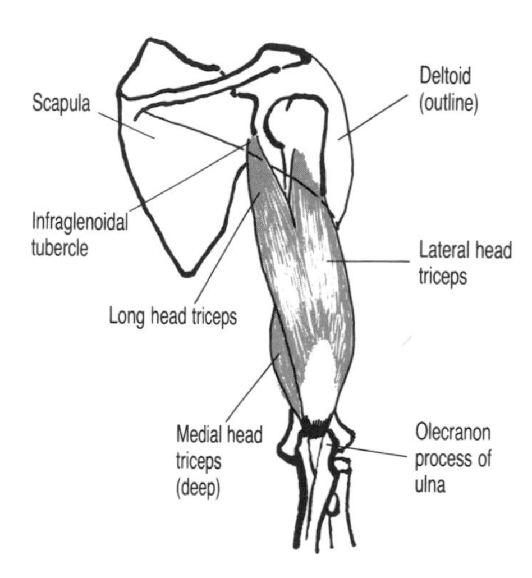

Fig. 5.17. Posterior view of the right scapula and arm to show the position of triceps brachii.

All extension movements involve the medial head; the other two heads are recruited when acting against resistance. It is the lateral head that becomes more obvious in the powerful triceps developed by the gymnast, weight lifter and wheelchair athlete. The triceps provides all the power of the elbow in extension movements to reach above the head, to push forwards and to the side. Figure 5.18 shows the use of the triceps with the pectoralis major to lift the body up from the sitting position if the muscles of the lower limb are weak.

Anconeus

This is the other posterior muscle which extends the elbow. This muscle is small and blends with the lower end of the triceps at the back of the elbow joint. The fibres of anconeus originate on the lateral epicondyle of the humerus, and insert distal to the triceps on the olecranon of the ulna. Anconeus adds little to the total strength of elbow extension, but does contribute to the stability of the elbow joint.

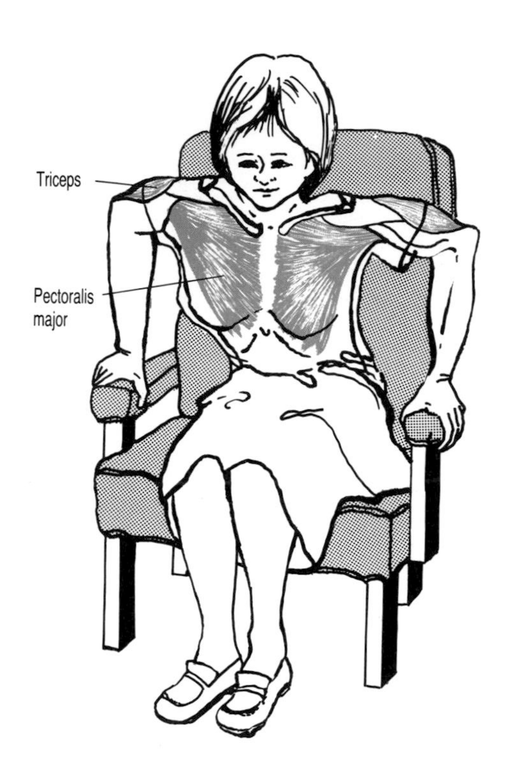

Fig. 5.18. Use of the triceps with the pectoralis major and latissimus dorsi to raise the body from sitting when the legs are weak.

5.7.3 Forearm muscles in flexion of the elbow

Brachioradialis

The brachioradialis is the most superficial muscle on the radial side of the forearm.

• *MOVE the forearm to be at a right angle to the arm.*
Turn the hand to face medially, i.e. mid prone position.
Flex the elbow and offer resistance with the other hand.
• *PALPATE the brachioradialis in position parallel to the long axis of the radius.*

The muscle originates from the ridge above the lateral epicondyle of the humerus. The fibres pass down the lateral side of the forearm, and the tendon inserts into the radius just above the styloid process at the wrist (Fig. 5.19). In the anatomical position, the muscle can only pull the head of the radius closer to the capitulum of the humerus. When the radius is rotated to bring the styloid process in line with the middle of the elbow joint (the mid prone position), the brachioradialis is able to flex the elbow in a powerful way. The mid prone position is frequently adopted to

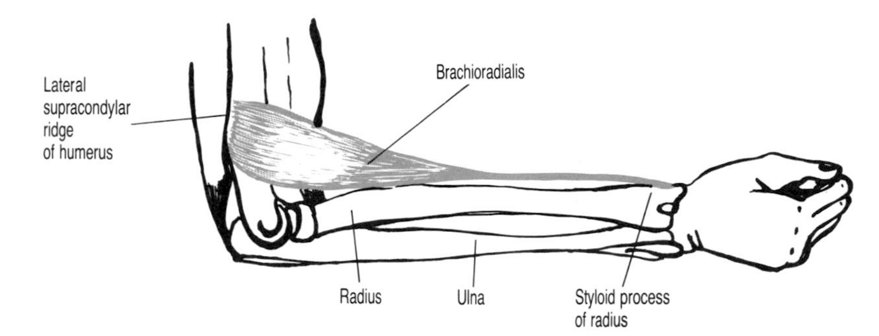

Fig. 5.19. Lateral aspect of the right forearm in the mid prone position to show the brachioradialis.

Fig. 5.20. Functional use of the brachioradialis — lift and hold a baby.

allow the strong leverage to aid flexion, e.g. using a hammer or saw, lifting a baby (Fig. 5.20), or heavy boxes. Working statically, the brachioradialis holds the elbow in flexion to support books or the handle of a bag over the forearm (for example).

Pronator teres

The pronator teres is another forearm muscle that helps in flexion of the elbow. It arises from the medial epicondyle of the humerus with the wrist and finger flexors. (Note: the brachioradialis origin is above the lateral epicondyle of the humerus with the wrist extensors.) The fibres of the pronator teres cross obliquely below the elbow joint to be inserted into the lateral shaft of the radius about half way down. When the forearm is supinated, the pronator teres gives least power to the elbow flexors. When all the elbow flexors are in action, the pronator teres counteracts the tendency for the biceps to supinate the arm.

Figure 5.21 shows both the brachioradialis and pronator teres, they will be considered again in Chapter 6 with the forearm muscles.

5.8 Positioning movements

The positioning movements of the upper limb involve activity in combinations of muscle groups around the shoulder and the elbow. Particular combinations that frequently occur together are called movement *synergies*. For example, the flexors of the elbow combine with the flexors, adductors and medial rotators of the shoulder, and protractors of the scapula, to bring the hand to the mouth in eating. The same groups combine to position the hand in front of the body and in the central visual field for precision movements of the fingers. Extension movements of the elbow combine with flexion, abduction and lateral rotation of the shoulder, and retraction of the

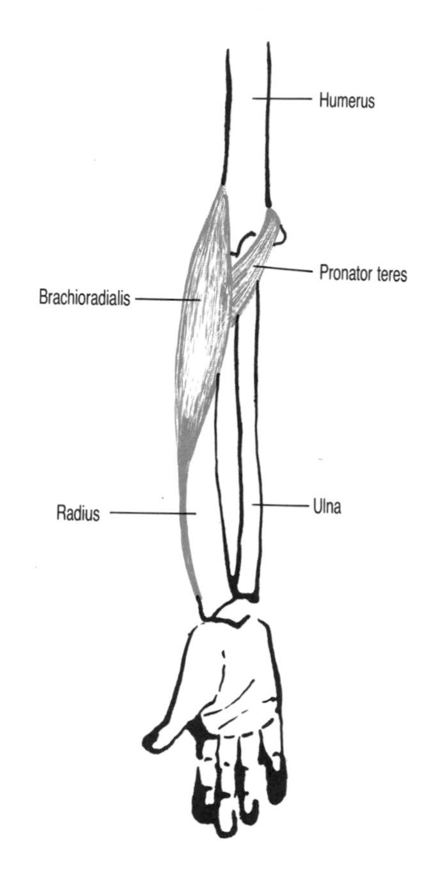

Fig. 5.21. Anterior view of the right forearm to show the position of the brachioradialis and pronator teres.

scapula, for example in reaching upwards and sideways to move the door of a high cupboard, or pull curtains across.

The wide variety of movements that can be performed by the shoulder and elbow is seen in a conductor of an orchestra moving the baton in all directions (Fig. 5.22).

Fig. 5.22. Positioning movements of the conductor's baton.

At the end of this chapter, you should be able to:

1 Outline the functions of the shoulder girdle, the shoulder joint and the elbow joint in movements of the upper limb.

2 Describe the position, attachments and actions of the muscles that:

(a) stabilize the glenohumeral joint;

(b) move the glenohumeral joint in flexion, extension, abduction, adduction, lateral and medial rotation;

(c) move the pectoral girdle; and

(d) move the elbow joint in flexion and extension.

3 Summarize the muscle cooperation around the shoulder and elbow in functional activities of the upper limb.

6 / Manipulative Movements. The Forearm, Wrist and Hand

6.1 Functions of the hand

The hand performs the most complex and intricate movements found in the body. The muscles controlling these movements originate partly in the hand itself (intrinsic muscles), and partly in the forearm (extrinsic muscles), passing over the wrist into the hand. It is the combined action of these forearm muscles and intrinsic muscles of the hand that is in all the movements of the fingers and thumb.

- *WATCH carefully the finger movements of a musician, keyboard operator or needlewoman.*
- *LOOK at your own hands making a cup of coffee, eating pasta, planting seeds.*

As well as these fine movements of the fingers and thumb used to operate small tools and keyboards, the hand is the mechanism to *grasp* handles and large tools while the upper limb moves them in space. In all gripping movements, the thumb is placed opposite to the fingers in different ways depending on the size and shape of the object. The wrist is also important in gripping by providing a stable base for the hand, and by directing the pull of the forearm muscle tendons acting on the fingers and thumb. Our complete ability to grasp includes 'letting go' or 'setting down' as well. These movements involve the opposing group of muscles to those that make the grip.

The hand is also a *sense organ*. The skin of the hand is richly supplied with receptors, and a large area of sensory cortex in the brain (see Chapter 3, Section 3.4.2, p. 59) interprets information from them. Trauma or pathological changes in the bones and joints of the wrist and carpus may damage sensory fibres in the nerves passing over them and affect hand sensation. Response from pain receptors in the skin of the hand is essential to protect it from injury. An essential part of all gripping activities is the continuous monitoring by the central nervous system of the activity in skin receptors in the hand. For example in writing, the pressure of the fingers on a pen and the hand on the paper, must be continually monitored for accurate formation of the letters.

- *TRY writing with a pen whilst wearing a thin pair of rubber gloves.*

The ability to 'recognize' objects held in the hand without seeing them is the result of the processing of information from all the receptors (see Chapter 3, Section 3.4.2).

Finally, the hand is used in *communication*, and expression of feelings. Watch how people use their hands as they greet each other, or chat in a group. Hands are used to compliment and reinforce the spoken word in a conscious way, or may be used unconsciously in 'body language'.

To summarize, the main functions of the hand are : (i) dexterity; (ii) grasp; (iii) sensation; and (iv) communication.

6.2 Pronation and supination of the forearm in hand function

Movements between the radius and ulna are important to position the whole hand on the forearm so that gripping and precision movements of the fingers and thumb can occur in a particular direction. Pronation and supination have already been considered in relation to forearm position during flexion and extension movements of the elbow in Chapter 5.

- *FIND handles and rails in different positions, i.e. vertical, horizontal, at an angle. Grip each one and notice how the position of the forearm changes in each position to allow the hand to grip.*
- *GRIP the vertical handle of a teapot or jug and then tip to pour out the contents. Note how the grip remains the same while the tipping is done by pronation and supination of the forearm (see Fig. 6.1).*
- *TURN a tap or a round door knob. The fingers and thumb exert pressure on the tap, while the forearm movement provides the power to turn it.*

The importance of the forearm in the use of the hand should now be appreciated. Fracture of the wrist or forearm bones usually results in loss of full range of pronation and supination. Hand function in daily activities is then limited until full range is restored.

In the anatomical position, the forearm is supinated and the radius and ulna lie parallel to one another. During pronation, the radius rotates and crosses over the ulna, carrying the hand with it. The movements occur at two synovial pivot joints given below.

1 The superior radioulnar joint lies between the head of the radius and the radial notch on the ulna. The joint lies inside the capsule of the elbow joint, but its movements are entirely independent. The anular ligament, (lined by a thin layer of cartilage) surrounds the head of the radius and is firmly attached to the margins of the radial notch on the ulna. The capsule of the elbow joint blends with the anular ligament so that the radius can rotate independently within this ring whatever the angulation of the elbow joint may be (see Appendix 2).

2 The inferior radioulnar joint. The lower end of the radius pivots round the head of the ulna, and is held in contact with it by a disc of fibrocartilage. This disc joins the styloid process of the ulna to the ulnar notch of the radius (see Appendix 2). The joint has a thin loose capsule, but the bones are held together by an interosseous membrane of dense fibrous tissue extending along the length of the shafts of the radius and ulna.

All the muscles involved in pronation and supination are inserted into the radius, which then moves around the fixed ulna. The supinators, inserted into the radius, can also assist other muscles to move the elbow, e.g. the biceps brachii is also an elbow flexor, and the supinator helps in extension of the elbow.

Strong pronation and supination movements are needed to use a screwdriver or a corkscrew (Fig. 6.1). Supination is more powerful than pronation, and so most screws have a right handed thread.

The **brachioradialis**, already described with the elbow flexors in Chapter 5, can move the forearm to the mid prone position from full pronation or full supination.

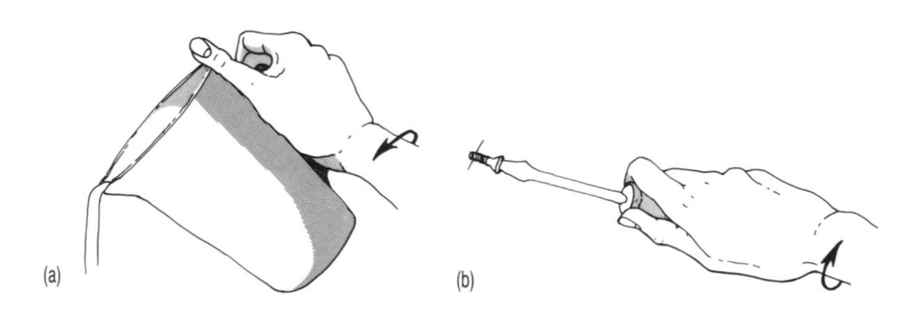

Fig. 6.1. (a) Pronation of the forearm — pouring from a jug; (b) supination of the forearm — turning a screw.

6.2.1 Muscles producing pronation

Two forearm muscles are active in pronation: the *pronator teres* and *pronator quadratus*.

The **pronator teres** (Fig. 6.2a) which crosses the anterior forearm from the medial side of the elbow to half way down the lateral shaft of the radius has already been described in Chapter 5, Section 5.7.3, p. 122.

The **pronator quadratus** (Fig. 6.2a) is a deep muscle of the forearm just above the wrist. Its fibres pass transversely between the lower anterior shafts of the radius and ulna, and the muscle is

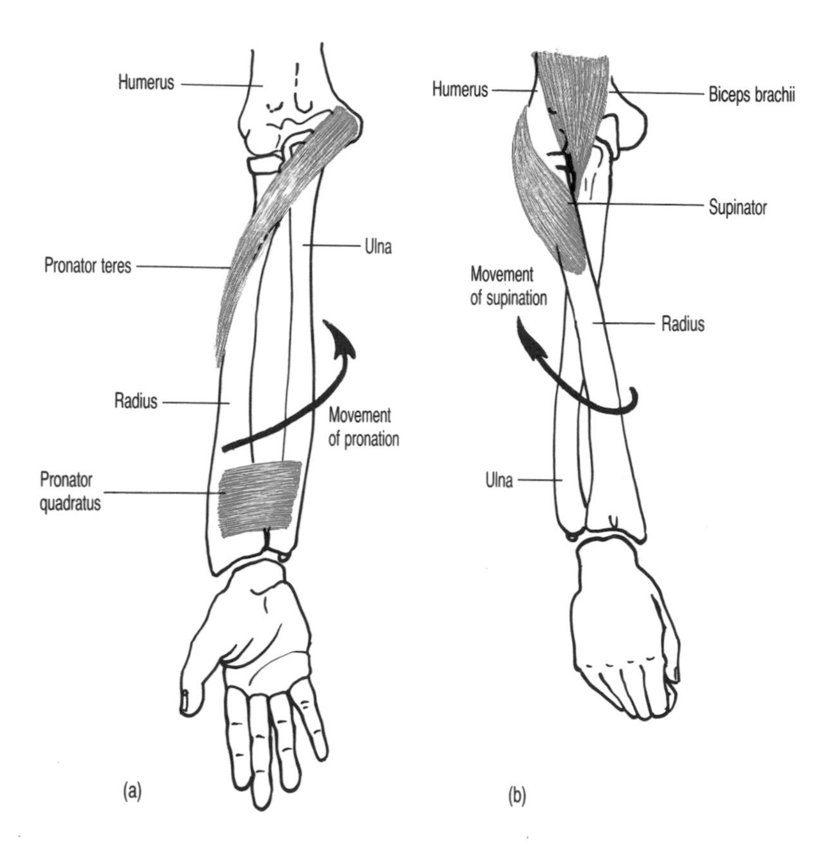

Fig. 6.2. Anterior view of right forearm and hand: (a) anatomical position, forearm supinated. Arrow shows direction of movement of the radius in pronation; and (b) forearm in pronation, hand turned to face backwards. Arrow shows direction of movement of the radius in supination.

deep to the flexor tendons which pass into the hand. When force is applied to the outstretched hand in pushing or falling, the pronator quadratus prevents separation of the radius and ulna. Many pronation movements are made with the pronator quadratus alone, with recruitment of the pronator teres for extra power against resistance.

6.2.2 Muscles producing supination

The two muscles active in supination are the *biceps brachii* and *supinator*.

The **biceps brachii** (Fig. 5.16) makes all supination movements against resistance. Its tendon pulls on the radial tuberosity just below the elbow to rotate the radius to a position parallel with the ulna. The attachments and action of the biceps have already been described in Chapter 5, Section 5.7.1, p. 118.

Slow, unopposed movements of supination, such as when the arm hangs by the side, are made by a deep posterior muscle of the forearm called the **supinator** (Fig. 6.2b) which is covered by the long extensors of the wrist and fingers. The origin of supinator is from the lateral epicondyle of the humerus and adjacent areas of the ulna. It is a short flat muscle, and its fibres wrap round the proximal end of the radius close to the bone, and insert into the upper end of the shaft.

6.3 Movements of the wrist

The wrist region is concerned with movements of the carpus of the hand on the radius and ulna of the forearm. The movements occur at the *radiocarpal* and *midcarpal joints*.

The **radiocarpal joint** is formed by the concave distal end of the radius and articular disc over the ulna, with the convexity formed by the three carpal bones in the proximal row, i.e. scaphoid, lunate and triangular (triquetral). The articular disc of fibrocartilage covers the distal end of the ulna and forms the medial part of the proximal joint surface (see Appendix 2).

The **midcarpal joint** lies between the proximal and distal row of carpals, i.e. distal surfaces of scaphoid, lunate and triquetral with proximal surfaces of trapezium, trapezoid, capitate and hamate. The joint cavity is continuous between the two rows of carpals and extends between the individual bones. (Note: the fourth bone in the proximal row, the pisiform, does not take part in either of the joints.)

The capsular ligament of the radiocarpal joint extends to cover the midcarpal joint, and the collateral ligaments continue over medial and lateral aspects of both joints.

The movements at the joints of the wrist are *flexion, extension, abduction* and *adduction*. There is no active rotation of the wrist about a longitudinal axis. Remember that rotation of the hand on the forearm occurs at the radioulnar joints of the forearm, i.e. pronation and supination movements.

Radiographs of the wrist in action show that all the carpals move as well as the radiocarpal articulation. In some movements, the scaphoid, for instance, may move as much as 1 cm. The radiocarpal joint contributes most to extension and adduction, whilst the midcarpal joint moves further in flexion and abduction. All the joints act together as a single mechanism for wrist movement.

- *PLACE the supinated hand (palm upwards) on a flat surface in a relaxed position. Notice the slight flexion and deviation to the ulnar side.*
- *LOOK at an articulated skeleton to see the shape of the lower end of*

the radius extending further on the dorsal side and laterally at the styloid process, which accounts for the position of the hand.

- *LIFT the hand and move the wrist into flexion, extension, abduction (radial deviation) and adduction (ulnar deviation). Note the range of each of these movements. You will see that the hands move further in flexion than extension, and more easily in ulnar deviation than radial deviation.*
- *COMPARE your own range of these wrist movements with those of other people. Notice the difference in range between individuals, but the relative amounts for each movement are usually the same.*

Since there is a variation in range of movement in normal subjects, the assessment of an injured wrist should be done by comparing it with the normal wrist of the same person and not with the 'average' wrist.

- *LOOK again at the wrist moving in flexion and extension. Notice that in flexion the hand moves slightly towards the ulnar side, and in extension towards the radial side.*

The reason for this difference is the shape of the lower end of the radius, as was noted in the rest position.

- *HOLD a mug of coffee or large tool, e.g. a hammer, in the hand. Note that the forearm is in the mid prone position and the weight of the mug or tool is tending to pull the wrist into ulnar deviation, so that the abductors of the wrist must work statically to hold the position.*

When the muscles are weak, the unsupported hand falls into flexion or ulnar deviation when holding a load.

6.3.1 Flexors of the wrist

The two main muscles which flex the wrist are the *flexor carpi ulnaris* and *flexor carpi radialis*. The palmaris longus is another wrist flexor which lies between the other two, but it is absent in 15% of people.

All three muscles have a common origin on the medial epicondyle of the humerus, and lie superficially on the anterior side of the forearm.

The **flexor carpi ulnaris** is attached to the pisiform bone and on to the base of the fifth metacarpal (Fig. 6.3).

The **flexor carpi radialis** lies deep to the muscles at the base of

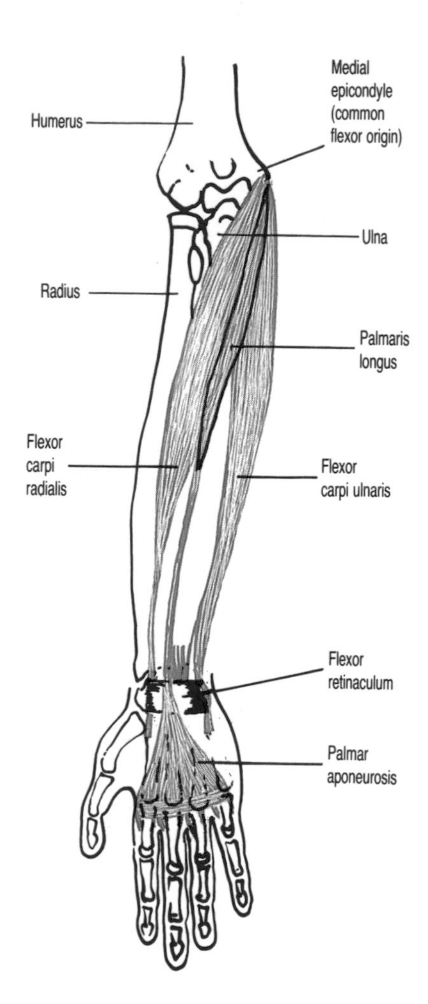

Fig. 6.3. Anterior view of the right forearm and hand to show the position of the three wrist flexors. Superficial layer.

Fig. 6.4. Combing the hair. The wrist flexes to draw the comb through the hair close to the scalp.

the thumb as it crosses the wrist and ends at the bases of metacarpals 2 and 3 (Fig. 6.3).

The palmaris longus has a long thin tendon which inserts into the palmar aponeurosis, a layer of dense fibrous tissue below the skin of the palm, considered in more detail in Section 6.4.1.

• *MAKE a fist and flex the wrist to see the flexor tendons appear on the anterior aspect. The palmaris longus is in the mid line with the flexor carpi ulnaris medial to it, attached to the pisiform. The flexor carpi radialis laterally may be more difficult to find.*

A functional use of the wrist flexors can be seen in Fig. 6.4 where they are used to counteract the resistance offered by hair on a comb.

6.3.2 Extensors of the wrist

The three muscles that extend the wrist are the *extensor carpi ulnaris*, and the *extensor carpi radialis longus* and *brevis* (Fig. 6.5). The long radial extensor takes origin on the ridge above the lateral epicondyle of the humerus with the, brachioradialis, already described in Chapter 5. The other two muscles are attached to the lateral epicondyle which is the common extensor origin. All three muscles pass down the posterior side of the forearm and insert at the wrist following the same pattern as the flexors.

1 The extensor carpi radialis longus into metacarpal 2.
2 The extensor carpi radialis brevis into metacarpal 3.
3 The extensor carpi ulnaris into metacarpal 5.

Note: the flexors insert into the anterior or palmar side, and the extensors insert into the posterior or dorsal side.

• *MAKE a fist and extend the wrist to see the extensor tendons on the posterior side. Extensor carpi radialis brevis is more central and may be difficult to feel, as it is crossed by tendons of muscles passing to the thumb.*

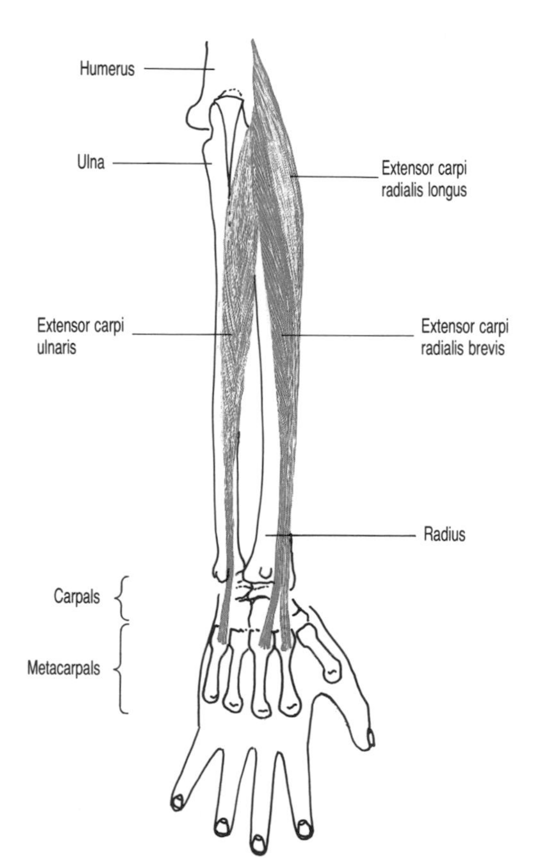

Fig. 6.5. Posterior view of the right forearm to show the position of the three wrist extensors.

In the use of the pronated hand, e.g. pressing the keys of a typewriter or piano (Fig. 6.6), the wrist extensors are active to lift the weight of the hand against gravity. Weakness of these muscles leads to 'wrist drop'. In strong gripping by the whole hand, the wrist extensors act as synergists to counteract flexion of the wrist by the long finger flexors.

Fig. 6.6. Position of the hand playing the keys of a piano. The wrist is held in extension to allow the fingers to move across the keys and press them down in rapid succession.

6.3.3 Abduction and adduction of the wrist

If the wrist is viewed in cross section, the flexor and extensor tendons involved in wrist movement can be seen around the oval shape of the carpus (Fig. 6.7). The tendons can combine in different ways, like the strings of a marionette, to produce each movement.

Contraction of the flexor and extensor tendons on the ulnar side will move the hand into adduction, often known as ulnar deviation. Similarly, contraction of flexor carpi radialis and extensor carpi radialis longus together will result in abduction of the wrist or radial deviation.

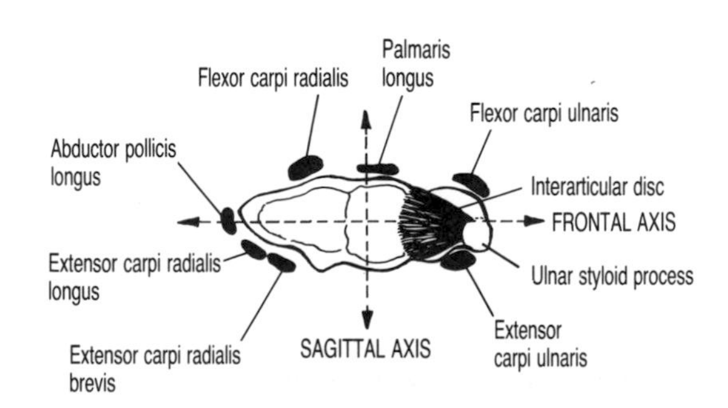

Fig. 6.7. Inferior aspect of the left radius and ulna to show the position of the tendons of the wrist flexors and extensors.

6.4 Movements of the hand

The hand performs complex and precision movements in the manipulation of utensils, tools and equipment in daily living. The increased use of electrically powered equipment in the home and in the work place has reduced the need for the hand to exert great power, but has introduced a greater variety of precision movements required to operate switches and controls.

A large number of muscles, originating in both the forearm and the hand are inserted into the fingers and thumb. Most of the tendons of these muscles pass over several joints, and the combinations of different directions of pull of the tendons allow the fingers to move in a variety of ways.

Before describing the muscles moving the hand, it is necessary to learn the terminology used in the description of the hand.

There are five digits numbered 1–5 from lateral (thumb) to medial. The fingers are usually identified by name: index finger, middle finger, ring finger, little finger.

The third metacarpal and the third finger form the central axis of the hand. When the fingers separate, the other fingers move away from the central axis.

The main joints of the hand are identified in Fig. 6.8, and illustrated in the Appendix. At the knuckles, the heads of the metacarpals articulate with the proximal phalanges at the metacarpophalangeal or MCP joints. In the fingers the joints between the phalanges are known as proximal interphalangeal joints or PIP, and distal interphalangeal or DIP joints (the thumb has an MCP joint, and only one interphalangeal — IP — joint). The movements at the MCP joints of the fingers are flexion, extension, abduction and adduction. The movements at the IP joints are flexion and extension only.

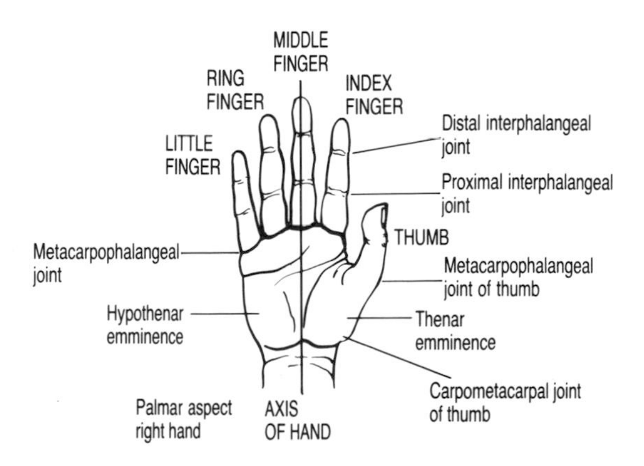

Fig. 6.8. Palmar view of the right hand.

The base of the first metacarpal bone in the thumb articulates with the trapezium, the most lateral bone in the distal row of carpals. The articulating surface on the trapezium is scooped out in two directions to form a unique saddle type joint which, combined with a loose capsule, allows the thumb considerable mobility. Movements of the MCP and IP joints are only in one plane, i.e. flexion and extension.

In the rest position of the thumb, the palmar surface faces inwards, since the first metacarpal is medially rotated at the first carpometacarpal joint. The planes of movement of the thumb are therefore at right angles to those of the fingers (Fig. 6.9).

Flexion of the thumb carries it across the palm in a plane at right angles to the thumb nail (Fig. 6.9a).

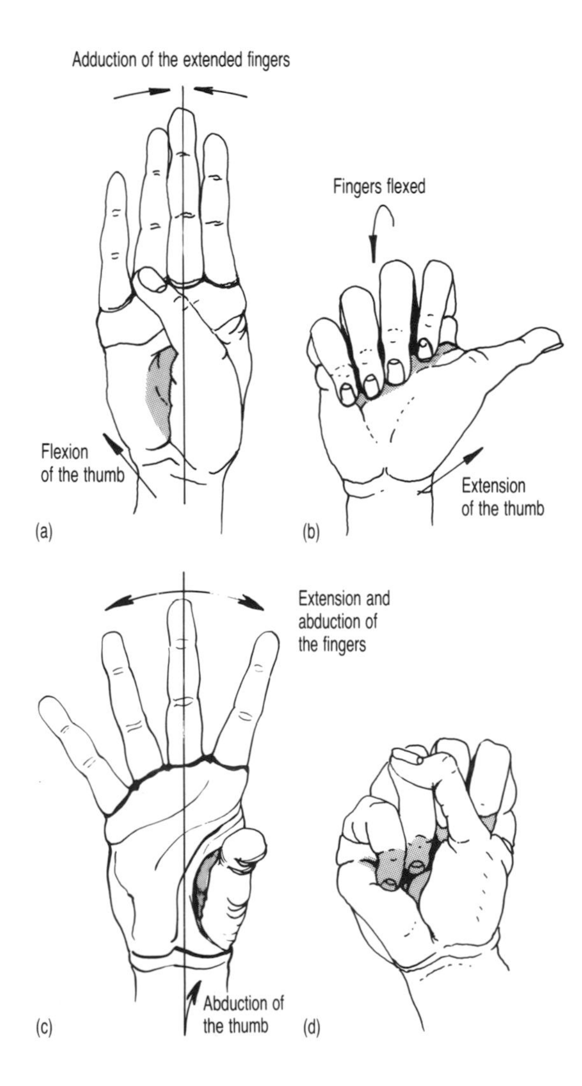

Fig. 6.9. Positions of the right hand seen in palmar view: (a) fingers extended and adducted, thumb flexed; (b) fingers flexed, thumb extended; (c) fingers extended and abducted, thumb abducted; and (d) fingers flexed, thumb in opposition.

Extension is the return movement from flexion and continues into the 'hitch a lift' position (Fig. 6.9b).

Abduction takes the thumb away from the palm of the hand and at right angles to it (Fig. 6.9c).

Adduction is the return movement from abduction, which pulls the thumb back towards the palm of the hand.

The first metacarpal is also able to rotate on the trapezium both medially and laterally. The combined movement of flexion, medial rotation and adduction, which brings the thumb into contact with the fingers, is known as **opposition** (Fig. 6.9d).

The thumb can be opposed to the fingers in a variety of ways (see Section 6.5).

During many functional activities, the hand closes round an object to hold or move it. In closing the hand, the fingers are flexed and adducted; the thumb is in opposition (Fig. 6.9d).

When the hand opens to release the object, the fingers and thumb are extended and abducted (Fig. 6.9c).

• *LOOK at your own hand. Starting at the base of the hand, notice the flexure line of the wrist, and then feel the shafts of the metacarpals on the back of the hand. Identify the MCP joints at the knuckles and check the movements that occur at these joints — flexion, extension, abduction and adduction.*
Identify the PIP and DIP joints and check the movements — flexion and extension only.
• *PALPATE the first metacarpal bone of the thumb, which moves independently of the other metacarpals.*
Move the thumb into flexion, extension, abduction, adduction and opposition.
• *OPEN and CLOSE the hand. Notice how the fingers and thumb abduct as they extend in opening the hand. The fingers and thumb adduct as they flex to close the hand.*

The muscles moving the hand will be described under three headings based on the functional use of the hand: (i) muscles that close the hand around an object to grasp it; (ii) muscles that open the hand in preparation for gripping; and (iii) muscles that move an object in a precise way.

6.4.1 Muscles closing the hand

The muscles closing the hand lie in the anterior side of the forearm deep to the wrist flexors, and in the palm of the hand. 'Digitorum' is included in the names of the muscles moving the fingers and 'pollicis' in those moving the thumb.

Forearm muscles

The forearm muscles that close the hand are : the *flexor digitorum superficialis*, the *flexor digitorum profundus* and the *flexor pollicis longus*.

The **flexor digitorum superficialis** originates at the medial side of the elbow with the wrist flexors, i.e. from the medial epicondyle of the humerus. The origin of the muscle continues diagonally across the forearm below the elbow, attached to the coronoid process of the ulna and the upper anterior shaft of the radius (Fig. 6.10a).

The **flexor digitorum profundus** lies deep to the superficialis and takes origin from the anterior and medial shaft of the ulna (Fig. 6.10b).

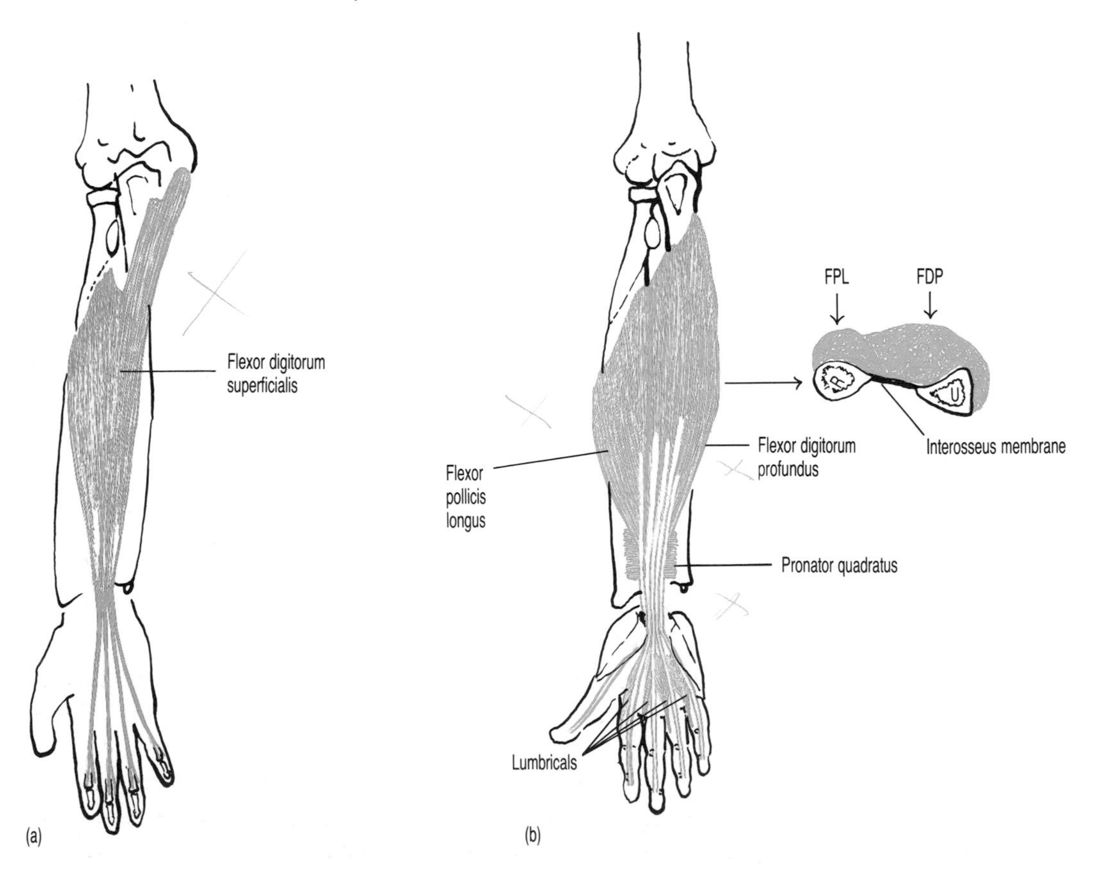

Fig. 6.10. Anterior view of right forearm and hand: (a) middle layer. Flexor digitorum superficialis; and (b) deep layer. Flexor digitorum profundus and flexor pollicis longus.

The **flexor pollicis longus** also lies deep to the superficialis and is attached to the anterior shaft of the radius (Fig. 6.10b). The flexor digitorum profundus and flexor pollicis longus appear as one muscle in the deep layer on the anterior side of the forearm covering the radius, the ulna and the interosseous membrane between them.

All three muscles pass down the anterior forearm to the wrist where the two muscles which insert into the fingers each divide into four tendons. Each of the tendons passes through the palm and over the palmar surface of each finger, where the flexor digitorum superficialis divides to insert into the sides of the middle phalanx, allowing the deeper flexor digitorum profundus tendon to pass onto the distal phalanx. Figure 6.17, p. 146 shows how these two muscles insert into each finger. The tendon of the flexor pollicis longus turns laterally to reach the thumb and insert into the base of the distal phalanx.

The three muscles together flex all the joints of the fingers and the thumb. The tendons of the index, ring and little fingers diverge from the axis of the hand from wrist to finger tip. This means that as the fingers flex, they also adduct towards each other.

Muscles of the hand

Five *intrinsic muscles* of the hand also assist the forearm muscles in closing the hand, acting on the thumb and little finger.

The **flexor digiti minimi** and **opponens digiti minimi** move the little finger and lie in the hypothenar eminence in the palm of the hand (Fig. 6.11a and b).

The **flexor pollicis brevis** and **opponens pollicis brevis** are comparable muscles in the thenar eminence below the thumb (Fig. 6.11a and b).

The **adductor pollicis** lies deep in the palm of the hand covered by the long flexor tendons and the thenar muscles (Fig. 6.11b).

A band of fibrous tissue, known as the flexor retinaculum crosses the palmar side of the carpal bones over the long flexor tendons. The thenar and hypothenar muscles originate from this retinaculum.

The flexor digiti minimi is inserted into the base of the proximal phalanx of the little finger, and the flexor pollicis brevis is attached to the proximal phalanx of the thumb. The opponens muscles are attached to the length of the shaft of the metacarpal bone of their corresponding little finger or thumb. During the opposition movement of the thumb, the shaft of the first metacarpal is rotated about its axis by the pull of the opponens pollicis. At the same time, the flexor draws the thumb across and towards the palm.

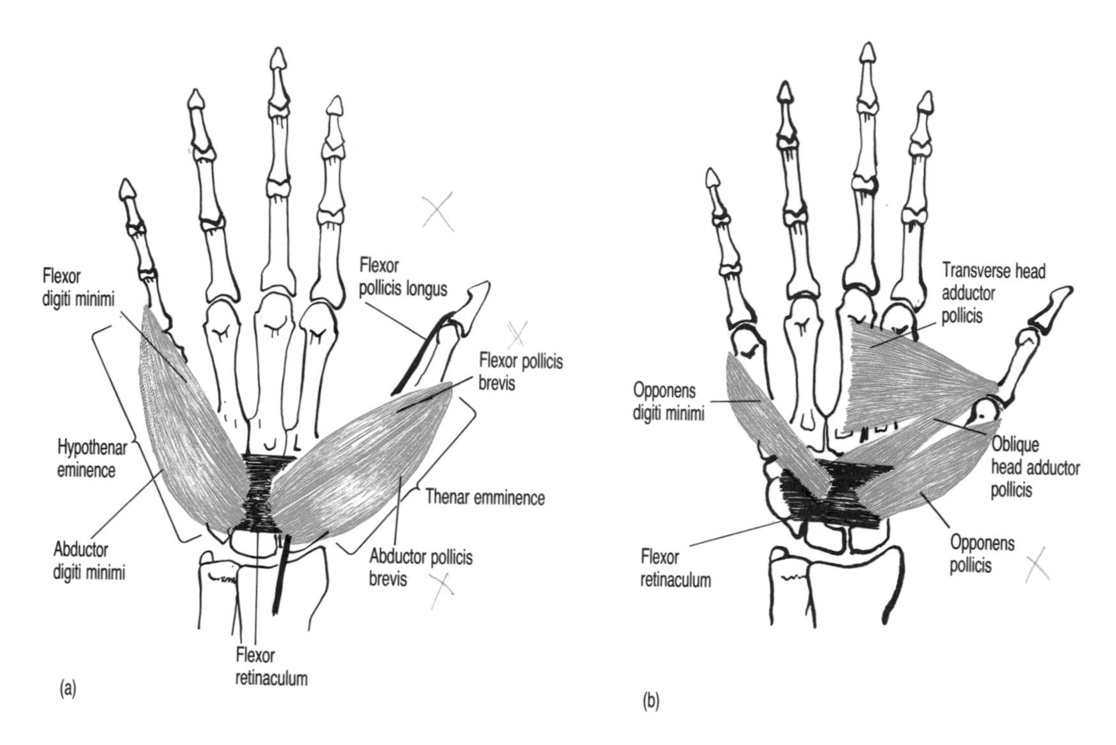

Fig. 6.11. Palmar view of the right hand: (a) flexor and abductor of the thumb and of the little finger; (b) opponens muscles and adductor pollicis.

The opponens digiti minimi increases the bulk of the medial border of the hand in a cupping movement used to grasp a round knob, such as the gear lever of a car.

The adductor pollicis is attached along a wide origin in the centre of the palm on the shaft of the third metacarpal and the capitate bone. This muscle forms the web of the thumb and inserts into the proximal phalanx of the thumb on the ulnar side (Fig. 6.11b). The adductor pollicis acts strongly to draw the thumb towards the hand in pinching movements between the thumb and index finger.

The connective tissues of the hand

The connective tissue in the *palm* of the *hand* plays an important role in the protection and binding of the muscles and tendons, so that smooth movement in the correct direction is achieved. Three particular sites of fibrous tissue are worthy of description: the *flexor retinaculum*, the *palmar aponeurosis* and the *flexor tendon sheaths*.

The flexor retinaculum of the wrist. The long finger flexors of the forearm enter the hand over the anterior side of the wrist. They are held in position by a band of fibrous tissue called the flexor retinaculum. This also provides a base for attachment of some of the thenar and hypothenar muscles (Fig. 6.11).

• *LOOK at the skeleton of the hand and note how the carpal bones form a trough on the palmar side for the long flexor tendons. Look at the arrangement of the carpal bones and find four raised bony points on either side of this trough. These are: the pisiform and hook of the hamate medially, and the tubercle of the scaphoid and crest of the trapezium laterally.*

The flexor retinaculum stretches across the carpal bones, converting the trough into a tunnel known as the *carpal tunnel* (Fig. 6.12). Note: the exact position of the flexor retinaculum is across the base of the hand, i.e. under the heel of the hand, and not in the position of a bracelet round the wrist.

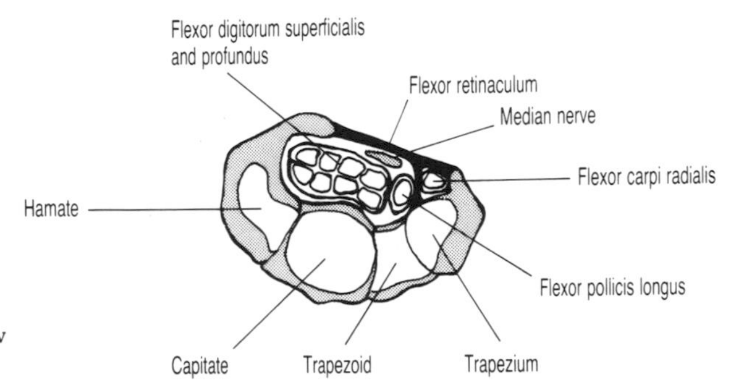

Fig. 6.12. Section through the carpus to show the carpal tunnel.

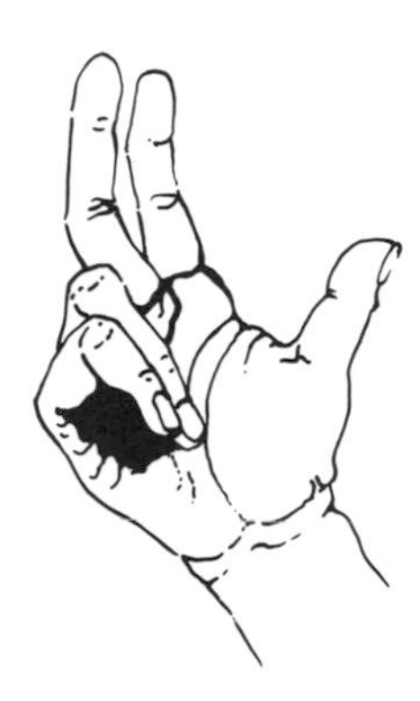

Fig. 6.13. Right hand with Dupuytren's contracture. Ring and little finger flexed into the palm by shrinkage of the palmar aponeurosis.

The palmar aponeurosis. The palmar aponeurosis is a triangular sheet of fibrous tissue covering all the long muscle tendons of the palm. The apex is joined to the flexor retinaculum at the wrist and receives the insertion of the palmaris longus (if this muscle is present) (see Fig. 6.3, p. 132). The sides of the triangle blend with the fascia covering the muscles of the thumb and little finger, and the sheet ends at the base of the fingers. The palmar aponeurosis is anchored to the metacarpals and to the deep transverse palmar ligament. The condition known as Dupuytren's contracture is when there is shrinkage of the fibrous tissue of the palmar aponeurosis, usually on the ulnar side. The little and ring fingers are pulled down so that they curl into the palm of the hand (Fig. 6.13).

The flexor tendon sheaths. As the long flexor tendons pass through the carpal tunnel and up over the palmar surface of each finger, they are wrapped in a double layer of synovial membrane known as a tendon sheath (Fig. 6.14). Each tendon sheath is held in

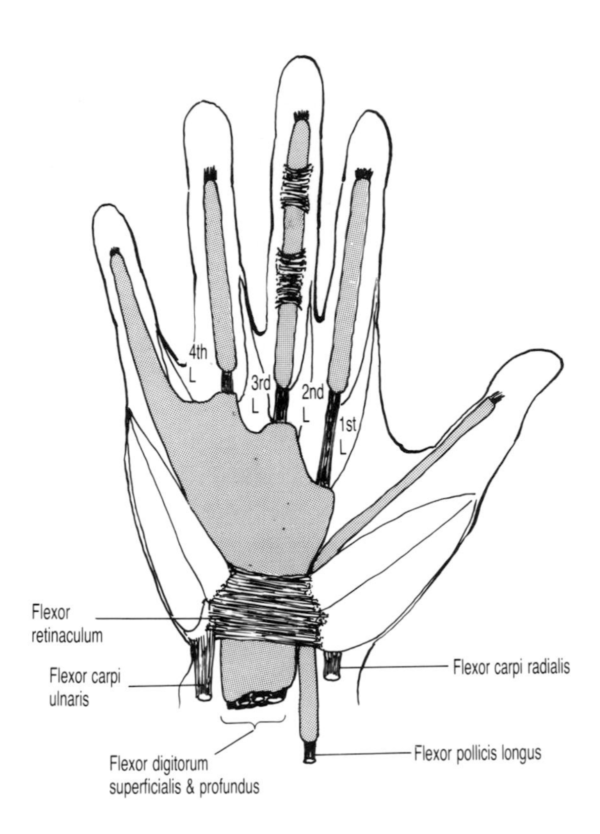

Fig. 6.14. Palmar view of right hand showing tendon sheaths of the long flexor tendons.

position on the palmar surface of the bones of the finger by fibrous bands forming tunnels. These fibrous bands are also joined to the palmar aponeurosis and are thin over the IP joints to allow flexibility of the fingers.

6.4.2 Muscles opening the hand

The muscles opening the hand lie in the posterior side of the forearm, and include one muscle in each of the thenar and hypothenar groups.

Forearm muscles involved in opening the fingers

The forearm muscles that open the fingers are the *extensor digitorum,* the *extensor indicis* and the *extensor digiti minimi.*

The **extensor digitorum** and the **extensor digiti minimi** originate with the wrist extensors from the lateral epicondyle of the humerous. The **extensor indicis**, a deep muscle, takes origin on the posterior border of the ulna. The tendons formed from these three muscles pass posteriorly over the wrist held down by a band of fibrous tissue, the extensor retinaculum. On the dorsal side of the hand, the extensor digitorum divides into four. The extensor indicis lies adjacent to the index finger tendon of the extensor digitorum and blends with it. The extensor digiti minimi lies medial to the other tendons and blends with the little finger tendon of the extensor digitorum (Fig. 6.15). The tendons of the muscles insert into the dorsal surface of the fingers via a complex arrangement of fibrous tissue known as the dorsal extensor expansion. This will be described in more detail at the end of Section 6.4.3.

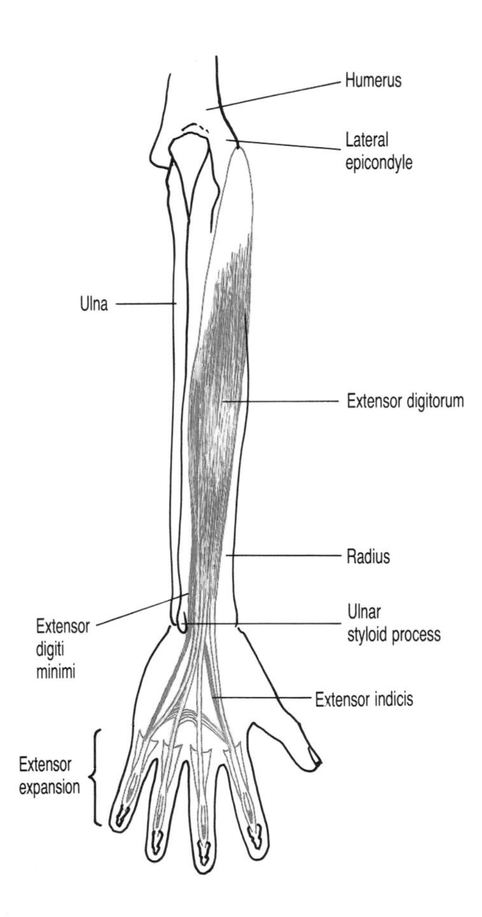

Fig. 6.15. Posterior view of the right forearm and hand showing the extensors of the fingers.

• *PALPATE the extensor tendons as they pass over the posterior side of the wrist and on to the back of the hand.*

• *OBSERVE how the long extensor tendons can be seen on the back of the hand when it is opened. Notice how the tendons are close together at the level of the wrist. The tendons of the index, ring and little fingers diverge away from the central axis of the hand to reach the fingers. The pull of the extensor tendons, therefore, abducts as well as extends these three fingers.*

Forearm muscles involved in opening the thumb

Three forearm muscles act in separating the thumb when opening the hand, the *abductor pollicis longus*, the *extensor pollicis longus* and the *extensor pollicis brevis.*

The three muscles originate from the posterior shaft of the radius and ulna as follows: (i) the abductor pollicis longus from the upper shaft of the radius and ulna; (ii) the extensor pollicis longus from the shaft of the ulna below; and (iii) the extensor pollicis brevis from the shaft of the radius below.

All three muscles pass deep to the extensor digitorum and become superficial on the lateral side of the wrist to reach the thumb (Fig. 6.16). At the base on the thumb they form the borders of the 'anatomical snuff box'. These long muscles of the thumb are called the 'deep outcropping muscles' of the forearm, since they begin deep in the posterior forearm and emerge near to the surface on the radial side at the wrist. Each muscle inserts into a different bone in the thumb: (i) the abductor pollicis longus inserts into the first metacarpal; (ii) the extensor pollicis brevis inserts into the proximal phalanx; and (iii) the extensor pollicis longus inserts into the distal phalanx

- *OBSERVE the 'anatomical snuff box' by extending the thumb with the wrist extended. A depression appears bounded by tendons below the thumb.*
- *PALPATE the abductor pollicis longus and extensor pollicis brevis lying together in the same boundary of the 'snuff box'. The other dorsal boundary is formed by the tendon of the extensor pollicis longus, which uses the dorsal tubercle of the radius to change direction at the wrist.*

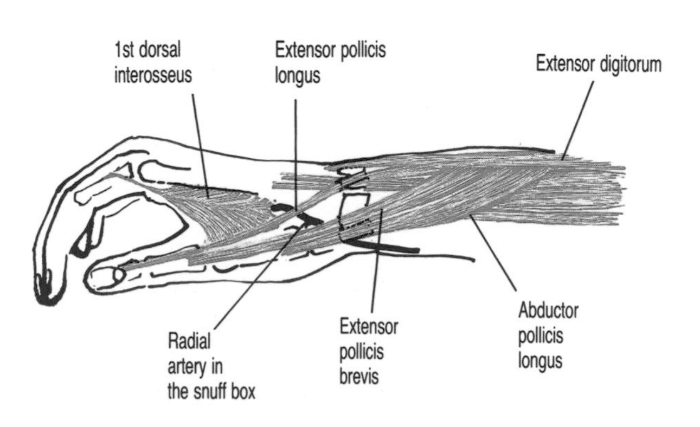

Fig. 6.16. Radial side of the right wrist and hand showing the muscles of the 'anatomical snuff box'.

Intrinsic muscles involved in opening the hand

Two intrinsic muscles of the hand assist the forearm muscles in opening the hand, acting on the thumb and the little finger: (i) the *abductor pollicis brevis* lies in the thenar eminence; and (ii) the *abductor digiti minimi* lies in the hypothenar eminence. Both these muscles originate at the flexor retinaculum at the palmar side of the base of the hand.

The **abductor pollicis brevis** is inserted into the base of the proximal phalanx of the thumb on the lateral side (see Fig. 6.11a, p. 139). Note: the thumb faces inwards at right angles to the palm, so that when the abductor pollicis brevis contracts, it draws the thumb away from the palm by movement at the base of the first metacarpal (saddle joint of thumb).

The fibres of the **abductor digiti minimi** originate from the flexor retinaculum and pisiform bone, and insert into the base of the proximal phalanx of the little finger on the medial side (see Fig. 6.11a).

The action of opening the hand is important in releasing a grip and in placing an object on a surface. A young baby can grasp a toy in the hand, but drops it randomly. At a later stage, when coordination between opposing groups of muscles has developed, the child can then put the toy down voluntarily and precisely as the hand opens.

6.4.3 Precision movements

The fingers and the thumb perform a variety of skilled movements: for example, alternate action of flexors and extensors at all the joints of the fingers is required to press the keys of a typewriter or a piano. While the fingers and thumb grip a pen or paintbrush, fine movements of flexion and extension of the distal joints carry the pen or brush over the paper.

Three sets of intrinsic muscles deep in the palm of the hand are important in precision movements: the *lumbricals*, the *dorsal interossei* and the *palmar interossei*. The lumbricals are four small muscles which originate from the tendons of the flexor digitorum profundus, the deepest long finger flexor in the palm (Fig. 6.17). Each muscle passes in front of the MCP joint of the corresponding finger, passes backwards on the radial side of this joint, and inserts into the dorsal surface of the finger on the radial side. The detail of the insertion will be considered later with the description of the dorsal extensor expansion of the fingers.

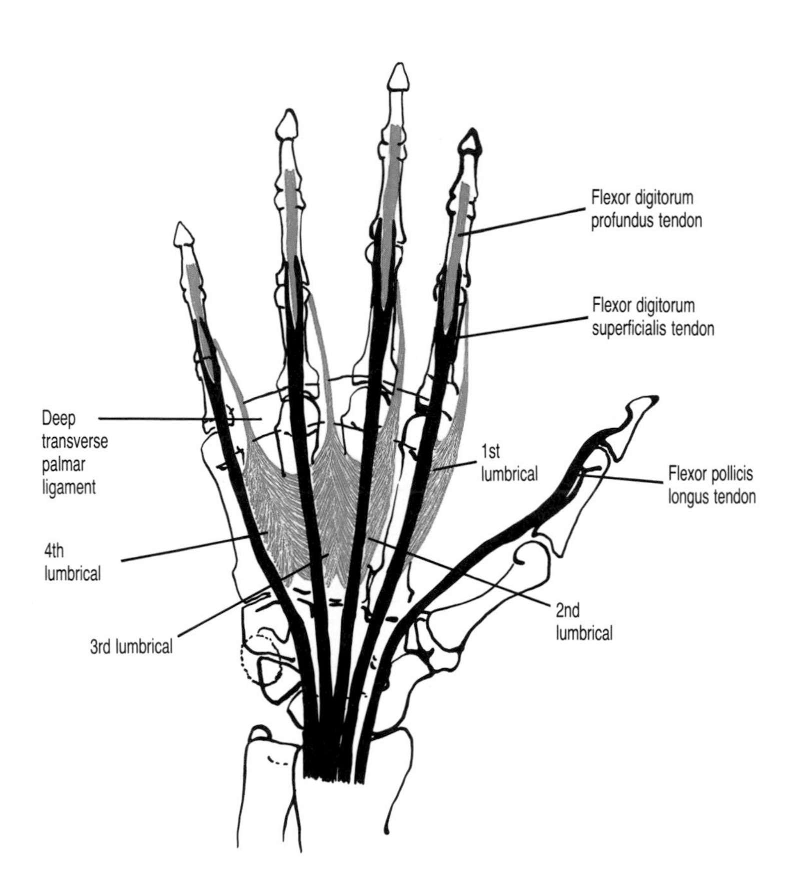

Fig. 6.17. Palmar view of the right hand to show the position of the lumbrical muscles.

• *LOOK carefully at a hand skeleton and your own hand. Work out how the lumbricals begin in the palm with the long flexor tendons, and end on the dorsal side of each finger, passing round the thumb side of the MCP joint.*

The **lumbricals** can act in flexion of the MCP joints and extension of the IP joints. Since they link the long flexor muscle to the long extensor muscle, their main function is to act as a bridge between the two, which balances the flexion and extension movements of the fingers.

There is evidence that the lumbricals are active in all fine movements of the fingers.

The interosseous muscles lie in the spaces between the metacarpal bones. There are four **dorsal interossei** originating from the sides of adjacent shafts of metacarpals 1–5, deep to the extensor tendons. Their insertions are best understood from a diagram, see Fig. 6.18a. Note: the two lateral dorsal interosseous muscles, i.e. on the thumb side, pass on the radial side of the metacarpal joints of the index and middle fingers. The medial two muscles pass on the

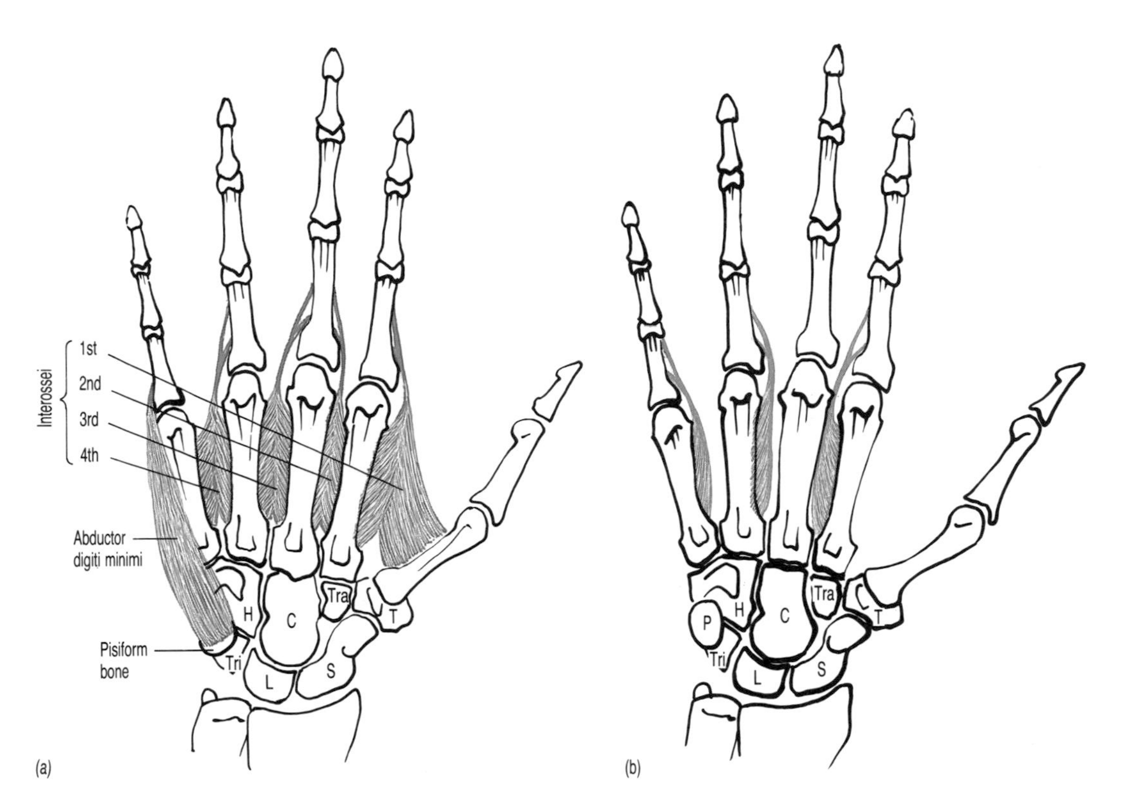

Fig. 6.18. Palmar view of the right hand: (a) dorsal interossei and abductor digiti minimi; and (b) palmar interossei.

ulnar side of the metacarpal joints of the ring and middle fingers. The tendons of all four muscles reach the dorsal surface of the fingers to blend with the outer bands of the extensor expansion in the index, middle and ring fingers, just beyond the level of the MCP joints. The thumb and little finger have abductors of their own which have already been described.

Action of all four dorsal interossei will spread the fingers away from the central axis of the hand. The middle finger has two dorsal interosseus muscles, and therefore can abduct from the central axis to either side. The attachment of each tendon into the dorsal surface of the finger means that each muscle will also assist in extension of the DIP joints.

The dorsal interossei can be palpated between the shafts of the metacarpal. When these muscles are wasted, due to nerve damage, the skin sinks between the metacarpals and the back of the hand looks like a skeleton.

• *PALPATE the first dorsal interosseous muscle by abducting the index finger whilst the thumb is in abduction.*

The three **palmar interossei** lie on the palmar side of the dorsal interossei (Fig. 6.18b shows their arrangement). Each is attached to one side of a metacarpal shaft, and is inserted into the outer band of the dorsal extensor expansion of the same finger. From their attachments it can be seen how they will draw the fingers together in adduction when they contract.

One way to remember the actions of the two sets of interossei is by the initials: Dorsal ABduct — DAB; Palmar ADduct — PAD.

Both the dorsal and the palmar interossei cooperate with the lumbricals in flexion of the MCP joint and extension of the IP joints.

Extensor (dorsal) expansion of the fingers

The mechanism for insertion of muscles onto the dorsal surface of the fingers will now be considered.

The extensor expansion is found on the dorsal surface of each proximal phalanx. The tendon of the extensor digitorum divides into three as it crosses the MCP joints. The middle band is inserted into the base of the middle phalanx, and the outer bands into the base of the distal phalanx (Fig. 6.19a). The outer bands receive the insertions of the lumbricals and interossei. The lumbricals are all inserted into the outer band on the radial side. Fine transverse fibres spread out from the middle band, covering the head of the metacarpal and are attached to the palmar ligament of the MCP joint. This 'extensor hood' prevents any bowstring of the extensor

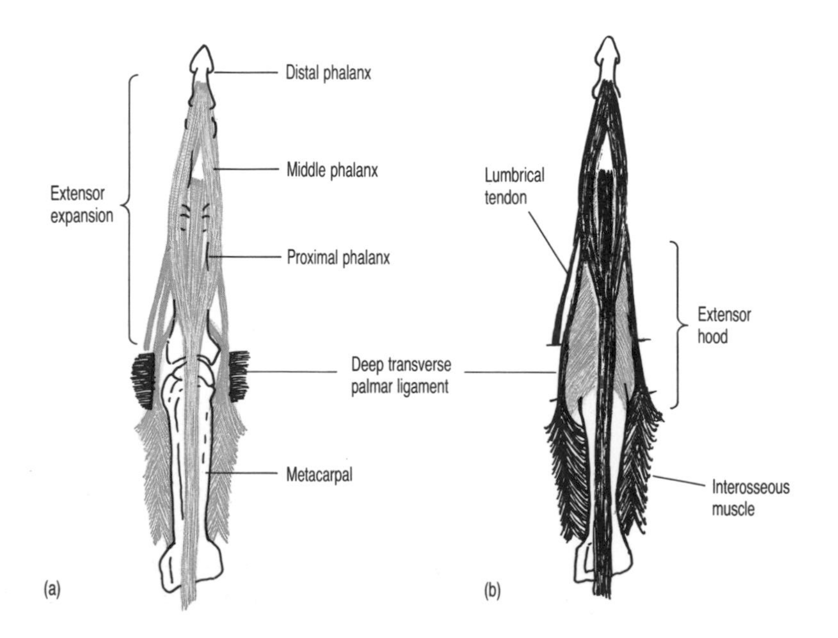

Fig. 6.19. Dorsal aspect of the middle finger (right): (a) the extensor expansion; and (b) the extensor hood.

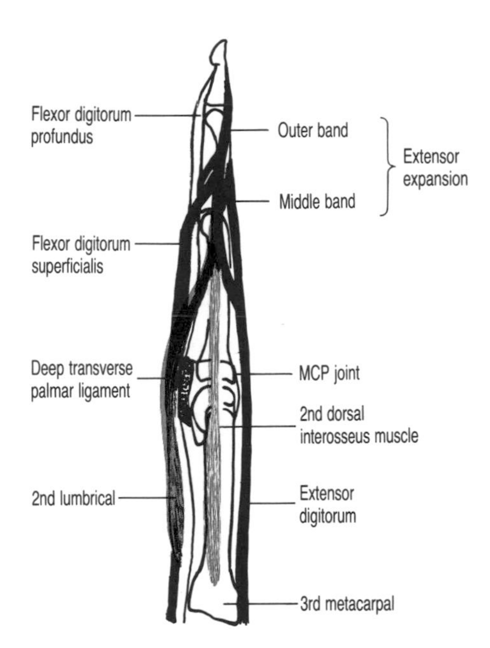

Fig. 6.20. Radial side of the right middle finger to show the relative positions of the insertions of the second lumbrical and second dorsal interosseous muscles.

tendon (Fig. 6.19b). The relative positions of the insertions of a lumbrical muscle and a dorsal interosseous muscle can be seen in the radial view of a finger seen in Fig. 6.20. The lumbrical lies on the palmar side of the metacarpal at first, and then crosses the MCP joint to insert into the outer band of extensor digitorum tendon. In this way the lumbricals can flex the MCP joint and extend the IP joints of the fingers. The interosseous muscles lie in parallel with the metacarpals, and are held down by the extensor hood at the MCP joint. The interossei pull on the outer band of the extensor expansion to produce abduction and adduction of the fingers.

The functional significance of the extensor expansion arrangement of the fingers is to allow a variety of combinations of movements at the three joints of the fingers to perform precision movements.

• *LOOK at the back of your own hand, and work out which lumbrical and interosseous muscles blend with the extensor expansion of each finger.*

6.5 Types of grip

The hand is used in a variety of ways to grasp and hold handles, tools, levers and so on. The different types of grip made by the hand in daily activities involve particular movements at the various joints of the hand, and the combination of activity in muscle groups in the forearm and hand. The ability to grip numerous objects is an important part of the assessment of the damaged hand.

• *OBSERVE the different ways that people use their hands to grip objects over a whole day, while dressing, cooking, eating, travelling, working and during leisure.*

The type of grip selected depends upon the shape of the object to be grasped, what we want to do with it and the texture of its surface. Naming all the different types of grip is a difficult task as the hand is used in such a variety of ways and individuals approach each method of grasp according to their own style of working.

There are two main varieties of grip: (i) power grips; and (ii) precision grips.

6.5.1 Power grips

In the power grips all the fingers are flexed around an object. The thumb is curled round in the opposite direction to press against, or meet the fingers around the object. All the muscles of the fingers and thumb that close the hand are active. Both the thenar and hypothenar muscles keep the hand in contact with the object grasped.

The hypothenar muscles are important to stabilize the medial side of the palm against the handle, and the muscles of the fingers and the thumb grip the object firmly. The wrist extensors are active to give a stable base for the gripping action, they increase the tension in the long finger flexors and prevent them from acting on the wrist as well. As the hand grips harder the wrist extensors increase their activity.

The power grips bring the maximum area of sensory surface of the fingers, thumb and palm into contact with the object being grasped, so that feedback from the sensory receptors in these areas ensures that exact pressure and control is being exerted on to the handle or tool.

The power grip is the most primitive grasping movement. One of the primary reflexes of the newborn baby is finger flexion in response to touching the palm. By 6 months, the whole hand can form a palmar grasp with the thumb in opposition. Exertion of power by the finger flexors requires the additional group action of the wrist extensors and elbow stability which does not develop until later. By the fifth year the child can grip strongly with each hand individually.

The unique feature of the power grip is to hold an object firmly so that it can be moved by the more proximal joints of the upper limb, such as the shoulder, elbow or radioulnar joints. For example, the hand grasps the door handle but it is the elbow and shoulder muscles that press it down, and the radioulnar joint that turn the knob. The hand moulds itself to the shape of the object grasped in the power grip before the power is exerted to move it.

1 The **cylinder grip** is used for handles that lie at right angles to the line of the forearm, such as a racquet, a jug handle or the hand brake of a car (Fig. 6.21a). The skin of the palmar surface of the fingers and the palm over the metacarpophalangeal joints curves round the handle, and the thumb lies in opposition over the finger tips.

2 The **ball grip** encompasses circular knobs, balls and the top of mugs or jam jars. The fingers and thumb adduct onto the object and very often the palm of the hand is not involved (Fig. 6.21b).

3 The **hook grip** is used for carrying a suitcase, bucket or shopping bag by the side of the body with a straight elbow and wrist. Only the flexed fingers are used in this grip, the thumb is not involved (Fig. 6.21c). Following a median nerve lesion when the thumb cannot be opposed, the hook grip is the only power grip possible.

4 Where a tool or object is being used in line with the forearm, such as a hammer, screwdriver or trowel, the fingers flex around the handle in a graded way — maximum degrees of flexion in the little finger and least in the index finger. The thumb either lies over the

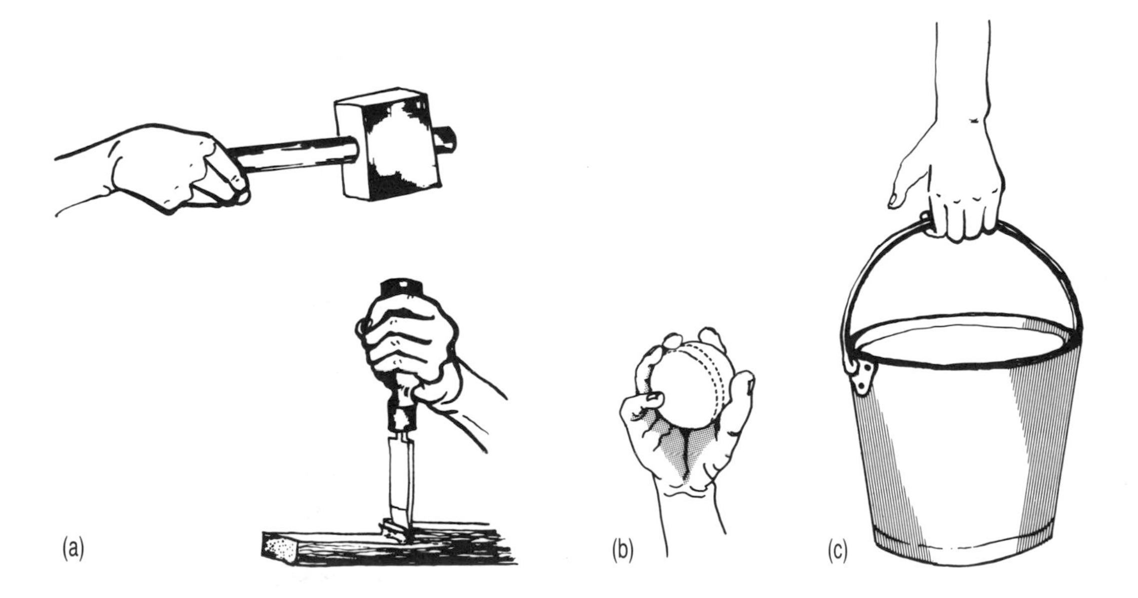

Fig. 6.21. Power grips: (a) cylinder; (b) ball; and (c) hook.

finger tips or lies along the handle of the tool being grasped. The wrist is ulnar deviated and the maximum area of the skin of the palm, thenar and hypothenar emminences is in contact with the handle of the tool. This is a grip giving considerable control, together with powerful manipulation of the tool (Fig. 6.21a).

6.5.2 Precision grips

The hand in the precision grip holds an object between the tips of the thumb and one, two or three fingers, e.g. holding a pencil or small tool. The intrinsic muscles of the hand are now involved in cooperation with the long flexors and extensors of the digits. The hand is positioned by the wrist and forearm, and the gripping is performed by the muscles acting on the joints of the fingers and thumb.

The precision grip is a more advanced manipulative movement than the power grip, appearing around 9 months of age in child development. Complex integration of the flexor–extensor mechanism of the fingers is essential for grasping a small object and moving it precisely.

The digits have serially arranged joints to perform these manipulative movements, the thumb has three: the first carpometacarpal (CMC) joint, the metacarpophalangeal (MCP) joint and the interphalangeal (IP) joint; each finger also has three: the metacarpophalangeal (MCP) joints; the proximal interphalangeal (PIP) joints and the distal interphalangeal (DIP) joints. It is the variety of movements at all these joints that combine to execute the different precision grips. The lumbrical and interosseus muscles form the balancing forces between the long finger flexors and extensors, and the intrinsic muscles of the thumb bring the pad of the thumb into opposition.

1 The **plate grip**. The MCP joints of the fingers are flexed with the IP joints extended, the thumb is opposed across the palmar surface of the fingers. This grip is used when holding a plate or other object that needs to be kept horizontal (Fig. 6.22a).

2 The **pinch grip**. The MCP and PIP joints of the index finger are flexed and the finger tip meets the opposed thumb. The DIP is pushed into extension in the finger and thumb. The pinch grip may include the middle finger (Fig. 6.22b).

3 The **key grip**. The extended thumb is held on the radial side of the index finger (Fig. 6.22c).

4 The **pincer grip**. All the joints of the index finger are flexed and the finger tip is brought into contact with the tip of the abducted thumb (Fig. 6.22d).

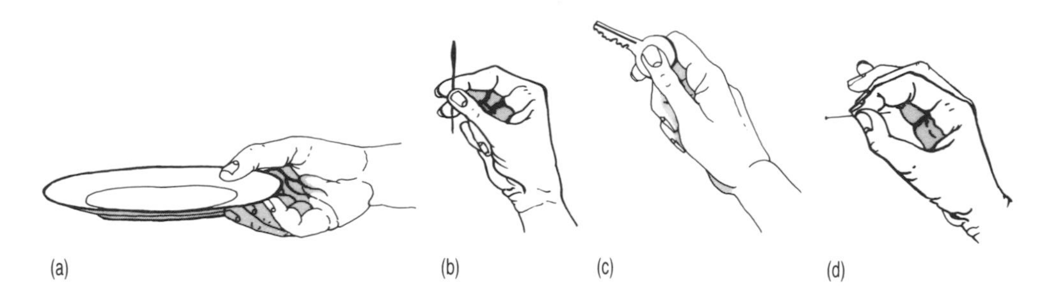

Fig. 6.22. Precision grips: (a) plate; (b) pinch; (c) key; and (d) pincer.

Alternative names for these grips are: (i) the lumbrical grip; (ii) the pad to pad grip; (iii) the lateral grip; and (iv) the tip to tip grip.

Manipulative movements

The bilateral activity in the two hands working together is an important feature of many manipulative tasks. The two hands may be performing similar movements, as in rolling pastry or pressing the keys of a keyboard, or one hand may provide stability while the other hand makes precise movements. The coordination of activity in the two hands is under the control of the nervous system. The hand is represented by large areas in both the sensory and motor cortex of the brain. Together with the motor centres of the brain stem and the cerebellum, there is the capacity to develop highly skilled symmetrical or asymmetrical movements in the two hands. For example the complex coordinated activity required to play many musical instruments (Fig. 6.23).

Fig. 6.23. Highly skilled bilateral movements of the hands.

6.6 Summary of forearm muscles and intrinsic muscles of the hand

The muscles of the forearm and hand have been described in three functional groups. For revision purposes, the muscles will now be grouped in their anatomical position with notes on common points of origin to assist learning of the attachments of the individual muscles.

The **forearm muscles** lie in the following positions:

Anterior:

1 Superficial layer:
Pronator teres, flexor carpi radialis, palmaris longus, flexor carpi ulnaris (common flexor origin is the medial epicondyle of the humerus).

2 Middle layer:
Flexor digitorum superficialis.
3 Deep layer:
Flexor digitorum profundus, flexor pollicis longus, pronator quadratus.

Posterior:
1 Superficial layer:
Brachioradialis, extensor carpi radialis longus and brevis, extensor digitorum, extensor digiti minimi, extensor carpi ulnaris, anconeus (common extensor origin is the lateral side of the elbow).
2 Deep layer:
Supinator, abductor pollicis longus, extensor pollicis longus and brevis, extensor indicis (origins from the posterior surface of the radius and ulna).

The twelve posterior muscles can be divided into the following:
1 3 act on elbow and radioulnar joints — brachioradialis, supinator and anconeus.
2 3 act to extend the wrist — extensor carpi ulnaris, extensor carpi radialis longus & brevis.
3 3 act to extend the fingers — extensor digitorum, extensor indicis and extensor digiti minimi.
4 3 act on the thumb — extensor pollicis longus and brevis, abductor pollicis longus.

The **intrinsic muscles of the hand** are arranged as follows:

Palmar view:
1 Thenar muscles — bulge of muscles found below the thumb:
Flexor pollicis brevis, abductor pollicis brevis and opponens pollicis.
2 Hypothenar muscles — bulge found below the little finger:
Flexor digiti minimi, abductor digiti minimi and opponens digiti minimi.

The six thenar and hypothenar muscles all originate on the flexor retinaculum at the base of the hand. The three thenar muscles are the mirror image of the three hypothenar muscles and vice versa. The opponens muscles of the two eminences are deep as they are inserted into the metacarpal shafts.

Deep muscles of the palm of the hand:
Lumbricals, palmar interossei, dorsal interossei, adductor pollicis.

At the end of this chapter, you should be able to:

1 List the functions of the hand.

2 Describe the movements of the forearm, the wrist and the hand, and discuss their importance in daily activities.

3 Describe the position, attachments and actions of the muscles that:

(a) pronate and supinate the forearm;

(b) move the wrist in flexion, extension, abduction and adduction;

(c) move the fingers and thumb.

4 Describe and give examples of power and precision grips.

7 / The Nerves of the Upper Limb

7.1 Introduction

The lower spinal nerves in the neck (C5, C6, C7, C8, T1) provide the nerve supply to the whole of the upper limb. Movement in the limb as a whole depends on activity in these five spinal nerves which form the roots of the *brachial plexus*. The nerves branch and join in a complex manner as they pass under the clavicle and over the first rib to reach the axilla.

Five terminal branches of the plexus are formed in the axilla. Each nerve is concerned with certain upper limb movements.

1 **Axillary nerve:** shoulder movement.
2 **Radial nerve:** extensors of the elbow, wrist and fingers.
3 **Musculocutaneous nerve:** flexors of the elbow.
4 **Median nerve:** flexors of wrist and fingers, grip of thumb.
5 **Ulnar nerve:** fine movements of the fingers.

Injury to nerves in the neck can have a widespread effect on upper limb movements. In the embryo, as the upper limb grows out from the sides of the trunk, the nerve from the central segment, C7, grows down towards the end of the limb (Fig. 7.1). The spinal nerves from segments C5 and C6 join to form the upper trunk of the plexus and supply the lateral border of the limb. C8 and T1 unite to form the lower trunk and its branches supply the medial border of the limb. The dermatomes lie in order down the lateral side of the limb, across the hand and up the medial side (Fig. 7.1b). In general, the nerves supplying the muscles of the shoulder originate from the upper segments (C5 and C6), and those concerned with movements of the fingers, from the lower segments (C8 and T1).

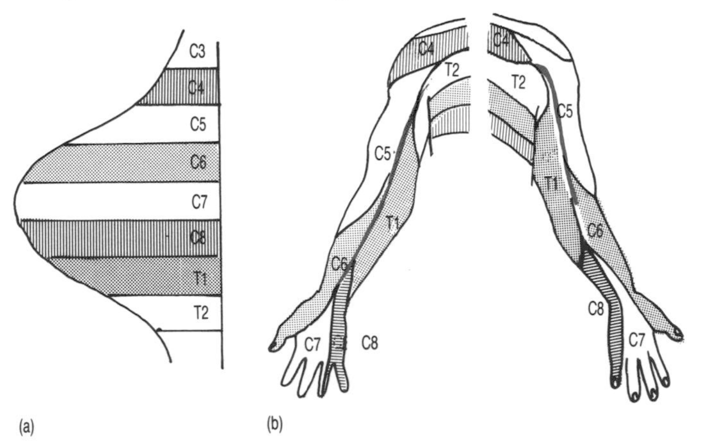

Fig. 7.1. Distribution of spinal segments C5 to T1 to the skin of the upper limb: (a) limb bud in the embryo, showing segmental distribution; and (b) anterior and posterior view of dermatomes in the adult showing mature organization of the spinal segments.

7.2 The brachial plexus (Fig. 7.2)

The *roots* of the brachial plexus are the anterior primary rami of the spinal nerves C5, C6, C7, C8 and T1.

Three *trunks* are formed from the five roots. The upper two roots

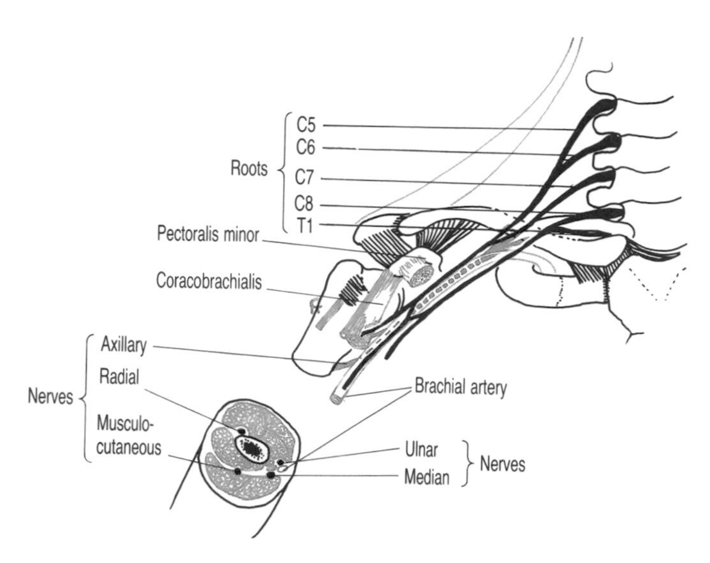

Fig. 7.2. Position of the right brachial plexus in relation to the clavicle, the first rib and the axilla. Position of the terminal branches in cross section of the arm.

join, the lower two roots join, and the middle root continues. These three trunks pass downwards and laterally between two muscles of the neck — the scalenus anterior and medius. The trunks meet the subclavian artery and continue with it behind the clavicle and over the first rib. Each trunk then divides into *anterior* and *posterior divisions*, the posterior divisions forming the nerves to the posterior muscles of the limb; and the anterior divisions, the nerves to the anterior muscles. The six divisions formed in this way continue to the axilla, then combine to form three cords.

The three cords are formed in the following way:

Three posterior divisions combine to form the *posterior cord.*
Two anterior divisions from upper and middle trunks form the *lateral cord.*
One anterior division of lower trunk becomes the *medial cord.*

The cords lie in the axilla close to the axillary artery, which forms the blood supply to the upper limb. At the lower part of the axilla, the cords split into the named nerves which enter the arm.

The **posterior cord** represents the *extensor* nerve of the upper limb.
The **medial** and **lateral cords** represent the *flexor* nerves of the limb.

• *LOOK at the articulated skeleton to identify the exact position of the brachial plexus, starting at the cervical vertebrae, passing over the first rib under the clavicle, to the axillary region below the shoulder joint.*
• *STUDY the plan of the brachial plexus shown in Fig. 7.3 to see the arrangement of the roots, trunks, divisions and cords.*
• *IDENTIFY: (i) the branches leaving the plexus which supply most of*

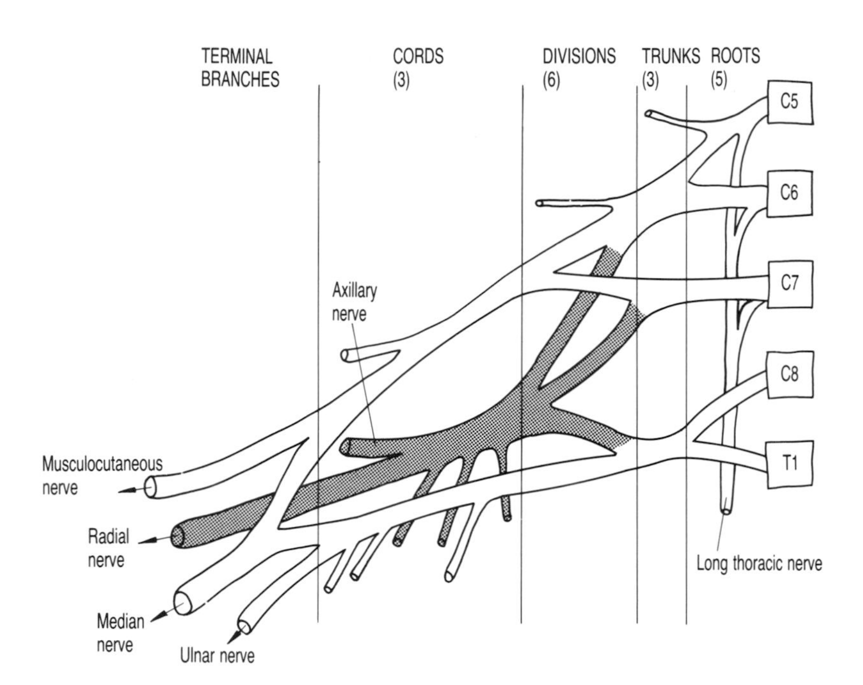

Fig. 7.3. Plan of the brachial plexus and its main branches.

the muscles of the shoulder region; and (ii) the terminal branches that enter the arm.

Brachial plexus lesions

The brachial plexus may be damaged in a variety of ways.
1 At birth.
2 By traction injuries to the neck.
3 Traction on the outstretched hand.
4 Through compression in the axilla.

The resulting loss of function is variable. The upper roots C5 and C6 may be damaged in **1**, **2**, or **3**, when there is loss of function in the abductors and flexors of the shoulder, and flexors and extensors of the elbow. The arm cannot be lifted from the side, and hangs in a position of adduction, medial rotation, pronation and finger flexion — waiter's tip position, known as Erb's paralysis.

Damage to the lower roots produces weakness of the intrinsic muscles of the hand, especially on the medial side, which is the ulnar or 'power' side. This is known as Klumpke's paralysis.

7.3 Terminal branches of the brachial plexus

Five terminal branches are formed from the three cords in the axilla, and enter the arm. One of the branches can be considered as two parts, one from each of the lateral and medial cords. Then each of the cords has two terminal branches.

The **posterior cord** forms the *radial* and *axillary nerves.*
The **medial cord** forms the *ulnar* and medial half of *median nerve.*
The **lateral cord** forms the *musculocutaneous* and lateral half of *median nerve.*

- *Trace from the roots to the formation of the five nerves in Fig. 7.3.*

Note: the ulnar nerve originates from the lower roots of the plexus. The radial nerve has fibres from all the roots.

7.3.1 Shoulder movement — the axillary nerve

The axillary nerve, a branch of the posterior cord, is important in all movements that lift the arm from the side, since it supplies the deltoid muscle and teres minor (Fig. 7.4). From the posterior cord, the axillary nerve branches backwards under the capsule of the shoulder joint, and winds round the surgical neck of the humerus to supply the whole of the deltoid muscle. A branch to the teres minor continues as a cutaneous nerve supplying the skin over the deltoid.

The other muscles moving the shoulder are mainly supplied by branches of the roots, the upper trunk and the three cords (see Section 4).

Injuries of the axillary nerve

These may occur in fractures of the humerus neck, or subluxation of the shoulder joint. The resulting loss of function is the inability to make movements that lift the arm away from the body.

7.3.2 Extensor nerve of the upper limb — the radial nerve

The radial nerve is the largest branch of the brachial plexus, formed as the continuation of the posterior cord (Fig. 7.4). In the arm, the radial nerve supplies the whole of the triceps muscle. The nerve is essential for extension movement of the elbow, since the triceps is the only muscle capable of this movement with any power (see Chapter 5).

In front of the lateral epicondyle of the humerus at the elbow, the nerve divides into two.

The *superficial terminal branch* continues along the lateral side of the forearm under the brachioradialis. Just above the wrist, the

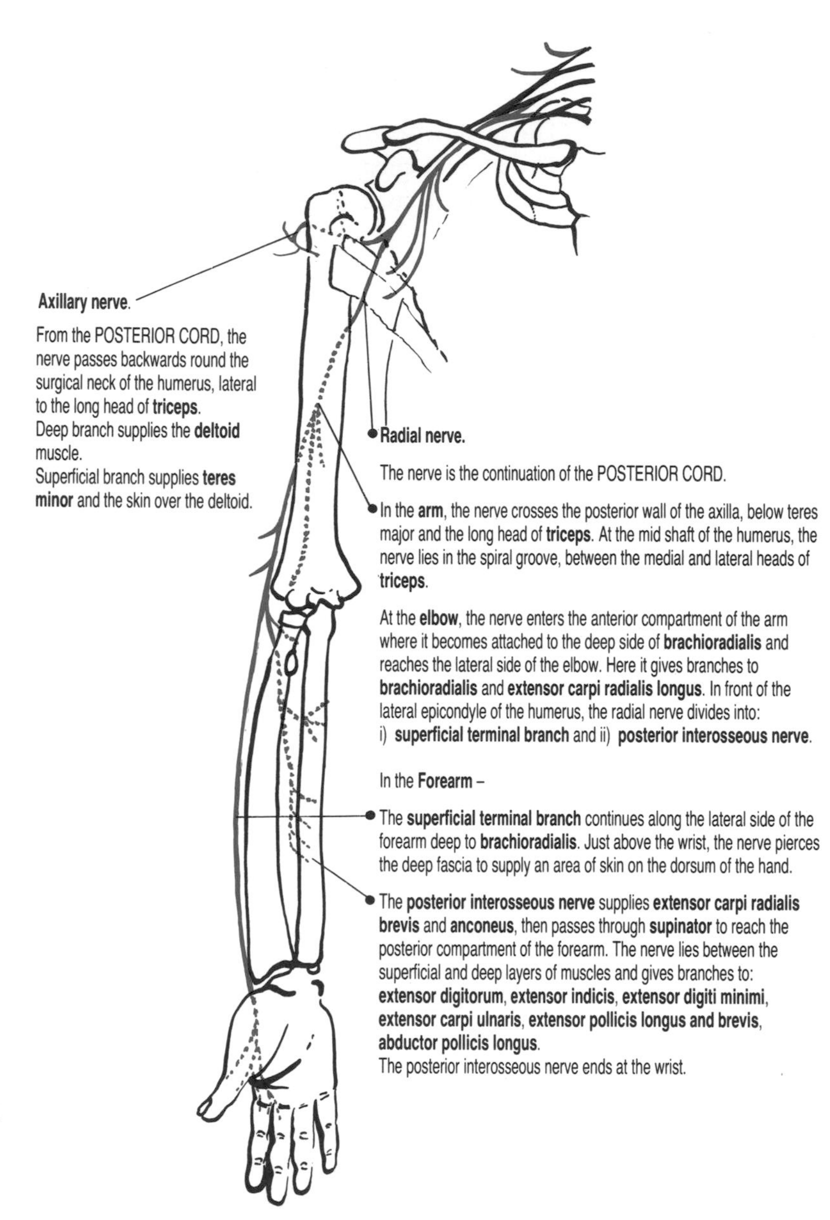

Fig. 7.4. Right axillary nerve and radial nerve; course and distribution, anterior view.

nerve pierces the deep fascia to supply a variable area of skin over the dorsal surface of the hand on the thumb side (Fig. 7.5).

The *posterior interosseous nerve* supplies the extensor muscles in the forearm, ending at the wrist, where it supplies all the joints of the wrist.

Note: the radial nerve as a whole supplies all the *extensor muscles* of the upper limb, but does not supply the intrinsic muscles of the hand that insert into the extensor expansion of the fingers.

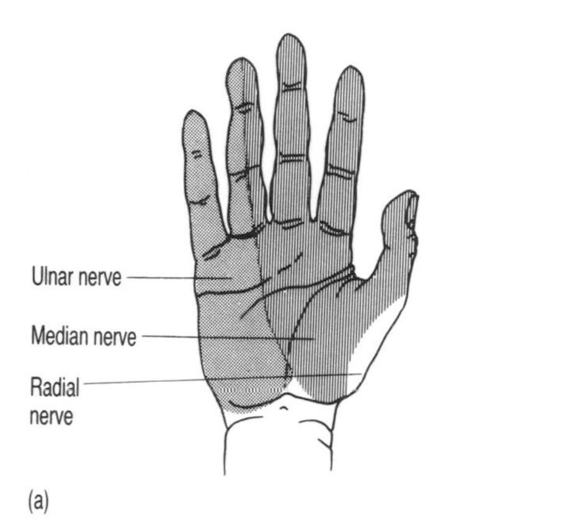

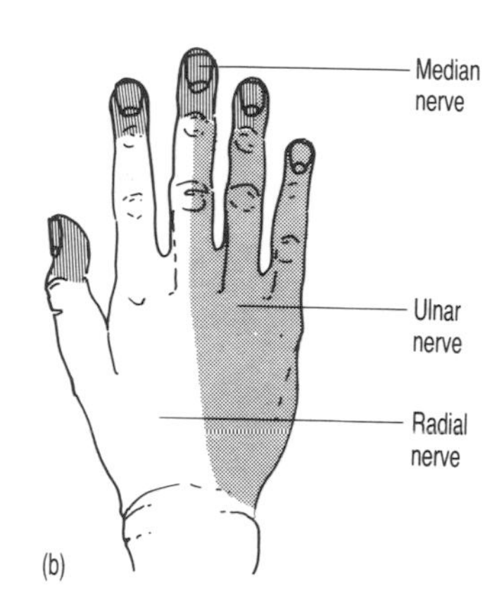

Fig. 7.5. The right hand. Areas of skin supplied by the radial, median and ulnar nerves: (a) palmar; and (b) dorsal.

Injuries to the radial nerve

In the **arm** — usually as a complication of a fracture of the middle of the humerus shaft. Loss of the triceps extension action leads to inability to push against resistance, e.g. a door, or to reach a high shelf. The wrist cannot be lifted against gravity to a functional position.

At the **elbow** — as a complication of supracondylar fracture of the humerus, the thumb cannot be opened and finger extension will be weak, particularly at the MCP joints.

At the **wrist** — which only results in sensory loss of a small triangular area on the dorsum of the hand over the first dorsal interosseus muscle.

Extension of the wrist is important in maintaining the functional position of the hand (Fig. 7.6a) for all movements of the fingers and thumb. Radial nerve injury results in 'wrist drop' (Fig 7.6b), the hand cannot be lifted against gravity and the power grip is weak.

• *WATCH the hand and forearm of a partner doing daily activities such as making a cup of tea and eating with a knife and fork. Note the position of the wrist during the movements. If the wrist could not be held in extension, the hand would drop under its own weight and the weight of any object held in it.*

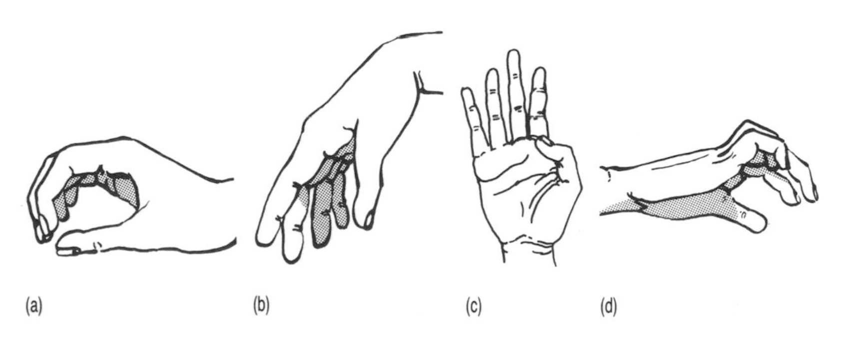

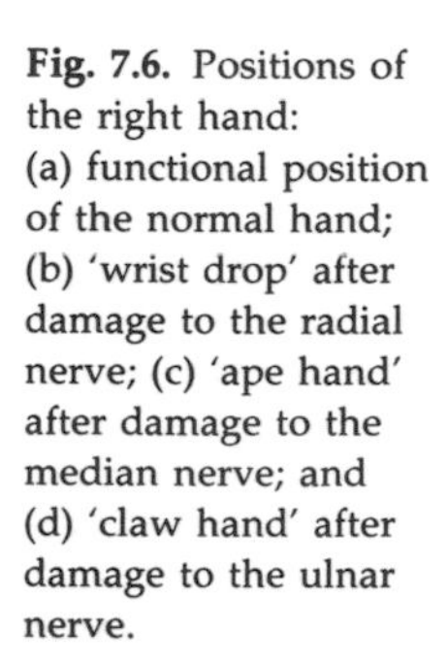

Fig. 7.6. Positions of the right hand: (a) functional position of the normal hand; (b) 'wrist drop' after damage to the radial nerve; (c) 'ape hand' after damage to the median nerve; and (d) 'claw hand' after damage to the ulnar nerve.

7.3.3 Flexor nerves of upper limb — musculocutaneous and median nerves

There are two terminal branches of the lateral cord of the brachial plexus important for flexion movements of the upper limb. The musculocutaneous is the nerve supplying the elbow flexors; and the median nerve supplies the wrist, fingers and thumb flexors, working in cooperation with the ulnar nerve.

The musculocutaneous nerve

This nerve pierces the coracobrachialis, and then passes down the arm between the biceps and brachialis. The nerve supplies these three muscles, which can be remembered by the initials — B B C.

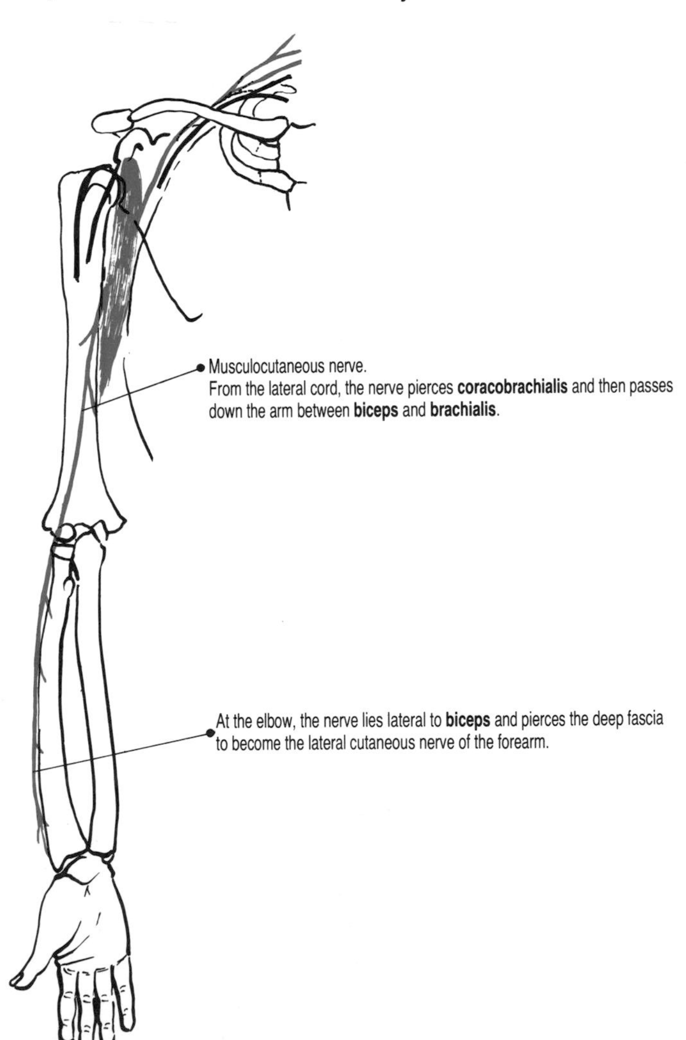

Fig. 7.7. Right musculocutaneous nerve; course and distribution, anterior view.

At the elbow, the nerve becomes cutaneous at the lateral side of the tendon of biceps, to become the nerve to the skin on the lateral side of the forearm (Fig. 7.7).

The median nerve

This is formed from the lateral and medial cords of the brachial plexus.

• *LOOK at Fig. 7.8 to see the course and distribution of the median nerve.*

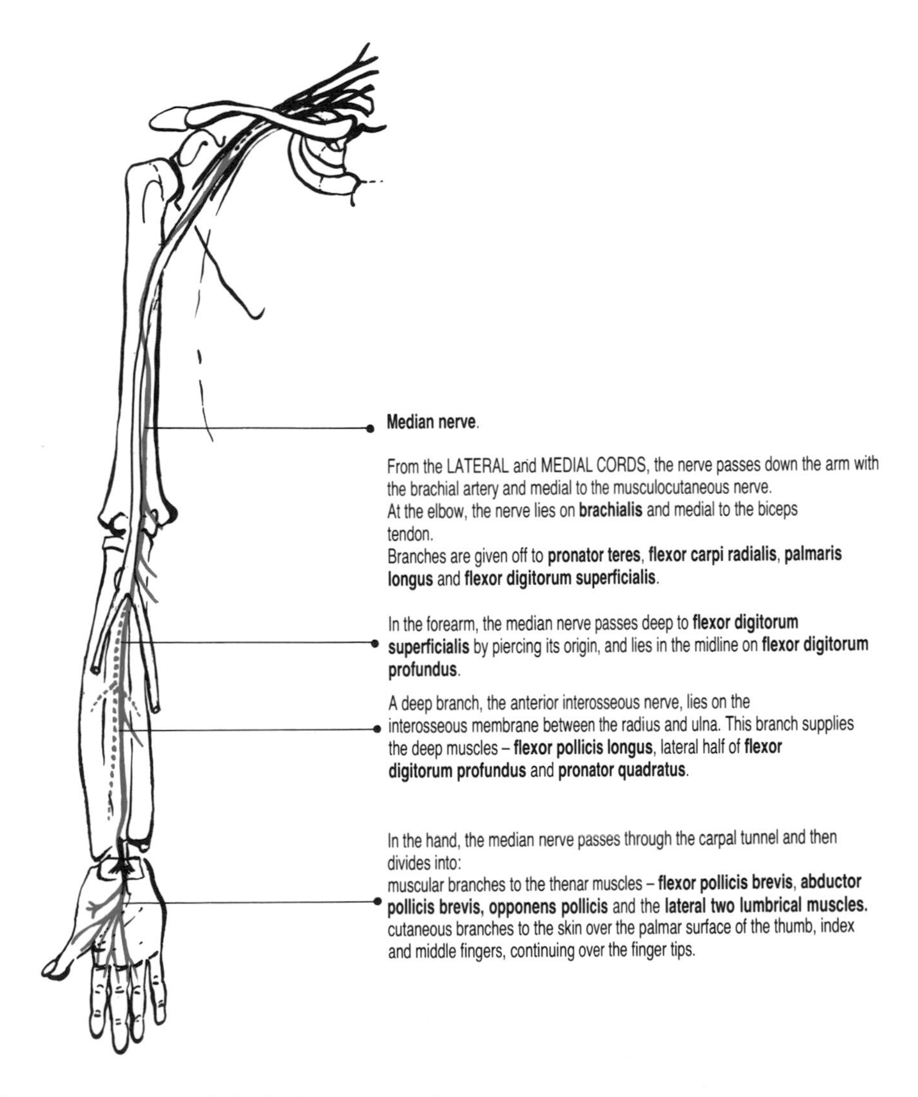

Fig. 7.8. Right median nerve; course and distribution, anterior view.

There are no branches of the median nerve in the arm, it is a nerve of the forearm and hand only. A communicating branch with the musculocutaneous nerve in the arm is present in some individuals.

At the *elbow*, the median nerve lies anteriorly and medial to the tendon of biceps.

In the *forearm*, branches are given off to the four muscles attached to the medial epicondyle of the humerus. A deep branch is given off, the *anterior interosseous nerve*, which lies on the interosseous membrane between the radius and ulna, and supplies the three muscles of the deep layer.

Note: the median nerve supplies all the flexors in the forearm except the flexor carpi ulnaris and the medial half of the flexor digitorum profundus (to the ring and little finger).

The median nerve enters the *hand* in the midline, passing underneath the flexor retinaculum, i.e. through the carpal tunnel (see Chapter 6). In the hand there are two main branches: (i) to the three thenar muscles and first two lumbricals; and (ii) a cutaneous branch to the skin over the palmar surface of the thumb, index and middle fingers, continuing over the finger tips to the dorsal side (see Fig. 7.5, p. 161).

The appearance of the hand in **median nerve lesions** is often called Ape or Monkey hand. The thenar eminence is wasted and the thumb is drawn backwards in line with the fingers, due to unopposed action of the extensor pollicis longus (see Fig. 7.6c). The loss of function of the lateral two lumbricals leads to flattening of the lateral side of the palm; the MCP joints are drawn into extension and the IP joints into slight flexion.

The most usual site of damage is at the wrist. Then the thumb is unable to oppose, and this, together with the loss of sensation from the finger tips makes many gripping movements difficult. If the nerve is damaged at the elbow, there is added loss of finger flexion particularly in the index and middle fingers which also affects gripping. It is the *precision grips* that are most affected by median nerve damage.

The median nerve may also be compressed in the carpal tunnel at the wrist (carpal tunnel syndrome), by an increase in pressure from the swelling of flexor tendon sheaths or carpal joints. The loss of sensation and muscle weakness leads to clumsiness and 'dropping things'.

• *PULL your thumb back and to the side of the palm of your dominant hand by winding a bandage round the wrist and round the thumb. Now try to use your hand in everyday activities to experience the problems when the thumb cannot be opposed to the fingers.*

• *WEAR a thin plastic glove with the ring and little fingers cut away on your dominant hand during hand activities. You will then experience the effects of loss of skin sensation in median nerve injury.*

7.3.4 Fine movements of the fingers — ulnar (and median) nerve

The **ulnar nerve**, is a continuation of the medial cord of the brachial plexus. Figure 7.9 shows the course and distribution of the ulnar nerve.

There are no branches of the ulnar nerve in the arm. The course of the nerve in the forearm and hand is apparent when the inside of

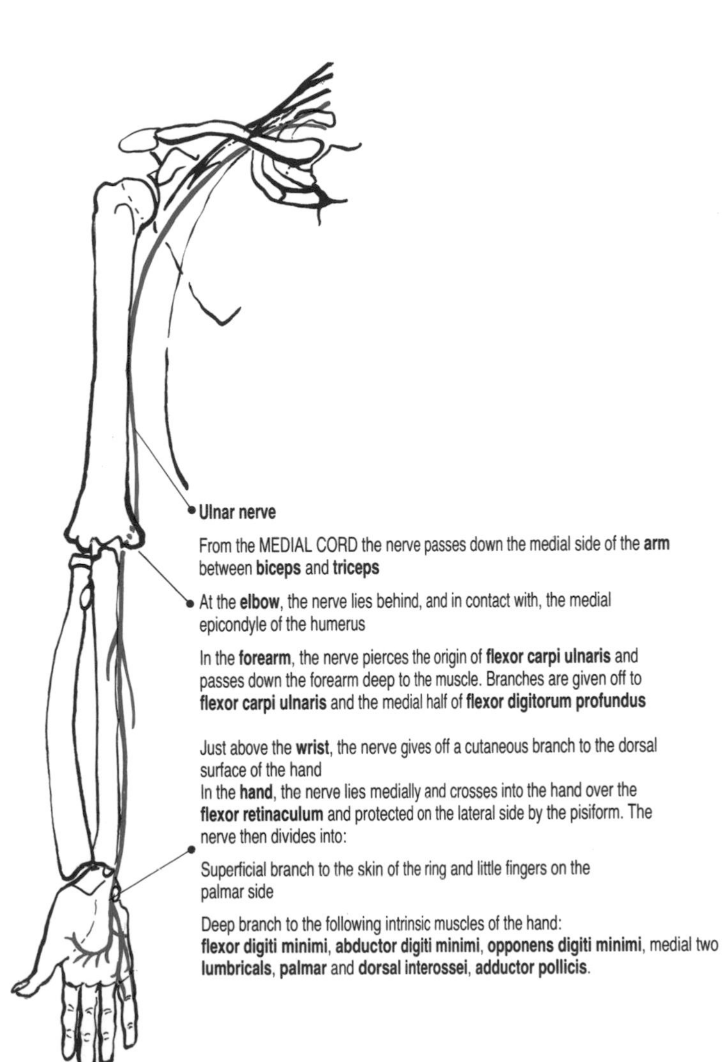

Fig. 7.9. Right ulnar nerve; course and distribution, anterior view.

the elbow is bumped. 'Banging the funny bone' gives a tingling sensation down the inside of the forearm and on to the little finger.

In the *forearm,* the ulnar nerve supplies the one and a half muscles not supplied by the median nerve.

At the *wrist,* the nerve lies medially and passes over the flexor retinaculum. Two cutaneous nerves are given off at, or above, the wrist to supply the skin over the palmar and dorsal sides of the hand medially, and the ring and little fingers (see Fig. 7.5, p. 161).

The terminal branches in the *hand* supply all the intrinsic muscles not supplied by the median nerve, which include all the muscles moving the ring and little fingers.

A condition known as 'claw hand' is the result of **ulnar nerve lesion** (Fig. 7.6d). The ring and little fingers curl in a flexion deformity, with hyperextension at the MCP joints, due to paralysis of the medial two lumbricals. Loss of the dorsal interossei means that the fingers cannot be separated. The web between the thumb and index finger, formed by the adductor pollicis and the first dorsal interosseous muscle, is wasted. The ulnar nerve is important for keyboard operators, musicians and all those who need fine coordinated movements of the fingers. The grips particularly dependent on the ulnar nerve are the *power grip* for stabilizing the medial side of the hand, and the *span grip* when the fingers must be separated to spread over a large object.

The **median nerve** cooperates with the ulnar nerve in all hand function. The loss of thumb opposition in median nerve damage can sometimes be compensated by use of the adductor pollicis, if the ulnar nerve is intact.

The **ulnar** and **median nerves** may be damaged together in severe laceration of the wrist. The result is impairment of total hand function, with loss of all grips.

- *REVISE the intrinsic muscles of the hand from Chapter 6.*
- *LIST all the muscles and their nerve supply.*

7.4 Outline of the direct branches from the brachial plexus

The five terminal branches of the brachial plexus supply all the muscles moving the elbow, forearm, wrist and hand. The deltoid and the teres minor are supplied by the axillary nerve, but the other muscles moving the shoulder region receive branches direct from the plexus.

Return to Fig. 7.3 (p.158) to identify the branches direct from the plexus.

1 Branches from the **roots** of the plexus are:

(a) C5 supplies the levator scapulae, the rhomboids and subclavius.

(b) C5, C6 and C7 form the *long thoracic nerve* to the serratus anterior on the floor of the axilla.

2 One branch from the **upper trunk** C5 and C6:

The *suprascapular nerve* leaves the upper trunk and passes over the upper border of the scapula at the suprascapular notch to reach the supraspinous fossa. The two posterior rotator cuff muscles, the supraspinatus and infraspinatus, are supplied by this branch.

3 Branches from the **cords**:

(a) The *posterior cord* has three branches that supply the muscles of the posterior wall of the axilla — the subscapularis, the teres major and the latissimus dorsi.

(b) The *lateral cord* gives a branch to the muscle of the anterior wall of the axilla — the pectoralis major.

(c) The *medial cord* has three branches. Two form separate cutaneous nerves to the skin on the medial side of the arm and forearm. The third branch supplies pectoralis minor and the lower fibres of pectoralis major.

The trapezius is the only muscle attached to the scapula that is not supplied by a branch of the brachial plexus. The spinal root of the spinal accessory nerve (cranial nerve XI) branches to the anterior of trapezius.

At the end of this chapter, you should be able to:

1 Outline the position and arrangement of the roots, trunks, divisions and cords of the brachial plexus.

2 Describe the course, distribution and functional importance of the *five* terminal branches of the brachial plexus: the axillary, radial, musculocutaneous, median, ulnar and radial nerves.

3 Outline the effects of injury to the nerves listed in **2**.

8 / Support and Propulsion. The Lower Limb

8.1 Functions of the lower limb

The lower limbs are the supporting pillars when we stand. A pillar must have strength and must not collapse under the weight above. The bones and muscles together convert the lower limb into a stable support. The pillar is divided into segments: the thigh, leg and foot. The segments are linked by joints: the hip, knee, ankle and foot joints, which can adjust to the changes that occur in the line of weight through the limbs as the head and trunk move above. The muscles around the joints counteract the effects of gravity and any external forces that disturb the body balance.

Locomotor movements require one limb to support the body weight while the other limb swings forward. In walking, running and climbing stairs, the lower limb has to keep its function as a support and also propel the body forwards or upwards. As we walk, the alternation of swing and support means that the limb as a whole must combine strength with mobility, and change its role at every step. The limb acts as a unit; unlike the upper limb where the shoulder is concerned with mobility, while the elbow and wrist provide stability for hand movements.

Daily activities such as getting out of bed, sitting down on a chair, getting up from a chair, using the toilet, all involve the lower limb. Weakness of muscles or loss of joint mobility make these transfer activities difficult and the upper limb then has to compensate. (This is discussed in Chapter 5.)

The sole of the foot is the contact area of the lower limb with the ground, and plays an important role in sensing the texture and friction properties of the supporting surface. Feedback from receptors in the skin and muscles of the sole of the foot is essential for an economical pattern of locomotion. The absense of this sensory information results in an abnormal gait.

The overall functions of the lower limb, in summary, are as follows.

1 Support in standing.
2 Swing and support in locomotion.
3 Transfer of the body from lying to sitting, to standing.
4 Provision of sensory information from supporting surfaces.

The hip joint is a stable ball and socket joint, whose articulating surfaces fit closely, with the capsule strengthened by ligaments on all sides. The knee is a modified hinge joint allowing movement in the sagittal plane with some rotation in full flexion and at the end of extension. The ankle is a true hinge joint, with collateral ligaments and no lateral or rotatory movement. The joints of the foot allow the sole to turn inwards and outwards by movement at the joints between the bones of the foot — the subtalar and mid-tarsal joints in particular.

• *LOOK at Appendix 2 to understand the structure of these major joints of the lower limb, and contrast them with the upper limb joints whose stability is less, but range of movement is greater.*

The muscles of the lower limb are as active in stabilizing the joints and converting the limb into a lively pillar of support, as in movement. The attachments of the muscles are often anchored in sheets of dense fibrous tissue. An example of this fibrous tissue is the fascia on the lateral side of the thigh, known as the iliotibial tract, which passes over the hip and knee (Fig. 8.1a). Two muscles keep the hip and knee extended by producing tension in the iliotibial tract. Other muscles are attached to bony points and make precise movements, e.g. the calf muscles pull on the heel to raise the foot onto the toes.

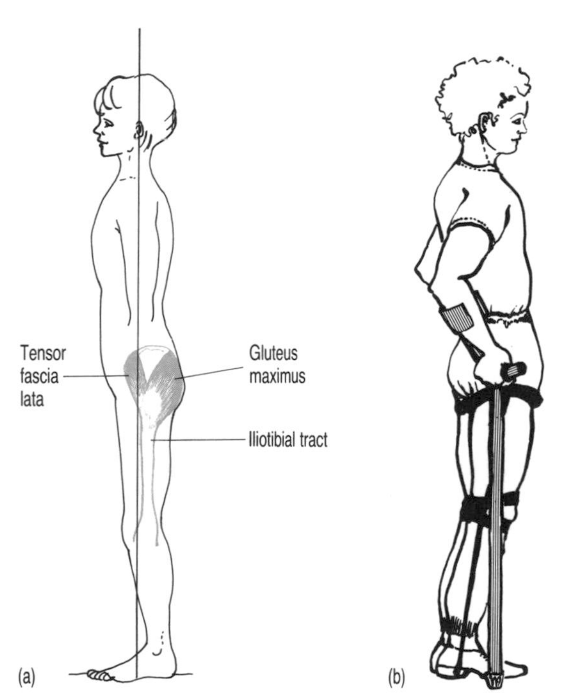

Fig. 8.1. Upright standing viewed from the side: (a) double support showing the line of weight, the gluteus maximus and the tensor fascia lata pull on the iliotibial tract to extend the knee; and (b) paraplegic standing, knee and ankle supported by an external brace, hip held in extension supported by the iliofemoral ligament of the hip joint.

8.2 Support

8.2.1 Double support

There is a remarkable economy of muscle activity involved in standing upright on two legs. The joints of the lower limb are in a close packed position when standing, and stability depends largely on the tension of the ligaments around the joints. Two particular structures are important.

1 The anterior ligament of the hip joint, the **iliofemoral ligament**, is important in resisting the tendency for the trunk to fall backwards on the lower limbs when the line of the body weight falls behind the hip joint. Little activity is required in the hip flexors and extensors. The paraplegic with paralysed hip muscles learns to place the hips well in front of the line of gravity and relies entirely on the tension in the iliofemoral ligament for stability at the hips in standing (Fig. 8.1b).
2 The **iliotibial tract** is a band of dense fascia which extends across the hip and knee on the lateral side of the thigh. In standing, the tension in a small muscle, known as the *tensor fascia lata*, which originates on the anterior superior spine of the ilium and inserts into the iliotibial tract, keeps the hip and knee extended with the help of the *gluteus maximus*, the large superficial muscle of the buttock (Fig. 8.1a).

The attachments of gluteus maximus will be described in Section 8.4.

The line of body weight lies in front of the ankle joints, so that some activity in the calf muscles is needed to keep balance over the foot base. (The calf muscles will also be described in detail in Section 8.4.)

We rarely stand to attention like the guardsmen on parade, but adopt changing positions of 'slack standing' with the knees slightly flexed and the weight shifting from one leg to the other.

• *WATCH people standing at a bus stop, queueing for tickets at a station, or talking in groups. Note the variety of lower limb positions. Shop assistants, teachers, nurses and surgeons spend long periods of time standing. The constant shifting of position reduces fatigue in any one muscle group, and also aids the return of blood to the heart by the pumping action of leg muscles.*

8.2.2 Change to single support

The change from standing on two legs to standing on one is the first stage in beginning to propel the body forwards. When one leg is lifted from the ground, muscles around the hip of the supporting leg are active to: (i) move the body weight over the supporting leg; and (ii) prevent the pelvis from dropping on the unsupported side.

The **adductor** group of muscles on the inside of the thigh contract to shift the pelvis over the supporting side. At the same time, the tendency for the pelvis to drop is counteracted by activity in the **abductors** of the hip in the supporting leg. Figure 8.2 shows the position of the abductors and adductors in the supporting leg.

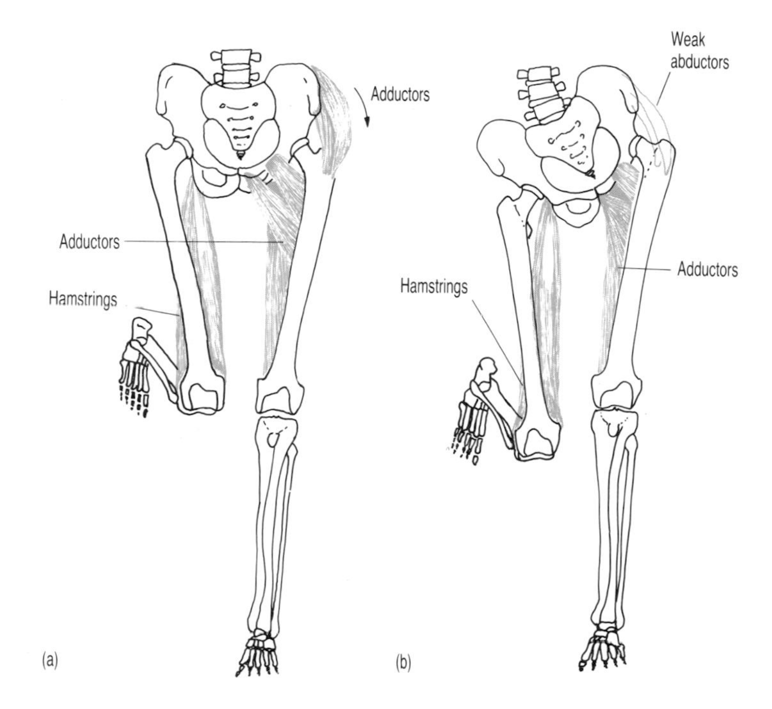

Fig. 8.2. (a) Action of the hip abductors and adductors to keep the pelvis level in single support; (b) Trendelenburg's sign. Abductors of supported side weak, the pelvis drops on the unsupported side.

You should be able to see how contraction of the abductors will pull on the pelvis and keep it level. Further tilt of the pelvis gives added clearance for the raised foot.

8.2.3 Hip muscles active in single support

Abductors of the hip

These abductors of the hip are the *gluteus medius* and the *gluteus minimus.*

Two fan shaped muscles lie deep to the gluteus maximus, the largest muscle of the buttock. The gluteus medius and minimus originate from the outer surface of the ilium, and both muscles insert into the greater trochanter of the femur (Fig. 8.3). When the leg is not acting as a support, the abductors are active in lifting the leg sideways.

• *STAND some distance from a long mirror, and take a few steps slowly. Note: if the right leg is off the ground, the right side of the pelvis is unsupported and could drop to the right, and so the left abductors must contract. For the next step, the opposite abductor muscles*

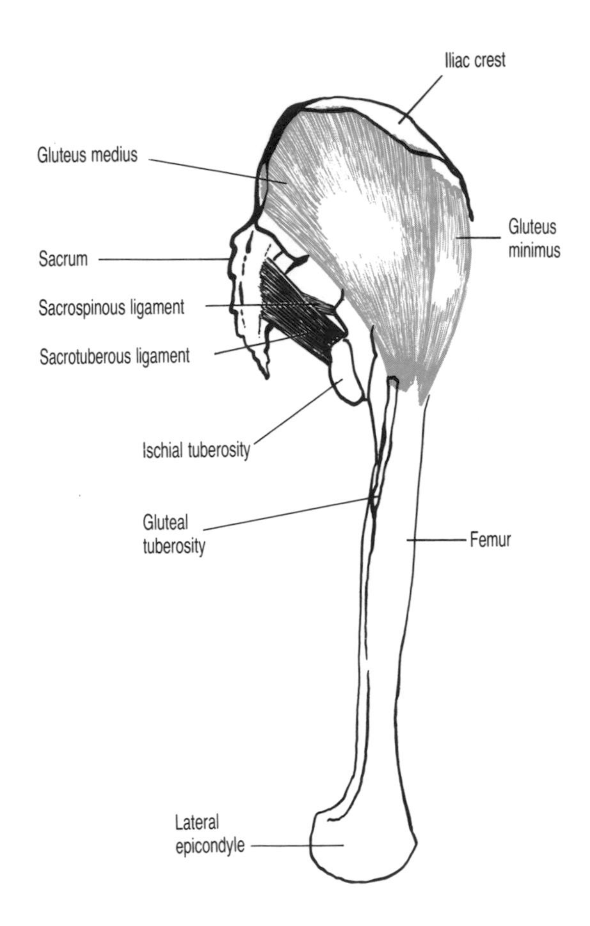

Fig. 8.3. Lateral aspect of the right pelvis and femur to show the gluteus medius and minimus.

contract. (There is also a muscle of the trunk involved, the quadratus lumborum, and this will be described in Chapter 10.)

The changes in hip position with each step can be seen if you walk behind someone wearing tight jeans. Notice how the sway varies in different people, at different speeds of walking, and also with mood.

Problems of the hip, e.g. congenital dislocation, fracture of the neck of the femur, or paralysis of hip abductors, produce an abnormal pattern of walking. The hip drops to the opposite side when weight is taken on the affected hip: this is known as *Trendelenburg's sign* (Fig. 8.2b).

Adductors of the hip

These are a group of five muscles lying on the inner side of the thigh (Fig. 8.4). In the various positions of the hip joint, the individ-

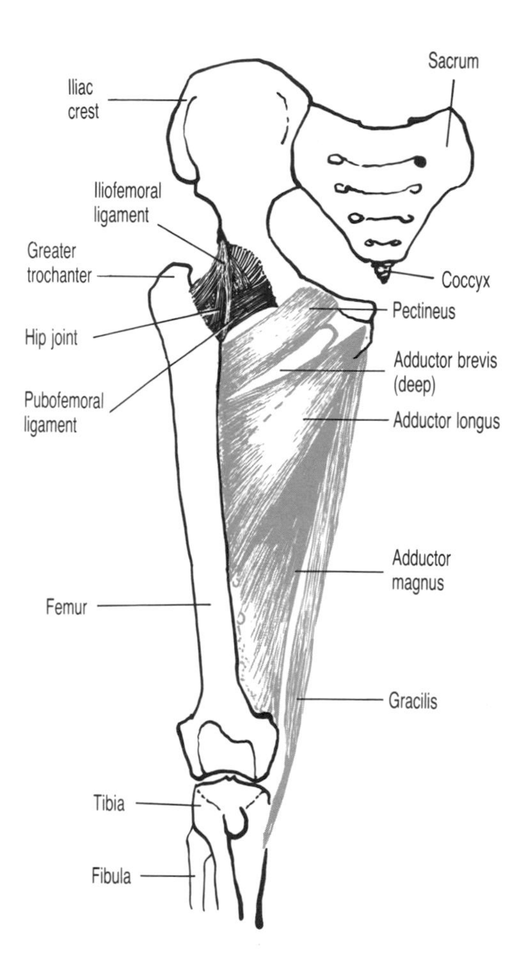

Fig. 8.4. Anterior view of the right pelvis and thigh to show the adductor group of muscles.

ual muscles of the adductor group can act as flexors, extensors and rotators. Strong adduction of the thigh is not very significant to everyday activities except when riding a bicycle or a horse, when contraction of the adductors keeps you on the saddle. When standing on an unstable platform, the adductors act with the abductors to keep the body weight over the feet.

The names of the adductors of the hip are *adductor magnus, adductor longus, adductor brevis, pectineus* and *gracilis.* The group of adductors originate from the anterior surface of the body of the pubis extending medially on to the superior and inferior ramus. The adductor magnus is the most posterior muscle of the group, and its origin extends back to the ischial tuberosity. From the small area of origin, the muscles fan out to insert into the full length of the posterior shaft of the femur. The posterior fibres of the adductor magnus pass vertically down to the adductor tubercle, just above the medial side of the knee. The gracilis is a strap muscle lying medially in the group, and ends below the knee.

The shift of body weight over the supporting leg demands more stability at the knee. The large muscle on the front of the thigh, the quadriceps femoris, acts to keep the knee joint extended. The quadriceps muscle will be described in more detail in Section 8.4 (see Fig. 8.10, p. 178).

The muscles around the ankle are important to keep the lateral balance when the body weight is supported on one foot. It is the muscles that turn the foot inwards and outwards (and are known as invertors and evertors respectively), that provide this stability. These muscles will be described in Section 8.5.

• *WATCH a partner in bare feet stand on one leg. Notice any changes in the level of the pelvis, and the side to side movement of the foot taking place just below the ankle joint.*

During walking, the relative amount of time spent in single support at each step depends on the speed of walking. At slow speeds, the swinging leg is only off the ground for a short time, and most of the cycle is spent in double support. As the speed increases, single support occupies a relatively longer time, so that patients with hip problems, weak quadriceps and inability to balance, can only walk at slow speeds.

8.3 Swing

8.3.1 Leg swing in daily activities

Leg swing can occur when one leg is free to move while the opposite leg is supporting the body weight. The movements of the free leg swing the limb to place the foot forwards, upwards or to the side.

• *STAND on one leg and swing the free leg in all directions. Think about the daily activities that use these movements.*
How does the leg swing in: (i) walking; (ii) climbing stairs; (iii) stepping into the bath; (iv) getting into a car; and (v) getting on to a bicycle?

8.3.2 Active muscles in the swing phase

The muscles involved in swinging the leg forwards are found in: (i) The *hip* (flexors combined with abductors and rotators); (ii) the *knee* (flexors); and (iii) the *ankle* (dorsiflexors — to raise the toes clear of the ground).

Hip flexors

The muscles that flex the hip are as follows.

1 The **iliacus** and **psoas** usually grouped together and called **iliopsoas**, assisted by the sartorius, the rectus femoris and the tensor fascia lata. The iliopsoas originates in the abdomen. The fibres of the psoas are attached to the transverse processes, bodies and discs of the lumbar vertebrae. The iliacus takes origin from the inner surface of the ilium on the iliac fossa. The two muscles leave the abdomen together, passing under the inguinal ligament, over the front of the hip joint, and insert into the lesser trochanter of the femur (Fig. 8.5).

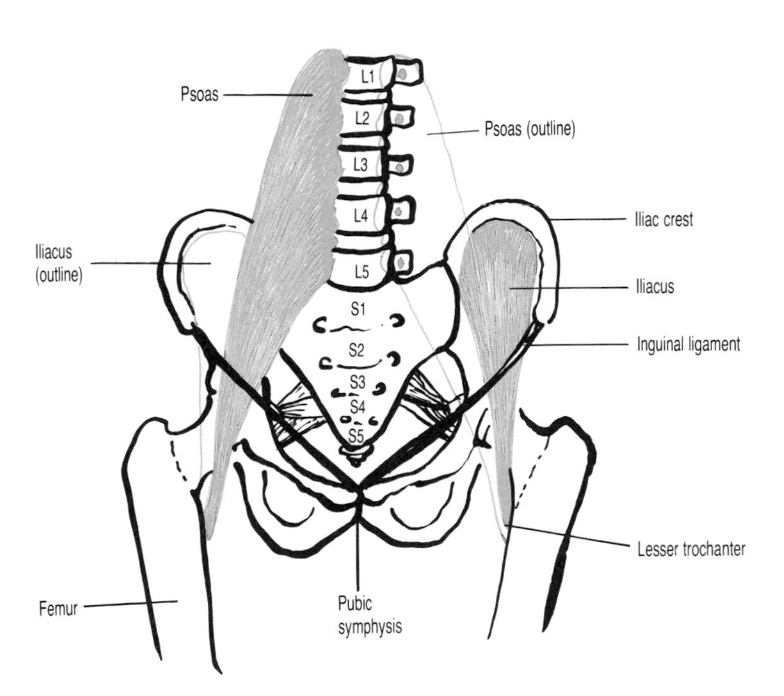

Fig. 8.5. Anterior view of the pelvis and hip joints to show the iliacus and psoas.

• *SIT with the trunk slightly forwards. Place one hand at the waist between the lower ribs and iliac crest, with the fingers across the lower back. Raise the foot off the ground and feel the activity in the psoas just lateral to the vertebral column. The bulk of the muscle you are feeling lies posteriorly, but remember its tendon passes over the anterior side of the hip joint to reach the femur.*

The iliopsoas is active in *walking, climbing stairs, sitting* and *sitting up*.

In **walking** the iliopsoas is used to start the leg swinging forwards. On level ground the leg then moves like a pendulum to

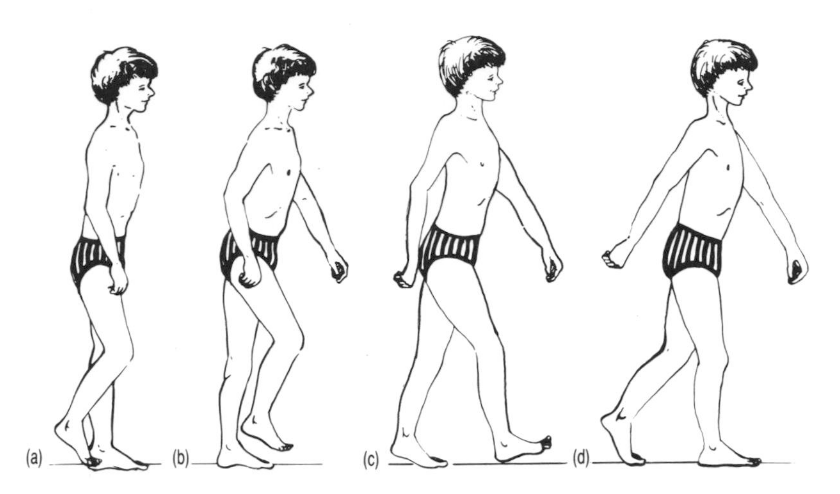

Fig. 8.6. Walking: (a) start of swing phase, raise stepping leg; (b) swing phase, bring stepping leg through; (c) heel strike, stepping leg comes to the ground; and (d) propulsion and support phase, weight transferred to stepping leg with forward thrust from back leg.

complete the swing phase. Greater activity in the hip flexors is needed in walking up hill, and running. Figure 8.6a and b show the hip flexion during swing.

In **climbing stairs** the iliopsoas lifts the leg and puts the foot on the step above. Figure 8.7a shows the flexion in the left hip.

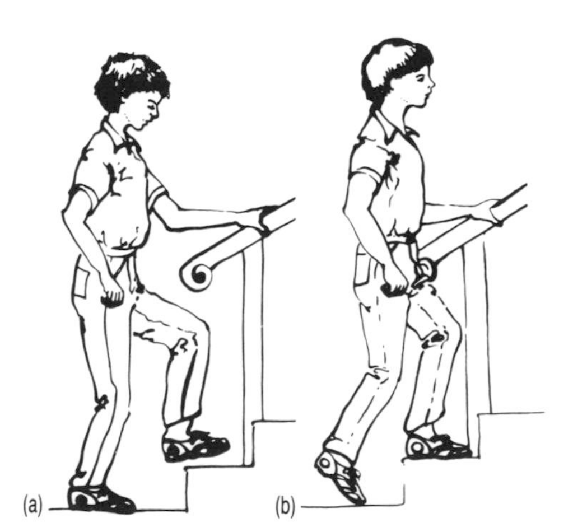

Fig. 8.7. Climb stairs: (a) lift left leg on to step, hip and knee flexion, ankle dorsiflexion; and (b) lift body weight on to the step, left hip and knee extend, right ankle plantar flexed.

In **sitting** and preparing for standing, the iliopsoas is used to pull the trunk forwards, i.e. the femur is fixed (Fig. 8.8a). As the trunk leans forwards, the centre of gravity of the trunk moves over the feet before standing upright (Fig. 8.8b).

In **sitting up** from lying (Fig. 8.9), the iliopsoas acts by pulling on the ilium of the pelvis and the lower vertebrae. The abdominal muscles (described in Chapter 10) start the movement, and the iliopsoas is active at the end of the movement to pull the trunk upright.

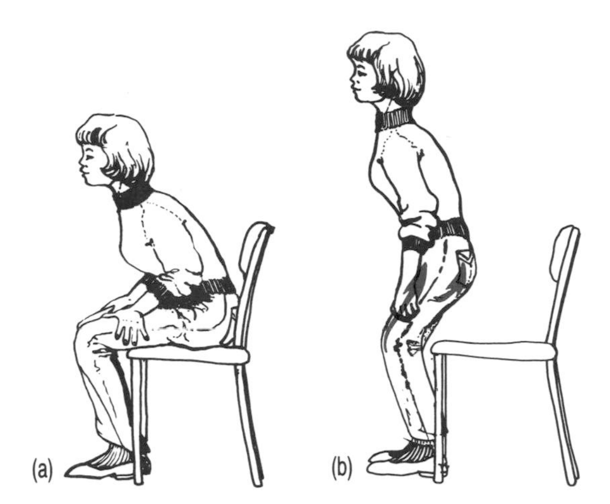

Fig. 8.8. Stand up from sitting: (a) lean forwards before standing; and (b) stand up, extension of knees, hips and trunk.

Fig. 8.9. Functional use of the hip flexors, sit up from lying.

2 The **sartorius** is a long thin strap like muscle that crosses the anterior thigh (Fig. 8.10). The overall actions of the sartorius put the limb into the cross legged position, adopted by tailors (hence the name) and in Yoga. Together with the tensor fascia lata, the sartorius assists the iliopsoas in hip flexion.

3 Hip abductors (gluteus medius and minimus) also help to swing the leg sideways.

4 Hip rotators. Many muscles around the hip have rotatory actions. The particular active muscles varies with the position of the femur in relation to the pelvis. There are six small *lateral rotators* arranged close to the joint, similar to the rotator cuff muscles of the shoulder. The six muscles lie across the posterior side of the hip joint deep to gluteus medius. The swinging leg carries the pelvis forwards and the limb tends to turn inwards. Activity in the lateral rotators will keep the foot pointing forwards in walking. Detailed attachments of these muscles is not important and can be found in standard anatomy textbooks.

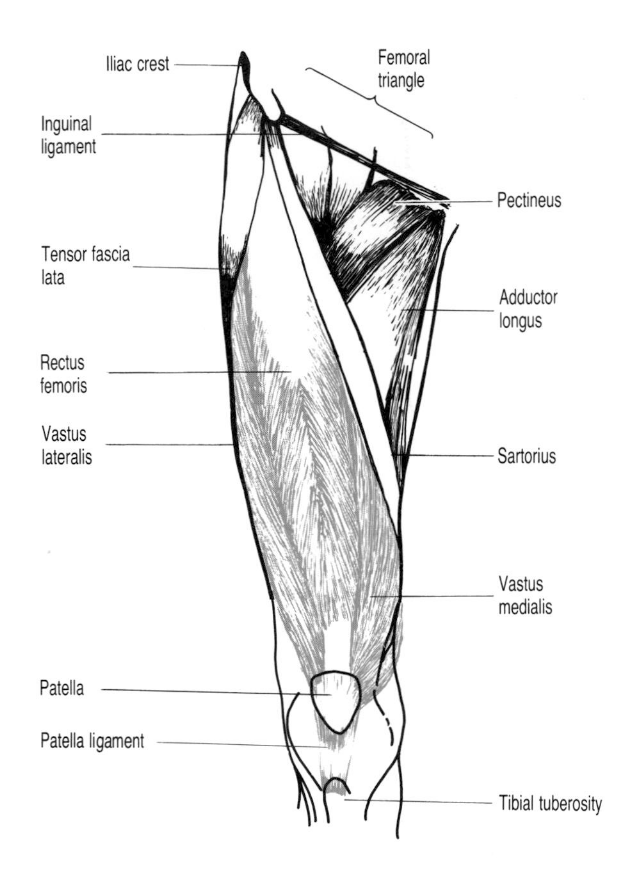

Fig. 8.10. Anterior view of the right thigh to show quadriceps femoris and sartorius.

Knee flexors

When the lower limb swings, the knee flexes to lift the foot clear of the ground. Figure 8.6a shows the knee flexion during swing. The muscles active in flexion of the knee are the **hamstring** group at the back of the thigh, and the medial part of the adductor magnus.

The three hamstring muscles are the *biceps femoris,* the *semimembranosus,* and the *semitendinosus.*

• *FEEL the tendons of the hamstrings in the fold of the knee in the sitting position. Two tendons lie medially (the semimembranosus and semitendinosus), and one tendon can be felt laterally (the biceps femoris).*

All three hamstrings originate on the ischial tuberosity of the pelvis (Fig. 8.11). The biceps femoris also has a short head of origin from the linea aspera of the femur, and passing laterally to the knee, both heads are inserted into the head of the fibula. The semimembra-

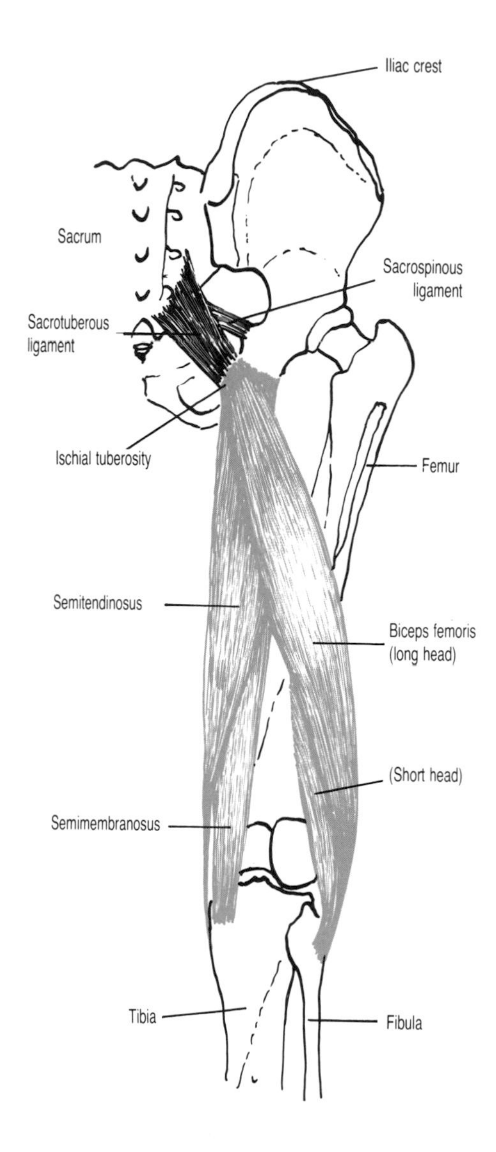

Fig. 8.11. Posterior view of right pelvis and thigh to show the hamstring group of muscles.

nosus begins as a flat tendon which forms a third of its length, and the muscle fibres then insert by a thick tendon behind the medial condyle of the tibia. The semitendinosus begins as muscle fibres and becomes tendinous two thirds of the way down the thigh, to insert into the tibia on the medial side below the knee.

All the hamstrings flex the knee to lift the leg towards the thigh. When the trunk leans forwards, the ischial tuberosities (the origin of the hamstrings) are carried upwards and backwards in relation to the hip, and the muscles can be felt stretching in the thigh. Contraction of the hamstrings then extends the hip, and the trunk is raised to the upright position. The action of the hamstrings in extension of the hip will be discussed again in Section 8.4.

Ankle dorsiflexors

The weight of the foot in the swinging leg will tend to pull the toes downwards, and so drag the toes on the ground. To counteract this dropping of the foot, the muscles passing over the front of the ankle onto the dorsal surface of the foot contract. Look at the ankle movement during swing in Fig. 8.6a and b (p. 176). The angle between the leg and the foot decreases in the movement known as *dorsiflexion*.

The individual muscles in the group of dorsiflexors are the *tibialis anterior*, the *extensor hallucis longus*, and the *extensor digitorum longus*.

• *PALPATE the bulge of muscles below the knee and lateral to the shin. Lift the toes upwards by dorsiflexing the ankle and feel the group in action. The tibialis anterior is the most superficial muscle that can be felt.*

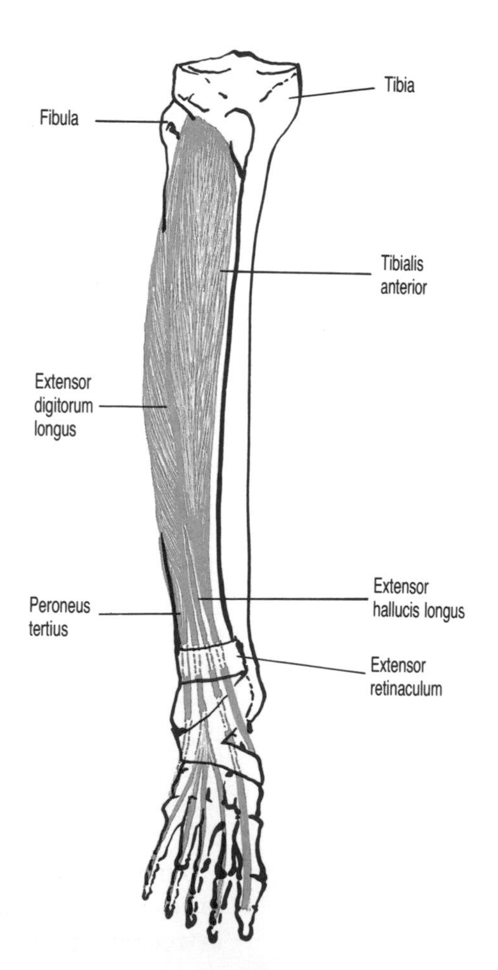

Fig. 8.12. Anterior view of the right leg and foot to show the anterior tibial group of muscles.

• *OBSERVE the three tendons passing over the front of the ankle, the tibialis anterior medially adjacent to the medial malleolus, then the extensor hallucis longus going to the big toe, and the extensor digitorum longus laterally. Check with Fig. 8.12.*

The **tibialis anterior** is attached to the anterolateral shaft of the tibia and inserts on the medial side of the foot into the medial cuneiform and the base of the first metatarsal.

The **extensor hallucis longus** originates on the shaft of the fibula and inserts into the distal phalanx of the big toe.

The **extensor digitorum longus** has fibres attached to the lateral condyle of the tibia and the medial shaft of the fibula and the interosseous membrane in between. The common tendon passes over the front of the ankle and divides into four, each inserting via an extensor expansion to the toes, in a similar way to the extensors of the fingers. A fifth tendon is sometimes present, which is the tendon of the *peroneus tertius,* inserting into the base of the fifth metatarsal.

Two transverse bands of fascia over the anterior side of the leg above the ankle hold the tendons of the dorsiflexors in position during movements of the ankle.

Loss of function of the the dorsiflexors leads to a condition known as 'foot drop', when the toes drag along the ground during the swing phase in walking.

8.4 Propulsion

So far the movements of the lower limb in support and swing have been considered. Next we shall look at the muscles which exert force against the ground to move the body forwards and upwards.

Muscle groups used in propulsion movements are found in: (i) the *hip* (extensors); (ii) the *knee* (extensors); (iii) the *ankle* (plantar flexors — raise the heel to 'push off'); and (iv) the *toes* (flexors — grip the floor and thrust forwards).

Hip extensors

The main extensor of the hip in propulsion movements is the gluteus maximus (Fig. 8.13). The most superficial muscle of the gluteal group, the gluteus maximus, is the largest muscle in the body. The gluteus maximus can be seen when lying prone or standing upright, forming the curve of the buttocks. Strong contractions can be felt in activities such as climbing stairs, running and jumping.

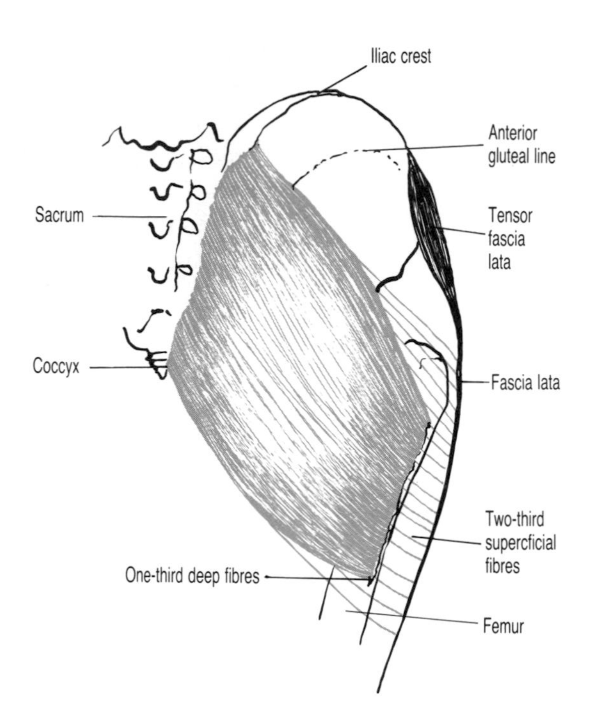

Fig. 8.13. Posterior view of the right hip to show gluteus maximus; superficial fibres insert into the fascia lata, deep fibres insert into the gluteal tuberosity of the femur.

The extensive origin of the gluteus maximus spreads from the posterior corner of the iliac crest across the posterior side of the sacrum and coccyx with some fibres attached to the fascia of the lower back (thoracolumbar fascia), and to the sacrotuberous ligament of the pelvis. All the fibres pass downwards and laterally over the posterior side of the hip joint. The main insertion of the muscle is into the iliotibial tract, with the remaining fibres passing deeply to attach to the gluteal tuberosity on the posterior shaft of the femur.

The action of the gluteus maximus is extension of the hip.

1 *Walking* up hill requires strong action of the gluteus maximus to propel the body onwards.

2 In *climbing stairs* the gluteus maximus is important in supporting the leg during swing, and in the stepping leg to lift the body onto the step (Fig. 8.7b, p. 176).

3 *Standing up* from sitting requires strong hip extension (Fig. 8.8b) as well as knee extension.

The muscle also works eccentrically in sitting down from standing and in the trailing leg when going down stairs.

The **hamstring** muscles also extend the hip, working with the gluteus maximus, for propulsion forwards in walking. The activity begins at heel strike and continues as the supporting hip moves over the foot (Fig. 8.6c and d).

Knee extensors

The muscles active in extension of the knee are the *quadriceps femoris* group. Four muscles on the anterior of the thigh form this group. The individual muscles are: (i) the *rectus femoris* the most superficial in the midline; (ii) the *vastus medialis* on the medial side; (iii) the *vastus lateralis* on the lateral side of rectus femoris; and (iv) the *vastus intermedius* which lies deep to rectus femoris. These muscles can be clearly seen in athletes and footballers when the vasti, in particular, become enlarged in response to weight training.

• *SIT on a chair and put your hands on the top of your thighs. Now stand up slowly and feel quadriceps in action. Pause in standing and feel the tension is less. Then sit down slowly to feel the quadriceps in action again. The muscle is working concentrically to extend the knee and lift the body upwards; then working eccentrically against gravity as the knee flexes and the body lowers to the seat of the chair again.*

The quadriceps group can be seen in anterior view in Fig. 8.10, with the exception of the vastus intermedius which lies deep to rectus femoris.

The **rectus femoris** is the only part of the quadriceps that passes over the hip joint. This muscle is attached by two heads, one from the anterior inferior iliac spine, and the other from just above the acetabulum.

The three vastus muscles surround the shaft of the femur.

The **vastus medialis** begins on the spiral line and down the medial side of the linea aspera. The fibres wrap round medially to approach the knee.

The **vastus lateralis** is attached posteriorly to the lateral side of the linea aspera and wraps round the lateral side of the femoral shaft.

The **vastus intermedius** originates on the anterior and lateral shaft of the femur.

Figure 8.14 shows how the muscles of the quadriceps group extend around three sides of the femur.

All four muscles meet at the patella on the front of the knee, and insert by a common tendon, the ligamentum patellae, to the anterior tubercle of the tibia. The ligamentum patellae provides extra stability for the knee joint on the anterior side where the capsule is absent. The lower horizontal fibres of the vastus medialis prevent lateral displacement of the patella at the end of the knee extension.

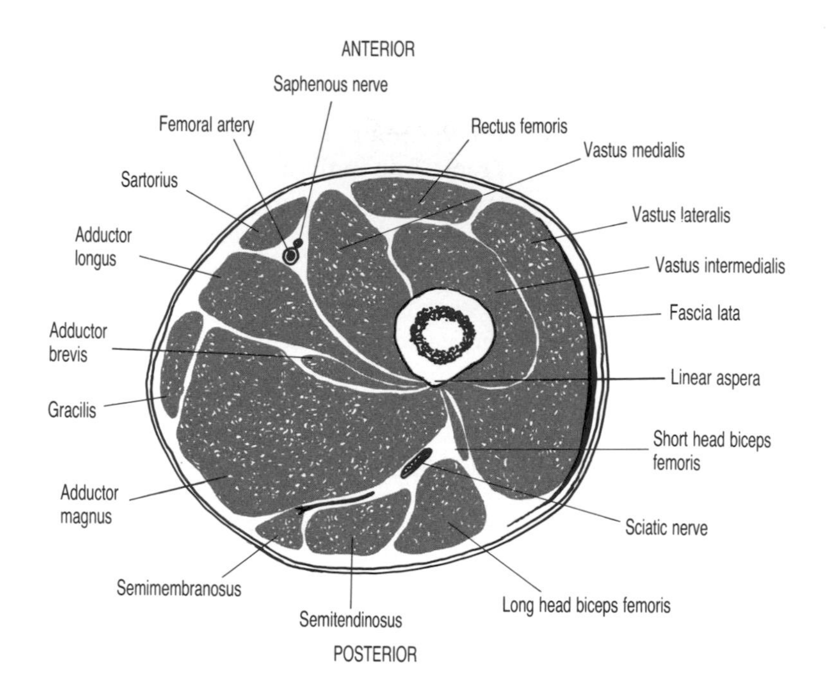

Fig. 8.14. Transverse section through the right thigh at the level of the upper third.

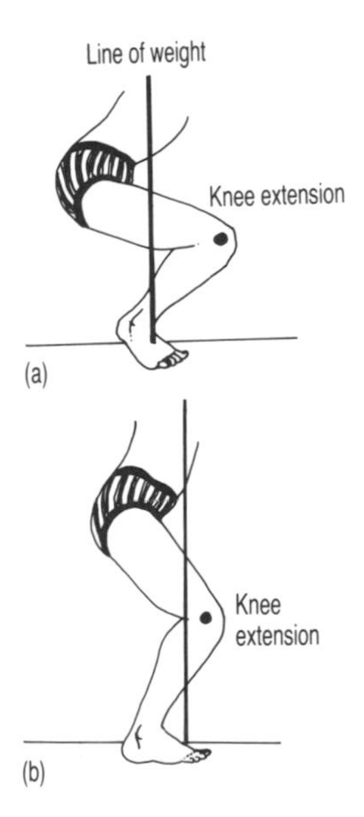

Fig. 8.15. Rise from a low squat position. Great force must be exerted by the quadriceps as the line of weight is a large distance from the knee joint.

The quadriceps group is a powerful extensor of the knee, while the rectus femoris alone has a weak flexor action on the hip. Acting with the gluteus maximus, the quadriceps raise the body from sitting and squatting. When the knees are flexed at an angle less than a right angle, the quadriceps has to develop a force of 4–5 times body weight to hold the position. The knees are under great stress when the body is raised from a low squat position since the line of body weight is some distance from the knee joint (Fig. 8.15) (see also Chapter 2).

When the knee is injured, or after a period of bed rest, the quadriceps wastes very rapidly. Strength of this muscle is restored by knee extension exercises, at first lifting the weight of the leg alone, and then by adding weights to the leg. Equipment operated by knee extension pressing a foot pedal or plate, e.g treadle lathe or sewing machine, can also be used.

Ankle plantar flexors

When the heel is raised from the ground, the body is lifted up or forwards in a 'push off' movement (Fig. 8.6d, p. 176). The calf muscles, attached by the Achilles tendon to the heel are active in this movement.

- *STAND on your toes and feel the calf muscles contracting. The*

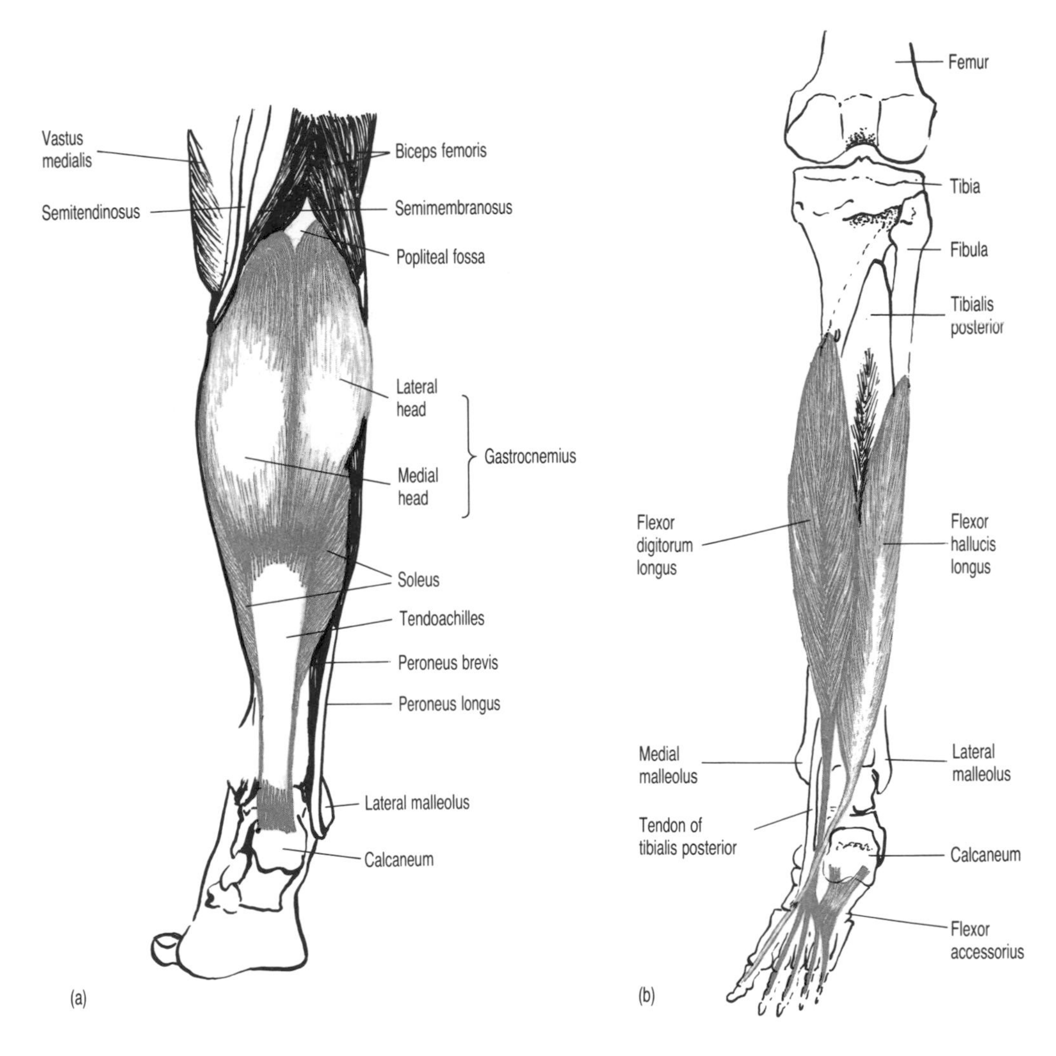

Fig. 8.16. Posterior view of: (a) the right leg to show the position of the gastrocnemius and soleus; and (b) the right leg and the sole of the foot to show the flexor muscles of the toes.

muscles are also working eccentrically when you lower your heels to the ground.

When the calf muscles are weak, the 'push off' in the trailing leg in walking is lost, and the body weight is transferred by trunk flexion; the thrust in jumping is also reduced.

The movement at the ankle to raise the heels is know as *plantar flexion.* The active muscles of the calf are the *gastrocnemius* and the *soleus* (Fig. 8.16a).

The **gastrocnemius** is attached by two heads to the posterior surface of the femur, one above each femoral condyle.

The **soleus** attaches below the knee, across the soleal line on the posterior shaft of the tibia, and to the head and shaft of the fibula.

Both muscles join to form the very strong Achilles tendon attached to the posterior surface of the calcaneum. The length of the calcaneum behind the axis of the ankle joint gives good leverage to the calf muscles.

There is a small muscle, the *plantaris*, lying between the gastrocnemius and the soleus. The short belly of this muscle originates above the lateral condyle of the femur, near to the lateral head of the gastrocnemius, and becomes a thin tendon just below the knee joint. The tendon passes down the length of the calf to insert onto the calcaneum in the Achilles tendon.

Another small muscle, the *popliteus*, is found at the back of the knee joint, deep to the gastrocnemius. It is attached to the triangular area above the soleal line of the tibia, and passes upwards to the lateral condyle of the femur.

Deep to the gastrocnemius and soleus are three muscles whose tendons pass round the medial side of the ankle to enter the sole of the foot. They assist the calf muscles in propulsion at the ankle, and are known as the deep plantar flexors. One of these muscles (the tibialis posterior) will be described with the foot in Section 8.5. The other two muscles are flexors of the toes as well as plantar flexors of the ankle.

Flexors of the toes

In propulsion movements, it is important to keep the toes firmly on the ground while the body moves above. At the end of the movement, some thrust is added by flexion of the toes, particularly the great toe in running.

The muscles involved in flexion of the toes are the *flexor hallucis longus* and the *flexor digitorum longus* (Fig. 8.16b).

The **flexor hallucis longus** is attached to the posterior shaft of the fibula below the soleus. The tendon crosses the lower end of the tibia to pass through a groove on the back of the talus, and under the sustentaculum tali of the calcaneum, to continue along the sole of the foot on the medial side to insert on the distal phalanx of the great toe.

The **flexor digitorum longus** also arises below the soleus on the posterior shaft of the tibia, and passes round the medial side of the ankle with the flexor hallucis longus. In the sole of the foot, the tendon divides into four to insert on the distal phalanx of the four lesser toes.

The tendons of both these mucles are held in position at the ankle by the flexor retinaculum, a band of fibrous tissue from the medial malleolus to the medial tubercle of the calcaneum. In the sole of the foot, the tendons are enclosed in synovial sheaths, similar to the flexors of the fingers in the hand.

8.5 The foot

8.5.1 Functions of the foot

The foot is a relatively small area to make contact between the body weight and the ground. The surface of the ground may be rough, smooth, hard, soft, level or sloping, and the sole of the foot has to be able to accommodate all these. The weight of the body above compresses the parts of the foot in different directions as the body moves.

The muscles acting on the foot originate in the leg and pass around the ankle to insert into the bones of the foot. Like the hand, the foot also has intrinsic muscles that begin and end in the foot. The arrangement of the intrinsic muscles of the foot is similar to those in the hand. Children with malformation of the upper limb may develop the muscles of the foot to take over the manipulative functions of the hand.

The most important function of all the foot muscles is to resist deformation of the foot by the ground, or any obstacle in contact with the foot. When the foot is off the ground (e.g. in the swing phase in walking) the muscles act to change the position of the foot in relation to the leg. For most of the time, however, the feet are in contact with the grounds, supporting the body weight in standing, and providing momentum for moving the body around. When the foot is firmly planted on the ground, the extrinsic muscles can move the leg on the foot, e.g. the dorsiflexors of the ankle pull the body forwards onto the leading leg in walking. The muscles are then acting in the opposite direction to pull the proximal attachments in the leg towards the foot.

8.5.2 Movements of the foot

Dorsiflexion and plantar flexion movements at the ankle have already been described under Sections 8.3 and 8.4 respectively.

Independent movements allow the foot to turn inwards and outwards (Fig. 8.17). Most of these movements occur at the joints of the foot, the subtalar and midtarsal joints in particular.

Inversion is the movement which turns the sole of the foot inwards when the foot is off the ground, or shifts the weight to the lateral side of the foot when weight bearing.

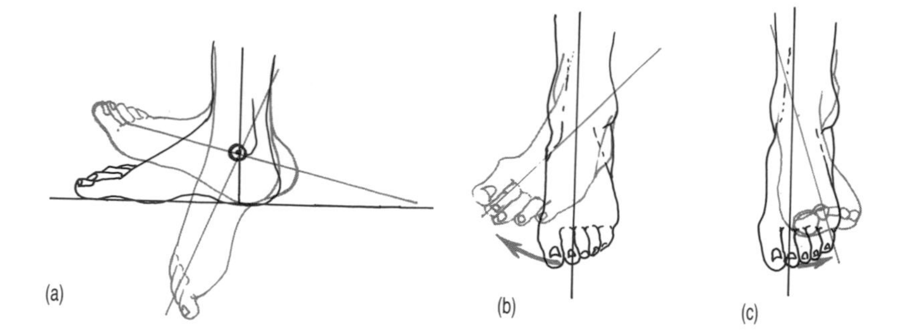

Fig. 8.17. Movements of the foot: (a) dorsiflexion (up) and plantar flexion (down); (b) inversion; and (c) eversion.

Eversion turns the sole of the foot outwards with the foot off the ground, or shifts the weight towards the medial side of the foot in weight bearing.

Invertors — *tibialis anterior* and *tibialis posterior.*
Evertors — *peroneus longus* and *peroneus brevis.*

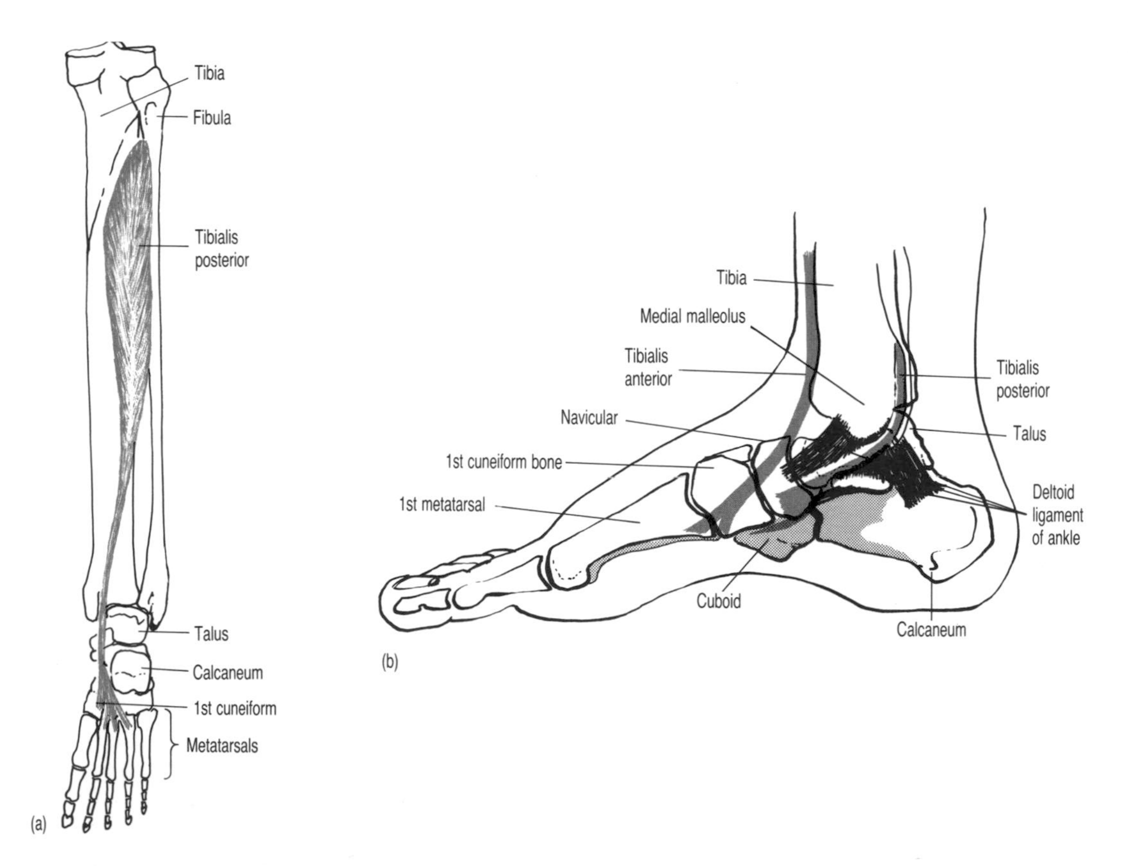

Fig. 8.18. (a) Posterior view of the right leg and the sole of the foot to show tibialis posterior; and (b) medial aspect of the right ankle showing the insertions of tibialis anterior and posterior.

Invertors

The **tibialis anterior** is one of the dorsiflexors already described in Section 8.3. Its tendon inserts on the medial side of the foot into the medial cuneiform and base of the first metatarsal, so that it lifts the medial border of the foot.

The **tibialis posterior** is the deepest muscle of the calf, originating on the posterior shaft of the tibia and fibula. The tendon of this muscle passes round the medial malleolus at the ankle and it inserts on the plantar surface of the navicular and adjacent tarsal bones (Fig. 8.18). Note the relationship between the tendons of the tibialis anterior and posterior on the medial side of the foot.

Evertors

The **peroneus longus** and **brevis** are attached to the lateral shaft of the fibula and their tendons pass round the lateral malleolus at the ankle. At the base of the fifth metatarsal, the peroneus brevis tendon ends, and the peroneus longus turns under, crossing the sole of the foot in a groove on the cuboid bone, to reach the medial cuneiform and base of the first metatarsal (Fig. 8.19).

When the evertors are weak, lateral stability of the ankle is lost, and the lateral ligament of the ankle is often torn.

When the foot is in contact with an uneven surface, the movements of inversion and eversion, together with the actions of the intrinsic muscles of the foot, allow the foot to adjust to the ground and stabilize the ankle. Shoes reduce the amount of adaptation required, but the foot and shoe still have to accommodate sloping ground and avoid slipping on wet or icy surfaces.

8.5.3 Resistance to deformation of the foot

The complex mechanism of the foot has resilience and spring to respond and adapt to the stress and strain of the body weight above.

- *REVISE the bones of the foot. Place an articulated skeleton of the foot on a flat surface. Note which bones are in contact with the table, and which bones are wholly or partly raised above the surface.*
- *STAND your feet in a tray of water soluble paint, then make footprints on a sheet of lining paper laid out on the floor: (i) sitting on a chair, i.e. non weight bearing; (ii) standing upright; and (iii) walking several steps. Note the variation in pattern of footprint in (i), (ii) and (iii). Compare your footprints with those of other students and note any individual differences.*

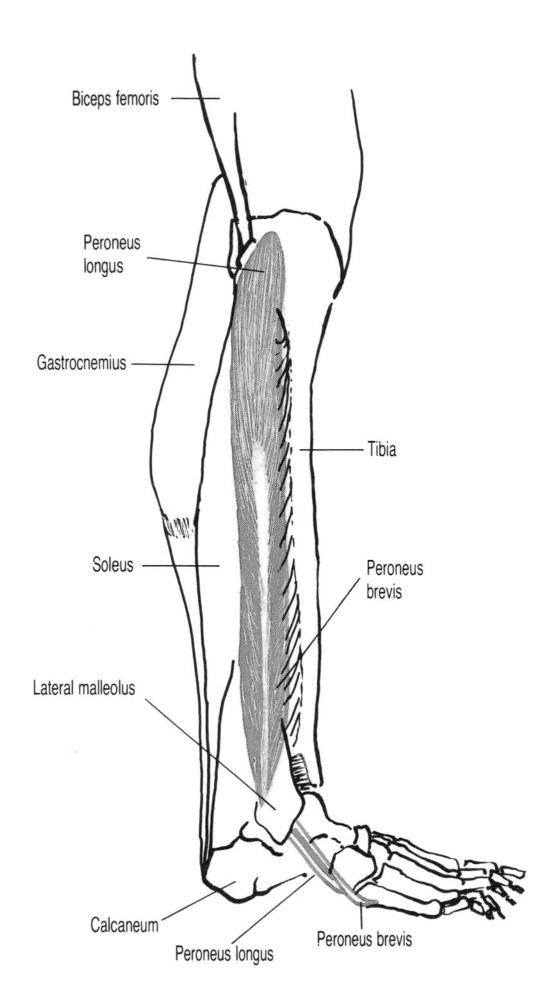

Fig. 8.19. Lateral aspect of the right leg and foot to show peroneus longus and brevis.

From (i) to (iii) above there will be an increase in the area of foot in contact with the ground. In all the prints, the heel and ball of the foot will be seen. The lateral border of the foot will be present when the foot is weight bearing. The medial border of the foot remains, absent, except when there is abnormal flattening of the foot.

Looking at the bones of the foot and the footprints, it can be seen that the foot is arched in different directions. A longitudinal arch from the heel to the ball of the foot is easy to recognize. The foot is also arched transversely across the distal row of tarsals and metatarsals (Fig. 8.20). Ligaments bind the bones of the foot together and provide the main factors supporting the arches in standing. During movement, it is the muscles of the leg acting as slings from above, and the intrinsic muscles of the foot acting as bow strings across the base of the arches, that maintain the arches.

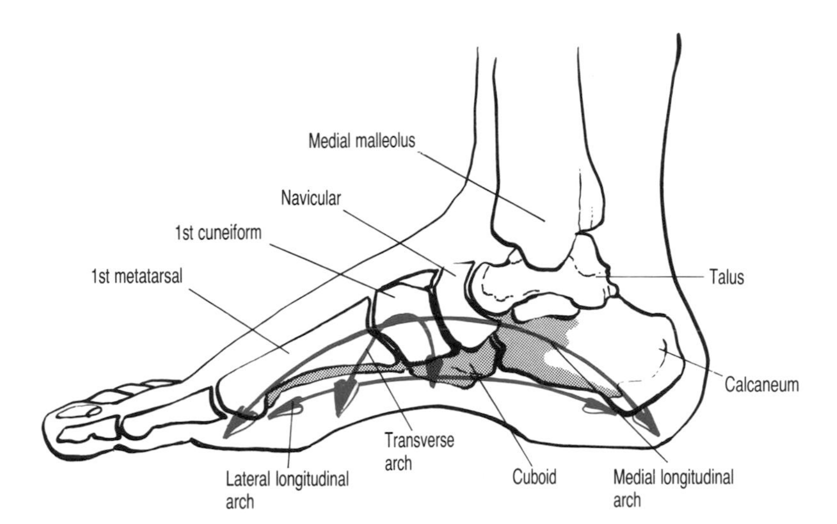

Fig. 8.20. Medial view of the right ankle to show the arches of the foot.

The height of the arches varies during different phases of locomotor movements, particularly the medial part of the longitudinal arch.

Bony components of the arches

1 Medial longitudinal arch. This arch is formed by the calcaneum, talus, navicular, three cuneiforms and metatarsals 1, 2, and 3. The highest part is the talus, which sits on the calcaneum supported by a shelf on the medial side known as the sustentaculum tali.

2 Lateral longitudinal arch. This arch also begins at the calcaneum and extends along the lateral side of the foot to the cuboid and metatarsals 4 and 5. There is considerable stress on this arch during running when the body weight is transferred along the lateral border of the foot and onto the big toe. The shoes of a marathon runner, which are often worn down on the outer border, show how high this stress can be .

3 Transverse arch. The foot is most arched in the transverse direction across the distal row of tarsals — the three cuneiforms and the cuboid. The metatarsals are also arched transversely, the region of the heads of the metatarsals is sometimes called the anterior arch. When the foot is stressed from above in standing, the anterior arch is flattened as the weight is taken by the heads of the metatarsals.

Ligaments bonding the arches

1 Spring ligament. A tough fibrous band extends from the sustentaculum tali of the calcaneum to the navicular, and supports the head of the talus. This ligament is very elastic and responds to compression of the medial longitudinal arch (Fig. 8.21).

2 Long and **short plantar ligaments.** These two ligaments bind the bones of the lateral longitudinal arch. The long plantar ligament stretches from the calcaneum to the ridge on the cuboid and the bases of the middle metatarsals. Deep to this, the short plantar ligament is attached to the anterior end of the calcaneum and to the cuboid (Fig. 8.21).

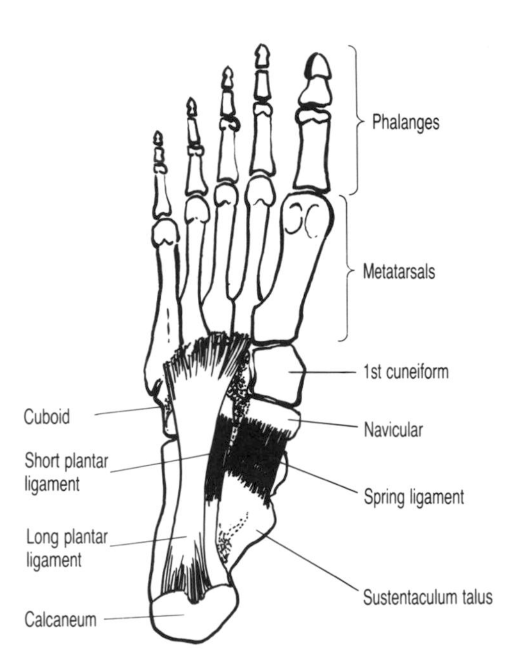

Fig. 8.21. Ligaments of the plantar surface of the right foot.

3 Plantar aponeurosis. A thick sheet of dense fibrous tissue is attached to the tuberosities of the calcaneum, and passes forwards over the muscles of the sole, to blend with the ligaments joining the heads of the metatarsals (deep transverse metatarsal ligaments). There are five main bands in the plantar aponeurosis, and each band blends with a fibrous sheath round a flexor tendon to a toe. The plantar aponeurosis joins the two ends of the medial and lateral longitudinal arches (Fig. 8.22).

Muscles supporting the arches

1 Muscles of the leg. Tibialis anterior inserts on the medial cuneiform and base of metatarsal 1 and the tendon of the tibialis posterior unites the plantar surfaces of the medial tarsals (Fig. 8.18,

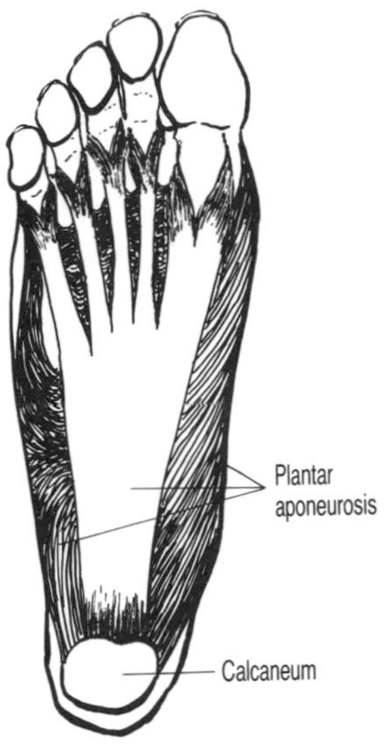

Fig. 8.22. Plantar aponeurosis of the right foot.

p. 188). These two muscles therefore support the medial longitudinal arch and resist compression of the inner side of the foot.

The tendon of the flexor hallucis longus acts as a bow string for the medial longitudinal arch. The flexor digitorum longus also has its tendon lying along the longitudinal arch and responds to compression from above (Fig. 8.16b, p. 185).

The peroneus longus has a tendon that crosses the foot from the cuboid laterally to insert on the plantar surface of the medial cuneiform and base of metatarsal 1 (Fig. 8.19). Crossing the foot in this way, the peroneus longus is the chief support for the transverse arch and also resists any compression of the outer side of the foot and protects the lateral ligament of the ankle.

2 Intrinsic muscles of the foot. Three muscles of the first layer of the foot originate on the calcaneum and insert into the toes. Like the plantar aponeurosis, these muscles (abductor hallucis, flexor digitorum brevis and abductor digiti minimi) join the two ends of the longitudinal arches. The muscles function by responding to depression of these arches by the weight above (Fig. 8.23).

The transverse head of the adductor hallucis, in the third layer of the foot, crosses the anterior end of the transverse arch, from the heads of metatarsals 3, 4 and 5 to the proximal phalanx of the big toe. This muscle is the main support for the anterior arch.

The interosseous muscles of the foot are arranged around the second metatarsal, which forms the axis of the foot. (The axis of the

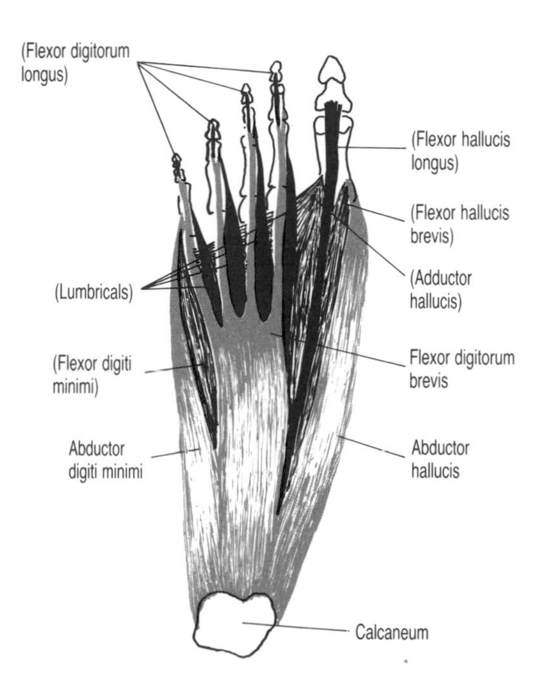

Fig. 8.23. Muscles of the first layer of sole of the foot, abductor hallucis, flexor digitorum brevis and abductor digiti minimi. Muscles of the second layer and third layers named in brackets.

hand is the third metacarpal.) Activity in the interossei and the flexors draws the metatarsals towards the axis of the foot, and resists compression of the transverse arch.

• *PLACE the foot flat on the ground whilst sitting in a chair. Try to pull up the centre of the foot (flexion) with the toes kept flat on the ground. Notice how the foot arches both longitudinally by some flexion at the tarsometatarsal joints, and transversely as the shafts of the metatarsals move towards the axis of the foot.*

Foot exercises are important for dancers who need to keep the flexibility of the foot for balance of the body on the whole or part of one foot.

Imbalance of the muscles acting on the foot leads to changes in the position of the foot which may persist as a deformity. If the muscles supporting the medial side of the foot are weak, the inner border becomes flattened. Neurological conditions affecting the lower limb often produce a flexion deformity when the longitudinal arch is raised and the forefoot drops. Weakness of the intrinsic muscles of the foot strains the plantar ligaments and pain is felt in standing.

While the bony arches respond to the compression of the foot in movement, the body weight is finally transmitted to the toes. The big toe takes most of the weight in standing, and provides the thrust in driving the body forwards in walking and running. Amputation or painful arthritic changes in the big toe drastically change the use of the foot in walking. Long distance runners commonly experience pain in the big toe as a result of the great stress on it.

Observation of the feet, particularly in the elderly, indicates how the complex mechanism of the foot adapts to its varied use and the constraints made upon it by footwear.

8.6 Summary of the lower limb muscles

1 *Muscles around the hip*
Gluteus maximus, medius and minimus, tensor fascia lata, iliacus and psoas (iliopsoas), six lateral rotators

2 *Anterior thigh*
Quadriceps femoris, sartorius

3 *Medial thigh*
Adductor magnus, longus and brevis, pectineus, gracilis

4 *Posterior thigh*
Hamstrings — biceps femoris, semitendinosus, semimembranosus

5 *Anterior leg*
Tibialis anterior, extensor hallucis longus, extensor digitorum longus

6 *Posterior leg*
Gastrocnemius, soleus, (plantaris and popliteus), flexor hallucis longus, flexor digitorum longus

7 *Lateral leg*
Peroneus longus and brevis

8 *Sole of the foot*

First layer:	Abductor hallucis, flexor digitorum brevis, abductor digiti minimi
Second layer:	Lumbricals, flexor digitorum accessorius
Third layer:	Flexor hallucis brevis, flexor digiti minimi brevis, adductor hallucis
Fourth layer:	Interossei — three plantar and four dorsal

At the end of this chapter, you should be able to:
1 Discuss the functions of the lower limb.
2 Describe the upright standing position in relation to orientation of the body segments and the factors maintaining the position.
3 Describe the changes from standing with double support to single support, including the muscles involved
4 Describe the muscle action of:
(a) swinging the lower limb;
(b) propelling the body forwards and upwards, with special reference to walking, climbing stairs and standing up from sitting.
5 Outline the functions of the foot.
6 Describe the movements of the foot and name the main muscles producing them.
7 Describe the factors maintaining the arches of the foot.

9 / The Nerves of the Lower Limb and some Observations of Gait

9.1 Introduction

The nerves of the lower limb are formed from the lumbar and sacral spinal nerves, which together form the lumbosacral plexus. Movement in the limb as a whole depends on activity in spinal nerves from the first lumbar down to the fourth sacral nerves. The spinal roots branch and join in the abdomen and pelvis before forming the peripheral nerves in the limb. The first four lumbar nerves form the **lumbar plexus** which lies embedded in psoas muscle in the posterior abdominal wall. A second plexus is formed from the fifth lumbar to the fourth sacral spinal nerves. These roots of the **sacral plexus** are part of the cauda equina (see Fig. 4.4, p.86) and enter the pelvis through the anterior foramina of the sacrum.

The **terminal branches** of the lumbar and sacral plexus which supply the large muscle groups of the lower limb are: (i) the *femoral nerve* — to the anterior muscles of the thigh; (ii) the *obturator nerve* — to the medial muscles of the thigh; (iii) the *gluteal nerves* — to the muscles of the buttock; and (iv) the *sciatic nerve* — to the posterior muscles of the thigh, this branches to form the *tibial* and *common peroneal nerve* — to all the muscles below the knee.

The lumbar and sacral plexi are less vulnerable to injury than the brachial plexus. The femoral, sciatic and common peroneal nerves may be damaged by trauma, and the obturator nerve may be compressed by any rise in pressure in the pelvis due to enlargement of pelvic organs. Injury to the back may compress the roots of the lumbar nerves, particularly L4 and L5.

In development, the lower limb bud grows out from the side of the embryo. The nerve from the central segment, S1, grows down the limb, to end along the outer side of the foot (see Fig. 4.7) in the same way that the middle segments of the brachial plexus supply the hand. Later in development, the lower limb rotates medially as it extends, so that the dermatomes of the upper segments (L2, L3, L4, L5) lie along the front of the limb, and those of the lower segments (S2, S3) lie on the back of the leg. The rotation also means that the anterior and posterior divisions of the spinal nerves forming the plexi do not pass to the corresponding side of the lower limb.

1 Lumbar plexus:

(a) Posterior divisions — anterior muscles of thigh.
(b) Anterior divisions — medial muscles of thigh.

2 Sacral plexus:

(a) Posterior divisions — muscles of the buttock, anterior and lateral muscles of leg.

(b) Anterior divisions — posterior muscles of thigh, leg and sole of foot.

Figure 9.1 shows the position of the lumbar and sacral plexi in relation to the lumbar spine, the sacrum and the hip. The course of the muscular branches will be described below, but the cutaneous branches will be mentioned in outline only.

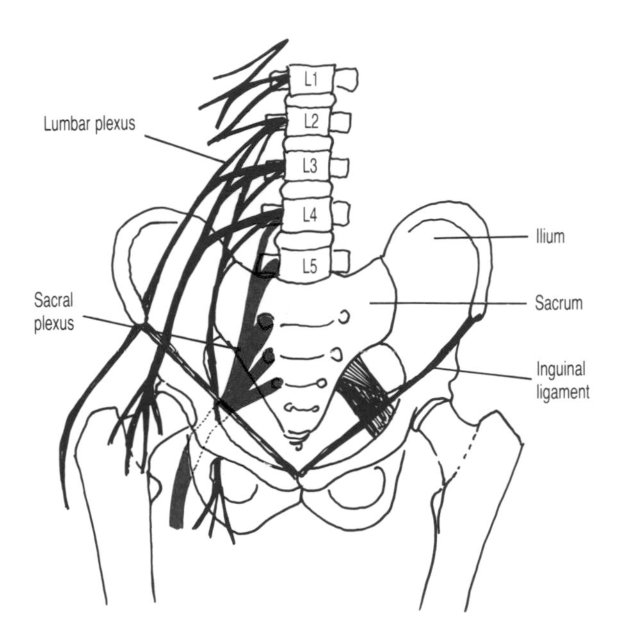

Fig. 9.1. Lumbar and sacral plexus, position and relations.

9.2 The lumbar plexus, position and formation

The lumbar plexus is formed by the anterior primary (ventral) rami of the upper four lumbar nerves. The psoas muscle lies alongside the lumbar vertebrae in the posterior abdominal wall, so that the lumbar plexus lies in this muscle, and the branches of the plexus emerge from it. Most of the fibres of L4 join the lumbar plexus, but the remainder join L5 to form the *lumbosacral trunk* which is part of the sacral plexus.

Branches direct from the plexus supply the psoas, iliacus, and the quadratus lumborum.

Figure 9.2 shows the five roots of the lumbar plexus and the formation of the main terminal branches.

9.3 Terminal branches of the lumbar plexus

There are three important nerves that are formed from the lumbar plexus: the femoral, the lateral cutaneous, and the obturator nerves.

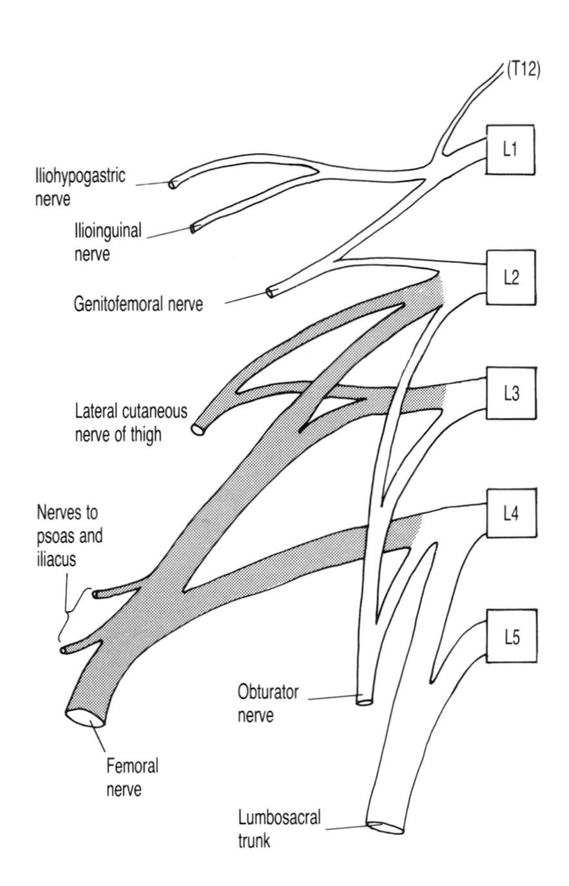

Fig. 9.2. Lumbar plexus, roots and main terminal branches.

Each nerve passes through the pelvis to enter the thigh: (i) the *femoral nerve* passes anteriorly to supply quadriceps and sartorius; (ii) the *lateral cutaneous nerve* passes laterally to the skin on the lateral side of the thigh; and (iii) the *obturator nerve* passes medially to supply the adductor group of muscles.

Figure 9.1 shows the way that the three nerves leave the pelvis.

• *LOOK at Figs 9.3 and 9.4 to follow the course and distribution of the femoral, lateral cutaneous and obturator nerves.*

The femoral nerve

This is the largest nerve of the lumbar plexus. It lies in the psoas muscle in the pelvis, and then emerges from the lateral border of the muscle to lie between the psoas and iliacus, leaving the pelvis anteriorly under the inguinal ligament. In the thigh, the femoral nerve branches to supply the quadriceps group of muscles. The

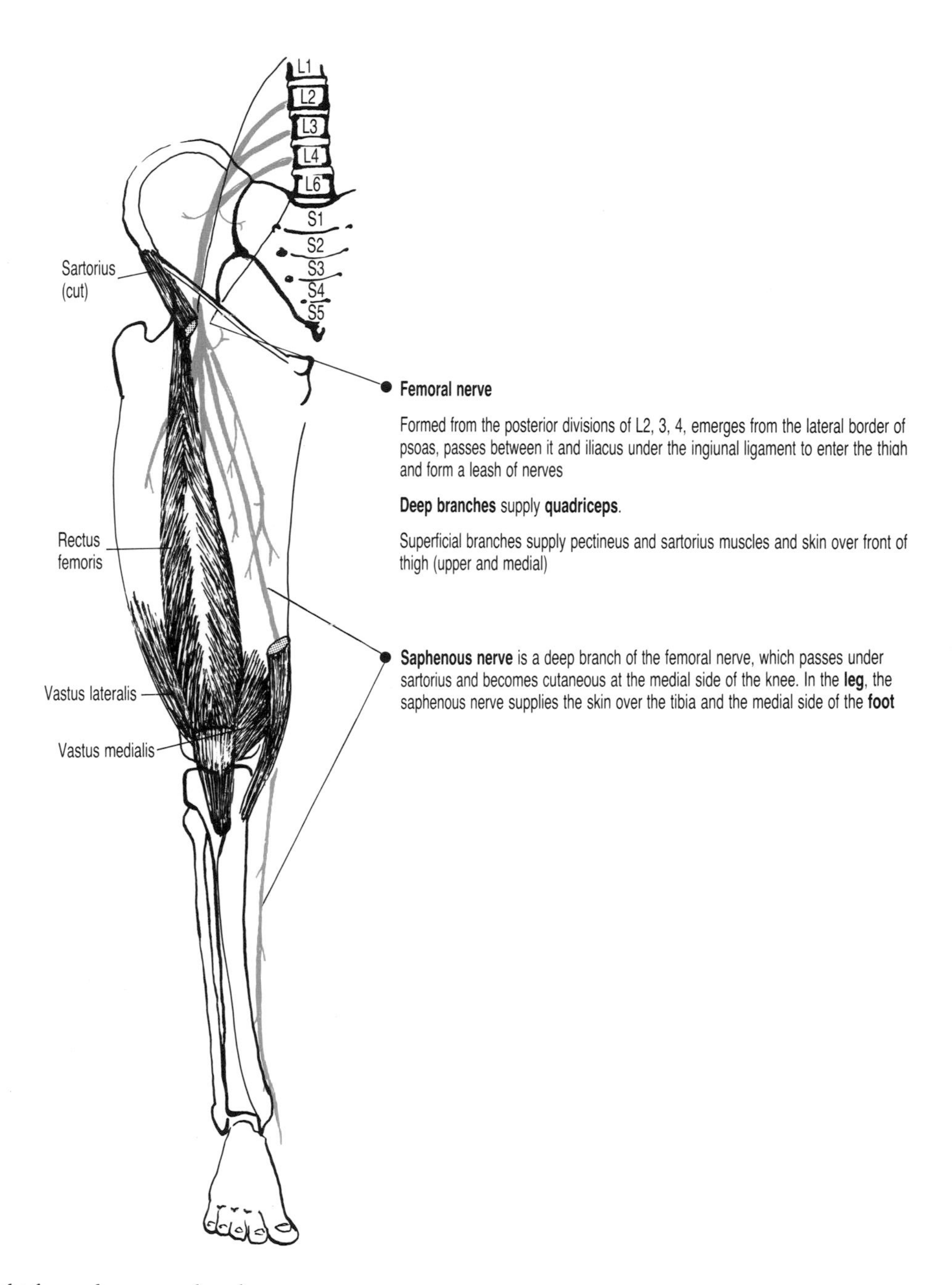

Fig. 9.3. Right femoral nerve and saphenous nerve.

saphenous nerve, a branch of the femoral nerve in the thigh becomes cutaneous at the medial side of the knee and continues onto the medial side of the ankle (Fig. 9.3).

The lateral cutaneous nerve

This nerve emerges from the lateral side of the psoas muscle and crosses the iliacus obliquely to the anterior superior spine of theilium. The nerve passes under the lateral end of the inguinal ligament and becomes cutaneous in the lateral thigh. Pressure on the nerve in the area of the iliac spine causes loss of sensation on the lateral side of the thigh (Fig. 9.4).

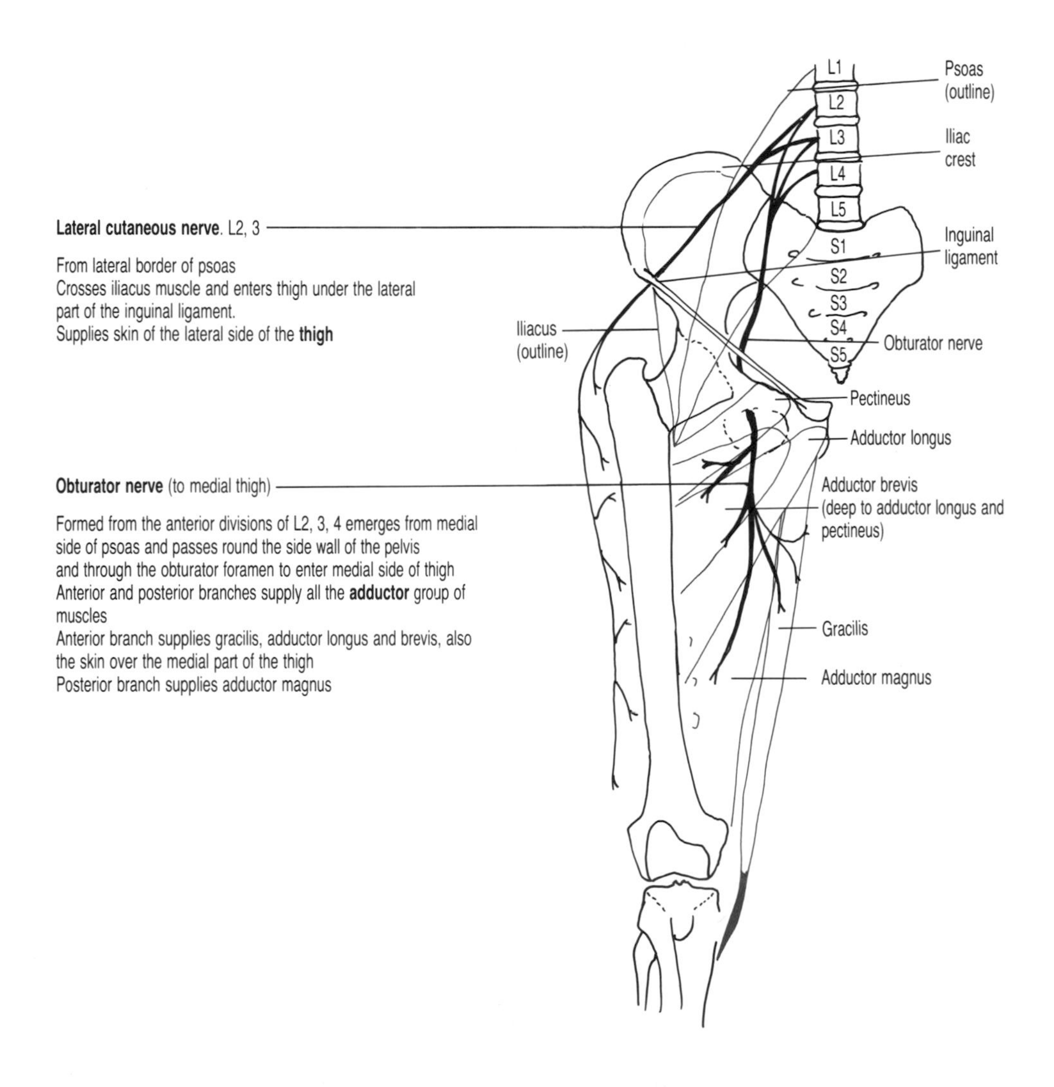

Fig. 9.4. Right obturator nerve and lateral cutaneous nerve.

The obturator nerve

This leaves the medial side of the psoas muscle at the brim of the pelvis and passes through the obturator foramen of the hip bone to reach the medial side of the thigh. In the medial compartment of the thigh, the obturator nerve supplies all the adductor group of muscles.

The functional importance of the femoral and obturator nerves

The femoral and obturator nerves are both important in **walking**. The quadriceps, supplied by the femoral nerve, stabilizes the knee during support, and propels the body upwards when walking uphill and climbing stairs. The adductors, supplied by the obturator nerve, are active in the supporting leg when the other leg is off the ground. If the adductors are weak due to damage of the obturator nerve, the leg swings outwards instead of forwards in the swing phase in walking.

Other nerves of the lumbar plexus

Three other cutaneous nerves are formed from the first two nerves of the lumbar plexus. The first lumbar nerve divides into two, the **iliohypogastric** and the **ilioinguinal** nerves, which supply the skin of the buttock and groin respectively. A third cutaneous nerve, the **genitofemoral**, is formed from L1 and L2, and supplies a small area of skin on the upper front part of the thigh (Fig. 9.2).

9.4 The sacral plexus, position and formation

The sacral plexus is formed in the pelvis from the joining of the anterior primary (ventral) rami of the lumbosacral trunk (L4, L5), and the first three and part of the fourth sacral nerves. The landmark to find the position of the sacral plexus in the pelvis is the *piriformis*, one of the six lateral rotator muscles of the hip. (The piriformis is attached to the anterior aspect of the second, third and fourth segments of the sacrum, and passes out of the pelvis into the thigh through the greater sciatic notch of the pelvis to attach to the apex of the greater trochanter of the femur. Figure 9.6 shows a posterior view of the piriformis muscle.) The main part of the plexus passes backwards with the piriformis to enter the posterior compartment of the thigh. (Figure 9.1 shows the emerging sacral nerves lying on the anterior surface of the sacrum.) Figure 9.5 shows the roots of the sacral plexus and the main terminal branches.

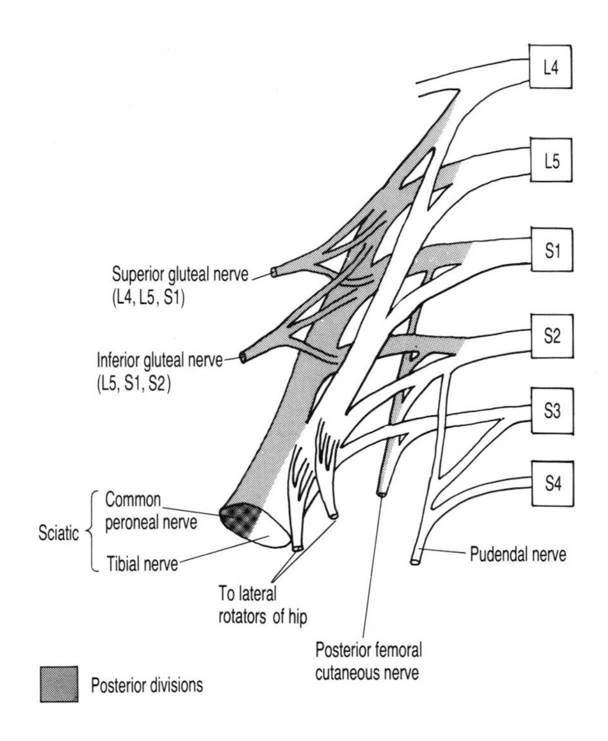

Fig. 9.5. Sacral plexus, roots and main terminal branches.

- *LOOK at an articulated skeleton and identify the greater sciatic notch of the pelvis, which is converted into the greater sciatic foramen by the sacrospinous ligament joining the sacrum to the ischial spine.*
- *LOOK at the anterior surface of the sacrum to see how spinal nerves originating inside the pelvis from the sacral foramina reach the back of the thigh by passing through the greater sciatic foramen. (Remember how the femoral nerve passed anteriorly under the inguinal ligament, and the obturator nerve passed medially through the obturator foramen.)*

You should now understand the three directions of exit of nerves to the thigh. Return to Fig. 8.14 (p. 184) and identify the sciatic and saphenous nerves in a transverse section of the thigh.

9.5 Terminal branches of the sacral plexus

9.5.1 Branches to the posterior thigh, leg and foot muscles

The three main nerves from the sacral plexus important for movement of the lower limb are the *superior gluteal nerve* which passes above piriformis to supply gluteus medius, gluteus minimus and tensor fascia lata; the *inferior gluteal nerve* which passes below the piriformis to supply the gluteus maximus; and, most importantly, the *sciatic nerve* which emerges below the piriformis to supply the hamstrings and branches to all the muscles below the knee.

Figure 9.6 shows the three nerves emerging through the greater sciatic notch and branching to supply the muscle groups.

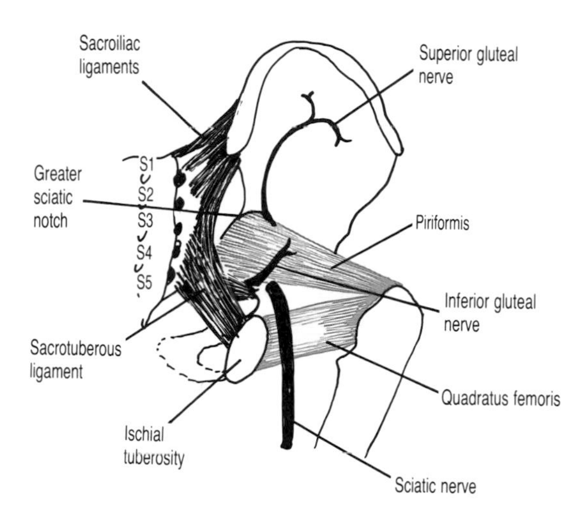

Fig. 9.6. Posterior view of right pelvis and hip to show nerves emerging through the greater sciatic foramen.

The sciatic nerve

This is the largest nerve in the body, about the same size as the thumb or 2 cm in diameter. The posterior divisions of L4, L5, S1 and S2 form the common peroneal component of the sciatic nerve. The anterior divisions of L4, L5, S1, S2 and S3 form the tibial component of the sciatic nerve (Fig. 9.5). These two components lie together in the thigh until they divide above the knee. Branches high in the thigh supply the hamstring group of muscles. Only the nerve to the short head of biceps comes from the common peroneal division, all other branches to the hamstrings are from the tibial component. The sciatic nerve also supplies the fibres of the adductor magnus that originate from the ischium.

Trauma to the sciatic nerve in the middle of the thigh does not usually affect the hamstrings since the branches to these muscles begin high in the thigh.

Figure 9.7 shows the course and distribution of the sciatic nerve. The *branches* of the sciatic nerve are as follows.

1 The **tibial nerve** (Fig. 9.7a). This is the nerve of the posterior muscles of the calf, supplying all the plantar flexors — the gastrocnemius, soleus, flexor hallucis longus, flexor digitorum longus and the tibialis posterior. The tibial nerve lies in the popliteal fossa at the back of the knee and continues on down the

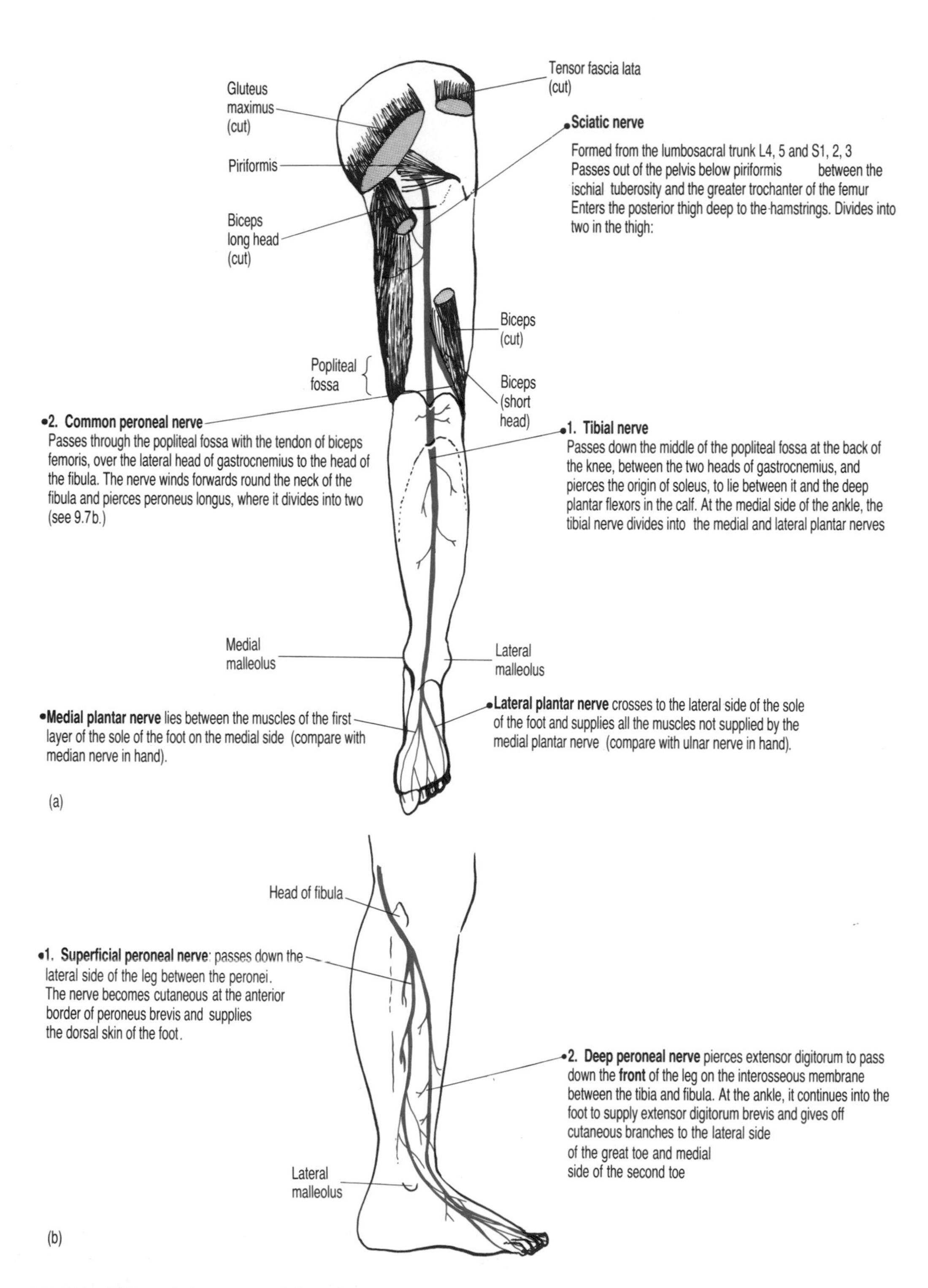

Fig. 9.7. Right sciatic nerve and branches, course and distribution: (a) posterior view; and (b) laterial view below the knee.

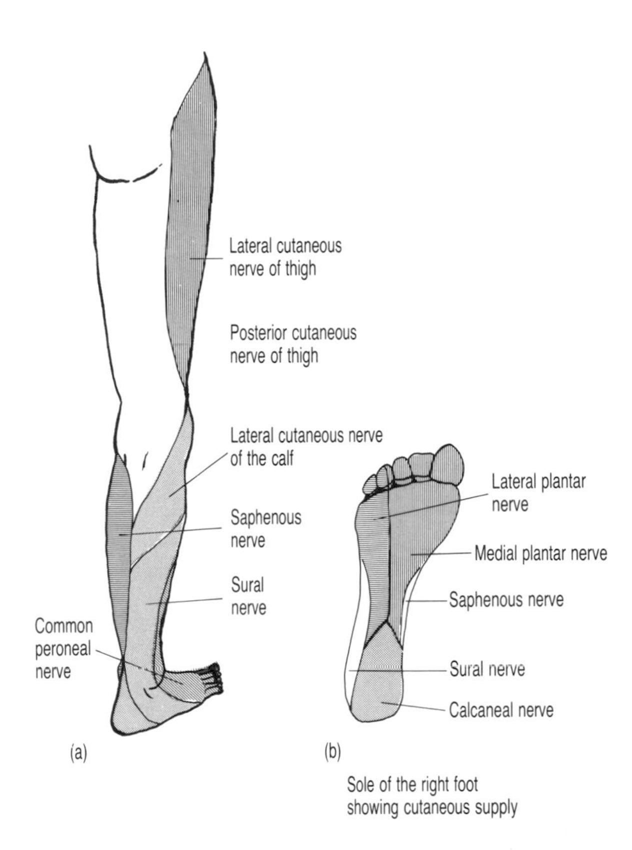

Fig. 9.8. Cutaneous nerve supply to the right lower limb: (a) posterior view; and (b) sole of foot.

leg between the muscles of the calf. The *sural nerve* is a cutaneous branch of the tibial nerve in the calf that supplies the skin on the lateral side of the leg and foot (Fig. 9.8). A branch of the tibial nerve at the ankle supplies the skin of the heel.

At the medial side of the ankle, the tibial nerve divides into two nerves that supply the muscles of the sole of the foot:

(a) The **medial plantar nerve** supplies the abductor hallucis, flexor hallucis brevis, flexor digitorum brevis and the first lumbrical.

(b) The **lateral plantar nerve** supplies the lateral three lumbricals, all the interossei, the abductor digiti minimi, the flexor digiti minimi and the adductor hallucis.

2 The **common peroneal nerve** (Fig. 9.7a). This is the lateral part of the sciatic nerve that divides in the upper part of the peroneus longus into the superficial and deep peroneal nerves.

(a) The **superficial peroneal nerve** which goes to the evertors, peroneus longus and brevis. This nerve lies deep to the peroneus longus in the leg and continues over the front of the ankle to supply the skin of the dorsum of the foot. Fractures of the shaft of

the fibula, common in skiing and skating falls, may injure the superficial peroneal nerve. The foot then tends to go into inversion and the ankle loses stability.
(b) The **deep peroneal nerve** goes to the dorsiflexors, the tibialis anterior, extensor hallucis longus, extensor digitorum longus and peroneus tertius. This nerve ends on the dorsum of the foot where it supplies extensor digitorum brevis and the skin over the first and second toes.

• *REVISE the muscles groups involved in support and swing described in Chapter 7. Record the nerve supply to each of the muscle groups. You should now be able to work out how the activation of the muscles used in walking depends on the following.*
1 *The superior and inferior gluteal nerves, and the femoral nerve keep the pelvis level and stabilize the hip and knee in support. The sciatic and deep peroneal nerves flex the knee and dorsiflex the ankle respectively during swing.*
2 *The tibial nerve initiates 'push off' by plantar flexing the ankle.*
3 *The superficial peroneal nerve stabilizes the ankle and protects its lateral ligament, particularly on rough ground.*
4 *The common peroneal nerve is vulnerable to damage at the lateral side of the knee, where it is subcutaneous as it winds round the neck of the fibula. The result of damage is loss of dorsiflexion and eversion of the ankle producing 'foot drop' when the leg is lifted off the ground.*

9.5.2 The pudendal nerve

The pudendal nerve (S2, S3, S4) (Fig. 9.5) passes out of the greater sciatic foramen medial to the sciatic nerve, and then turns below the ischial spine to supply the muscles of the pelvic floor.

9.5.3 Cutaneous nerves of the sacral plexus

Posterior femoral cutaneous nerve

Branches of the posterior femoral cutaneous nerve (S1, S2, S3) (Fig. 9.5) supply the skin over the posterior side from the buttock to the upper calf.

Sural nerve

The sural nerve is a branch of the tibial nerve at the back of the knee which supplies the skin of the lateral side of the lower calf and the lateral side of the foot. A branch of the tibial nerve at the ankle supplies the skin of the heel.

9.5.4 Summary of the cutaneous supply in the foot

The nerve supply to the skin of the sole of the foot is important for sensory information about the contact of the foot with the ground during standing and walking, and the distribution of the body weight over the feet. The major part of the skin of the sole of the foot is supplied by the medial and lateral plantar nerves (Fig. 9.8). The heel receives a branch of the tibial nerve, and the lateral border of the foot from the sural nerve.

The dorsal surface of the foot is largely supplied by branches of the superficial peroneal nerve, except a triangular area over the first and second toes which receives branches of the deep peroneal nerve. The saphenous nerve, a branch of the femoral nerve in the thigh, becomes the cutaneous nerve to the medial side of the lower leg and continues to the medial dorsal surface of the foot.

9.6 Some observations of gait

Walking is a pattern and sequence of movements performed by the lower limbs to propel the upright body in one direction, usually forwards. To keep the balance of the body, the head and trunk constantly adjust their position, so that the centre of gravity remains over the foot base. In fast walking and in running, the forward propulsion is aided by swinging the arms, but the upper limbs only play a minor role in walking at average speeds. The timing and coordination of all the active muscles is controlled by the central nervous system.

• *WATCH people walking to the shops, to the station and in the park; alone and in groups. Notice the variety of walking speed, length of stride, rate of stepping, position of the head and body, and amount of arm swing.*

Each person seems to have his or her own way of walking, which varies with mood, time of day and many other factors. If you try to 'walk in step' with someone else, it is always difficult, especially if you are not the same height. All the measurable parameters of gait, such as stride length and step frequency are related to stature. Tall people take along strides and make fewer steps per minute compared with short people walking at the same speed. Other features of gait are not easy to measure, but can be observed by the experienced eye of the therapist and athletics coach. The walking pattern should be stable, rhythmic and coordinated. The ability to change direction and speed, and to accommodate changing surfaces on the ground, are all part of the demands of normal walking. Knowledge of the features of walking patterns makes it easier to observe deviations from it in those who have difficulty in walking.

9.6.1 The walking cycle

The cycle of movements in walking are usually divided into: (i) the support or stance phase; and (ii) the swing phase. In Chapter 8 the lower limb actions in support and swinging have been considered. Figure 9.9 shows the stages in swing and support during walking, and the changes occurring during support and swing will be considered in more detail here.

Support phase

When the heel of the leading leg is lowered to the ground at 'heel strike' the support phase begins (Fig. 9.9c). The leading leg is now preparing to receive the body weight transferred from the trailing leg: contraction of the quadriceps occurs and the hamstrings stabilize the hip and the knee: the ankle is dorsiflexed at heel strike, and then plantar flexes (by eccentric action of the anterior tibial muscles) to place the whole foot flat on the ground.

The 'mid stance' phase is when the leading leg has become a support for the body weight.

At the end of the support phase, the heel is raised in plantar flexion at 'push off' to transfer the weight forwards to the opposite leg (Fig. 9.9d).

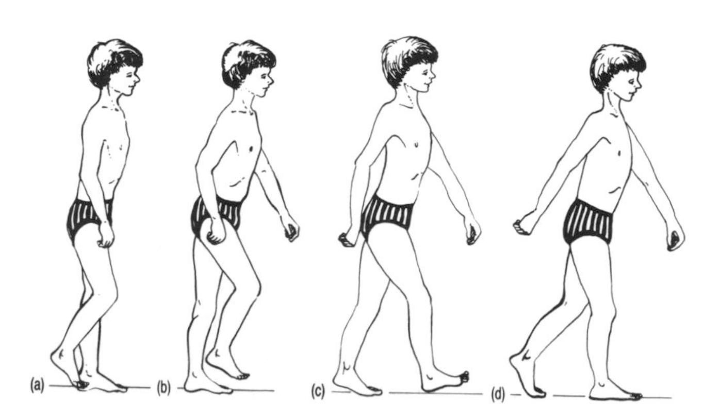

Fig. 9.9. Walking cycle: (a) start of swing; (b) swing phase; (c) heel strike; and (d) propulsion and support.

During support the hip and knee extensors stabilize the leg; the hip abductors and adductors stabilize the pelvis.

• *WATCH a partner walking slowly and notice the changes in the pattern of movements of the supporting leg from heel strike through to push off.*

Swing phase

After push off, the trailing leg is lifted from the ground and starts the forwards swing (Fig. 9.9a). The hip flexors start the swing, and in walking at slow speeds on level ground, the limb then swings like a pendulum. During the swing the knee flexes and the ankle dorsiflexes to lift the toes clear of the ground (Fig. 9.9b).

The swinging leg carries the pelvis forwards on that side. To counteract this rotation of the pelvis, the femur is rotated laterally by the hip adductors, which insert on the posterior shaft of the femur. In this way, the toes are kept pointing forwards. At the same time the trunk is carried forwards on the side of the swinging leg, so the trunk rotators are active to keep the shoulders facing forwards and the eyes looking ahead. At the end of the swing, the hip extensors halt the thigh, the knee extends and the heel is placed on the ground.

• *WATCH a partner walking and observe the changes in the pattern of the swinging leg at each step. The leg accelerates, swings and then deccelerates at each step.*

When one leg is in the swing phase the other leg is in the support phase. At most natural speeds of walking, there is a 'double support phase' in between when both feet are on the ground. With an increase in the speed of walking, there is a decrease in the time of double support. In running, double support disappears altogether.

9.6.2 Abnormal gait

An abnormal pattern of walking may have mechanical or neurological origin. Only one side of the body may be affected or both sides. Walking demands the ability to perform the swing and support movements of the lower limb, at the same time as maintaining the balance of the body. Coordination of the bilateral movements of the lower limbs, trunk and arms is essential. The postural reflexes, largely based on the brain stem, control the balance reactions.

Some causes of abnormal gait

1 Mechanical:

(a) Inability of skeletal structures to bear the body weight. For example, fracture or osteoporosis of bone, pain in lower limb joints.

(b) Weakness of muscles.

(c) Restricted range of movement at joints.

2 Neurological:

(a) Abnormal muscle tone— hypotonia, spasticity or rigidity.

(b) Presence of abnormal movement synergy in the lower limb.

(c) Disturbance of postural reflexes.

(d) Absense of sensory feedback from the sole of the foot and from proprioceptors in muscles and joints.

(e) Loss of body image when one side is ignored and the affected side is 'left behind'.

(f) Perceptual problems leading to difficulty in judging distances and depths and therefore where to put limbs.

Apparently minor problems such as a painful toe, or ill fitting shoes can produce marked changes in gait. Also, one particular feature of abnormal gait may occur for a variety of reasons. For example, the forefoot may drop during the swing and the toes drag on the ground due to damage to the common peroneal nerve, or general weakness of lower limb muscles, or increased tone in the

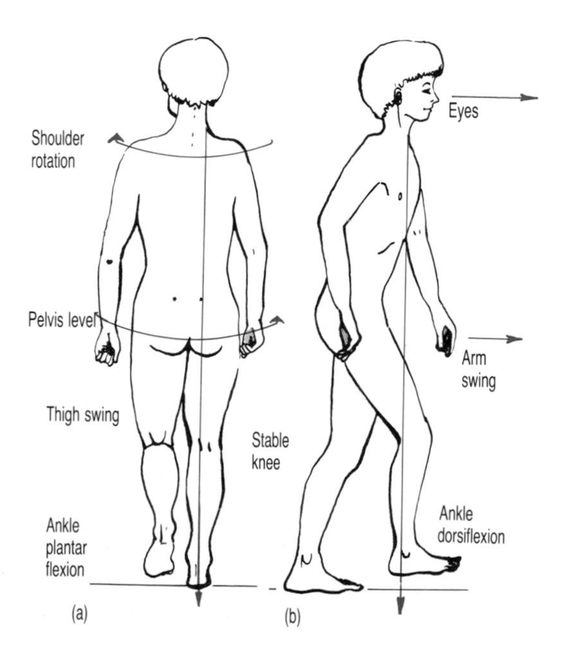

Fig. 9.10. Observation of gait.

plantar flexors. It is therefore important to *observe* the particular features in each individual (Fig. 9.10), including any abnormal movements of body segments and also the ability to: (i) balance; (ii) coordinate the two sides of the body; (iii) maintain walking rhythm; and (iv) change direction and speed of walking as required.

Some types of abnormal gait

1 Shuffling gait. The feet make short shuffling steps with rigidity in hip and knee extensors. There is little movement of the trunk and arms (Parkinson gait).
2 High stepping gait. During the swing phase, the foot is lifted high off the ground. This is due to loss of sensory information from the skin of the sole of the foot and from the proprioceptors in the muscles of the limb, known as *sensory ataxia*. If this gait is unchanged when the eyes are closed, then the cerebellum is involved and it is known as *cerebellar ataxia*.
3 Spastic gait (with hypertonia). The lower limb shows extensor synergy with overactive adductors, extensors and invertors. The thigh swings across the body during the swing (called *scissors gait*). There is difficulty in putting the heel down at the beginning of the support phase.
4 Hemiplegic gait (with hypotonia). The pelvis is lifted on the affected side by 'hip hitching', and the thigh swings out to that side due to weak adductors. The foot drops during the swing, and toes or lateral border of the foot are placed on the ground first.
5 Waddling gait. General weakness in the supporting muscles means that the pelvis tilts with every step and the trunk is thrown from side to side.

At the end of this chapter you should be able to:
1 Outline the position and arrangement of the lumbar plexus.
2 Describe the course and distribution of the main branches of the lumbar plexus: femoral, lateral cutaneous and obturator nerves.
3 Outline the position and arrangement of the sacral plexus.
4 Describe the course and distribution of the main branches of the sacral plexus: superior gluteal, inferior gluteal, sciatic, tibial and common peroneal nerves.
5 Summarize the nerve supply to the skin of the sole of the foot.
6 Describe the swing and support phases of the walking cycle.
7 List the main features of abnormal gait patterns.

10 / Posture and Breathing. The Trunk

10.1 Functions of the trunk

The trunk is the central axis of the body. The limbs use the trunk as a base on which to move. When the body is upright, the trunk supports the head and maintains the erect posture with minimal effort.

The trunk consists of the *thorax, abdomen* and *pelvis,* three cavities stacked one above the other (Fig. 10.1). Any change in the pressure inside one of the cavities affects the adjacent ones. The vertebral column links the three boxes posteriorly. The thoracic cavity extends from the clavicle and first rib above, to the muscular diaphragm below. The abdominal cavity has the dome of the diaphragm as the roof. The blade of the iliac bone of the pelvis lies in the abdomen and the cavity leads down into the pelvis below. The pelvic cavity is a bowl formed by the sacrum and the two innominate bones with a muscular floor.

The joints and muscles of the trunk combine to form a stable system when standing upright. The muscles act like guy ropes keeping the balance when external forces act on the trunk. If a

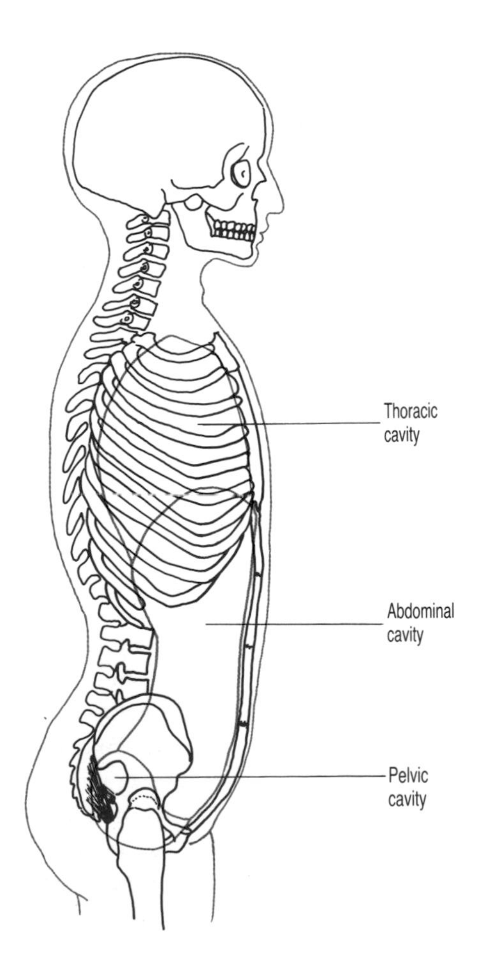

Fig. 10.1. Side view of the trunk. Outlines of the thoracic, abdominal and pelvic cavities.

group of muscles become weak, the trunk changes its position, in the same way as a tent will lean to one side if a guy rope is loosened.

The trunk has a protective function for the lungs, heart, digestive tract, kidneys and pelvic organs (bladder, rectum and reproductive organs). The spinal cord is also protected by being enclosed by the bones of the vertebral column, with pairs of spinal nerves emerging between adjacent vertebrae to be distributed to all parts of the body.

Ventilation of the lungs is the result of movements of the thorax, the anterior abdominal wall and the muscular diaphragm. The muscle action of the anterior abdominal wall exerts its effect either on the diaphragm above to expel air, or on the pelvic cavity below to expel urine or faeces.

Lifting, carrying, pushing and pulling heavy loads all involve the trunk, which counteracts the forces on the limbs, and adjusts the line of weight over the foot base. Carrying a heavy load of shopping in one hand requires muscle activity on the opposite side of the trunk to balance the weight. Activity in the anterior abdominal muscles is important to reduce the load on the back in lifting loads from the front.

To summarize, the overall functions of the trunk are as follows.

1 To maintain the upright posture.
2 To protect organs.
3 To aid ventilation of the lungs.
4 To adapt to changes in internal pressures when lifting loads.
5 To expel urine, faeces and the foetus at birth.

It is important to consider the three parts of the trunk as a functional unit, since changes in one part affects the other two. Most of the trunk movements are performed by large muscles arranged in sheets around the axial skeleton. The position of the muscles and direction of the fibres determine the ways in which each contributes to trunk movement.

10.2 Upright posture

The bones and ligaments of the *vertebral column* form a stable balanced support that requires little muscle activity when standing still. Any slight sway is counteracted by tension in the strong longitudinal ligaments joining the individual vertebrae. Each bone of the vertebral column articulates with the one above and the one below by a cartilaginous joint (intervertebral disc) and four synovial joints between the articular processes.

At birth, the vertebral column has a primary curve, concave forwards. As the baby learns to support the weight of the head and trunk in sitting and then standing, two secondary curves develop in the neck and lower back. From 2 years onwards, the vertebral column has four curves as follows:

7 cervical vertebrae	— convex forwards	— secondary
12 thoracic vertebrae	— concave forwards	— primary
5 lumbar vertebrae	— convex forwards	— secondary
5 sacral vertebrae (fused)	— concave forwards	— primary
3 coccygeal vertebrae		

The four curves provide an ideal way of combining support with flexibility and resilience (Fig. 10.2).

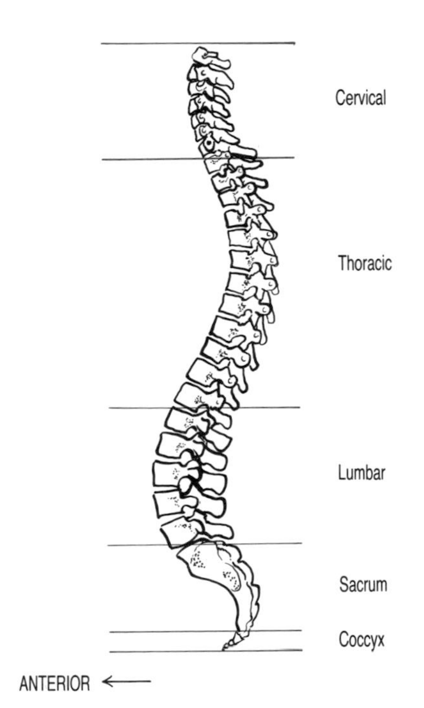

Fig. 10.2. Vertebral column viewed from the side. Cervical, thoracic, lumbar and sacral curves.

• *OBSERVE a partner standing upright. Look first from the side to imagine a line from the ear through the vertebral column to the hip and knee, ending just in front of the ankle. Move the trunk until the position looks balanced. Notice the curves of the back. Refer to an articulated skeleton to see the curves more easily. Next, look at your partner from the front to see if the shoulders and hips are level, i.e. no lateral curves.*
• *WATCH a person sitting at a keyboard and notice the shape of the back in relation to the shape of the back of the chair. Try raising and lowering the keyboard to see the effect on the working posture.*

• *LOOK at elderly people sitting in easy chairs. Think where a cushion should be placed to support the lumbar curve of the back.*

If an abnormal posture is adopted over long periods of time, a balanced position is progressively lost and muscle activity must be used to a greater extent. Examples of abnormal posture are:

Kyphosis: standing with rounded shoulders
Lordosis: standing with a hollow back
Scoliosis: lateral curvature to the spine, and tilting of the shoulders

Shoes with high heels throw the body weight forwards and the vertebral column adapts by increasing the lumbar curvature (lordosis). Problems with breathing may develop in scoliosis due to the effect on the shape of the thorax. Poor working posture increases the possibility of low back pain, even in the young. In the elderly, degenerative changes in the vertebrae and discs due to disease or ageing, coupled with the loss of the need and motivation to move about during the day, give general loss of mobility, and deformity develops which may become permanent.

10.3 Movements of the trunk

When the trunk moves in different directions, movement at the synovial joints between adjacent vertebrae is small, but the result of the combined movement of vertebrae at all levels results in a considerable range of movement. During movement of the vertebral bodies, the cartilaginous discs are compressed on one side. A sudden compression may result in tearing of the disc, and this allows the gelatinous centre to protrude (a prolapsed intervertebral disc) and press on the spinal cord or the roots of a spinal nerve. Severe pain then radiates down the path of the affected nerve.

The movements of the trunk, seen in Fig. 10.3, are described as follows:

Flexion: bend forwards, or sit up from lying. The ribs move towards the pelvis
Extension: straighten the back and bend backwards. The ribs move away from the pelvis
Lateral flexion: bend to the side, e.g. pick up a basket or case; the ribs move towards the pelvis on one side only
Rotation: the trunk twists, the head and shoulders are turned so that the eyes can look to the side or behind, either to the right or left.

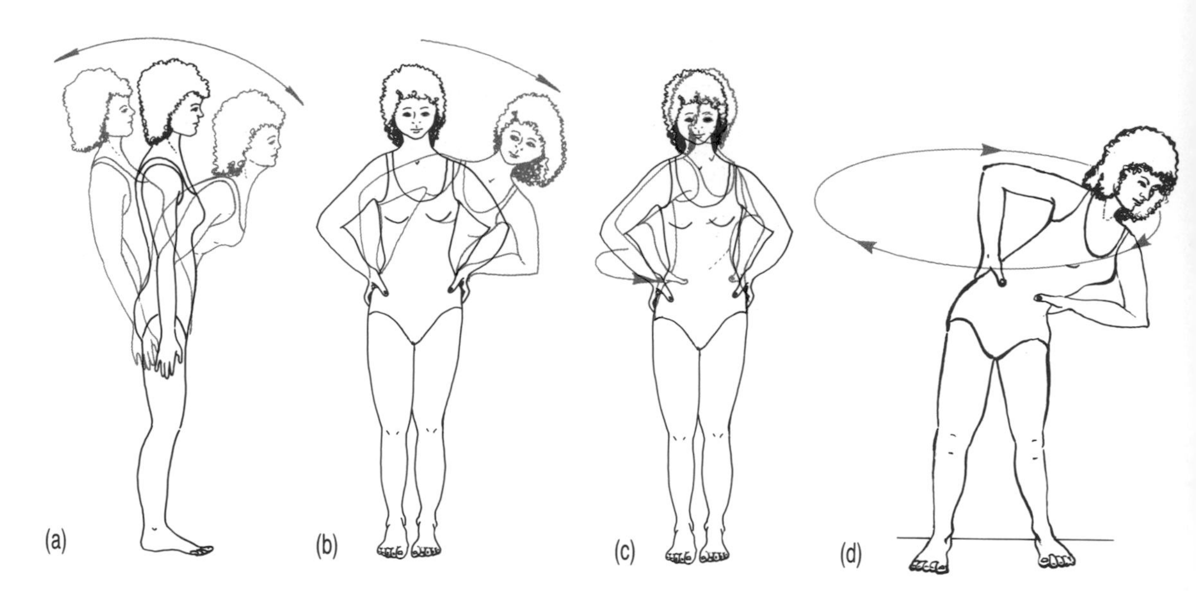

Fig. 10.3. Movements of the trunk: (a) flexion (forwards) and extension (backwards); (b) lateral flexion; (c) rotation; and (d) circumduction.

The 'trunk rolling' exercise shown in Fig. 10.3d is a combination of all these movements.

The range of the individual movements varies in different parts of the vertebral column, depending on the thickness of the intervertebral discs, the direction of the articular facets of the synovial joints, and the length and angulation of the spines. The regions with secondary curves have the greatest mobility. Movements of the cervical region are important for a large field of vision. Reversing a car becomes difficult when there is loss of mobility in the neck. The lumbar region has the greatest range for flexion and extension movements. The extreme bending movements of the acrobat and gymnast are made by continual exercises to stretch the intervertebral ligaments and increase the separation of the lumbar vertebrae. Conversely, the fusion of the lumbar vertebrae in some pathological changes of the spine will reduce the overall mobility of the trunk by a significant amount.

10.3.1 Muscles moving the trunk

Two systems of muscles collectively perform all the movements of the trunk: (i) the deep posterior muscles of the back; and (ii) the abdominal muscles.

Deep posterior muscles of the back

The posterior side of the vertebral column, from the sacrum to the skull, provides a long line of bony processes for the attachment of muscle fibres. Some of these muscle fibres are long, extending from the sacrum to the thorax, while others are short and only span one, two or three vertebrae. The vertical fibres pull the column into extension, those arranged obliquely can rotate one vertebra on the next, and the lateral fibres which are attached to the angles of the ribs can assist lateral flexion.

The largest muscle in this group of deep back muscles is the **erector spinae** (also known as *sacrospinalis*) which originates from the sacrum by a thick broad tendon. In the lumbar region, this muscle is thick and can be palpated in the lower back. Continuing upwards, the muscle is in three bands in the thoracic region, attached to the spines, transverse processes and ribs. The uppermost fibres in the cervical region end on the base of the skull.

The muscles connecting the trunk to the upper limb, for example the latissimus dorsi and trapezius (described in Chapter 5), are separated from the deep muscles of the back by a layer of deep fascia.

Figure 10.4 follows the line of the erector spinae on the right hand side of the vertebral column. Note how the muscle starts at the sacrum and climbs up the back to the head. Deep to the erector spinae another group of muscles is found (Fig. 10.4 — left hand side

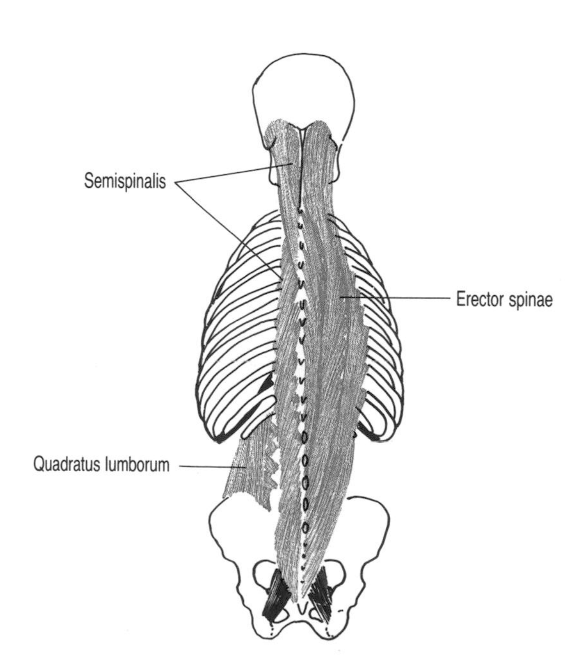

Fig. 10.4. Posterior view of the trunk. Erector spinae on right, semispinalis and quadratus lumborum on the left.

of vertebral column). Most of the fibres in this deeper group lie obliquely from the transverse process of one vertebra to the spine of the vertebra above, or they may span three or four vertebrae. The parts found in the thorax and neck are known as the **semispinalis**.

In movements of the trunk, the erector spinae acts strongly to raise the body from flexion to an upright position.

The erector spinae is important in stabilizing the bones of the vertebral column, and plays a role in lifting and carrying loads.

Lifting from a starting position of bending forwards with straight legs is shown in Fig. 10.5a. From this position, the erector spinae compresses the lumbar discs as it extends the back to raise the load of the trunk, the arms and the child. The centre of gravity of the total load is some distance from the fulcrum in the lower back, so that the load arm is long. The erector spinae, acting on a short lever arm, must develop considerable force to overcome the moment of force of the trunk. In lifting from the sitting position (Fig. 10.5b), the line of weight is even further from the fulcrum and the compression load on the discs is therefore much greater as the erector spinae extends the spine. People in wheelchairs should avoid lifting heavy loads, since the stress on the back will be greater than the same load lifted by someone who can stand close to the load. Lifting from a starting position with bent knees, and with the load as close to the body as possible (Fig. 10.5c) reduces the stress on the back by allowing the hip and knee extensors to contribute most of the power for the lift, and by shortening the lever arm of the trunk plus load.

Fig. 10.5. Lifting: (a) straight legs; (b) sitting; and (c) bent knees.

It should now be clear how bending the knees as well as the back before lifting puts less stress on the back.

Abdominal muscles

The **anterior abdominal wall** consists of flat sheets of muscle forming a four way corset or girdle between the ribs and the pelvis. The individual muscles are: (i) the *rectus abdominis*, which is a vertical panel down the centre of the abdomen; (ii) the *external* and *internal obliques*, which form diagonal fibres at the side of the trunk; and (iii) the *transversus abdominis*, which is a large horizontal waist-band.

Figure 10.6 shows the direction of the fibres of the abdominal muscles seen from the side. The fibres of the two oblique muscles and transversus abdominis blend into an aponeurosis (dense fibrous tissue) towards the midline, connecting with those from the opposite side to form a sheath round the rectus abdominis.

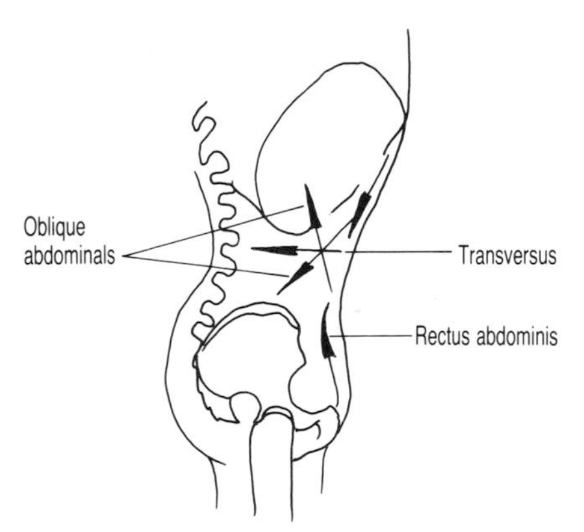

Fig. 10.6. Side view of the trunk to show the direction of the muscle fibres in the anterior abdominal wall.

The **rectus abdominis** (Fig. 10.7a) is a strap like muscle extending from the lower end of the sternum and costal cartilages of the 5th, 6th and 7th ribs to the pubis below. The muscle fibres are usually interrupted at three intervals by transverse bands of fibrous tissue. The four bulges of muscle fibres can be seen in men who have done weight training.

The rectus abdominis flexes the trunk by pulling the ribs towards the pelvis, so acting strongly in sitting up from lying. When the body is lifted off the ground, as in running and jumping, the rectus abdominis supports the front of the pelvis.

Fibres of the **two oblique abdominal muscles** lie at right angles to each other. The **external oblique** is attached to the outer surfaces of the lower eight ribs. The posterior fibres pass vertically to insert

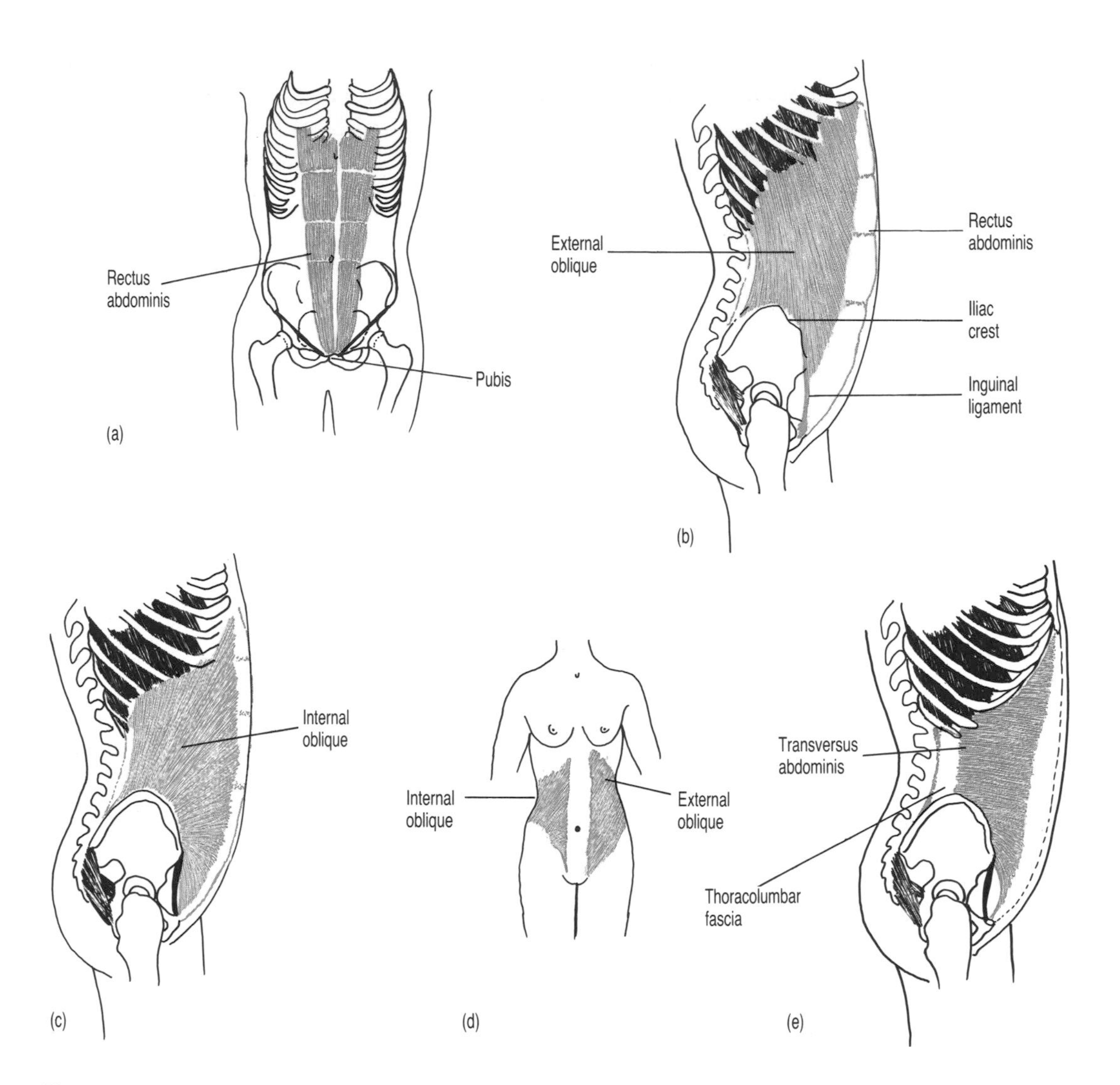

Fig. 10.7. Muscles of the anterior abdominal wall: (a) rectus abdominis, anterior view; (b) right external oblique abdominal, side view; (c) right internal oblique abdominal, side view; (d) oblique abdominals working together; and (e) right transversus abdominis.

on the posterior part of the iliac crest of the pelvis. All the other fibres lie in a direction downwards and forwards, i.e. like hands in a side pocket, to attach to the wide central aponeurosis (Fig. 10.6b). The lower margin of the muscle and aponeurosis is thickened to form the inguinal ligament, which extends from the anterior superior iliac spine to the pubic crest. The inguinal ligament acts as a retinaculum forming the division between the trunk and the thigh. The **internal oblique** is attached to the fascia of the lower back (thoracolumbar fascia), the anterior iliac crest (deep to the external oblique), and the inguinal ligament. The muscle fibres pass upwards and inwards, to attach to the lower ribs, and become a wide aponeurosis as far as the midline (Fig. 10.6c). The aponeurosis of the right and left obliques meet in the midline at the linea alba, a strip of fascia from the lower end of the sternum to the pubic symphysis.

The various ways in which the two layers of oblique abdominal muscles work in combination to produce movements of the trunk will now be considered.

External and internal obliques on *both sides together* **flex** the trunk.
External and internal oblique on *one side only* produces **lateral flexion** to the same side.
External oblique on *one side only* produces **rotation** of the ribs to the opposite side.
Internal oblique on *one side only* produces **rotation** of the ribs to the same side.
External oblique on *one side* with the internal oblique on the *opposite side* produces **rotation** of trunk to the side of internal oblique (Fig. 10.7d).
In standing and sitting, the oblique abdominals work with the neck muscles in turning to look to the side and behind.
In walking, the pelvis is carried forwards on the side of the leading leg and the trunk rotates to keep the eyes looking forwards. The amount of rotation increases with the stride length.

• *LIE down supine and feel the abdominal muscles working in:*
— *Sitting up from lying*
— *Lifting the head from lying. Feel the rectus abdominis working statically to fix the thorax so that the neck muscles can act on the head.*
— *Sitting up from lying while the trunk turns to the left at the same time. Think which abdominal muscles are working.*

The reasons why it is difficult to sit up from lying if the abdominal muscles cannot function, for example after abdominal surgery, fractured ribs, or in the late stages of pregnancy should now be clear. In these instances the movement can be performed by turning on to the side and pushing up with the arm to raise the trunk.

The **transversus abdominis** is the deepest abdominal muscle originating from the inner aspect of the costal margin, the thoracolumbar fascia, iliac crest and inguinal ligament. From this extensive posterior origin, the fibres pass transversely round the abdomen to form a central aponeurosis anteriorly. The muscles from each side meet in the midline at the linea alba. Figure 10.7e shows the right transversus viewed from the side.

The transversus has no action in moving the trunk. The tension in the transversus holds in the abdominal organs, and contraction increases the pressure inside the abdomen. This muscle will be discussed further in Section 4.3 on breathing.

Collective functions of the anterior abdominal wall. All the muscles support and protect the organs of the abdomen and pelvis. A blow to the abdomen produces reflex contraction of the anterior abdominal wall and a temporary cessation of breathing. When lifting loads with the back, contraction of the abdominal muscles can be used to reduce the pressure on the intervertebral discs of the lumbar region. The rise in intra-abdominal pressure during the lift is then distributed upwards and downwards, and this relieves the pressure on the lumbar vertebrae set up by the back muscles. Weight lifters learn to use the abdominal muscles to reduce stress on the back. A sudden or unexpected demand for lifting can produce back strain, and even simple everyday tasks, like making a bed, can cause back injury. Some of the lifting tasks used in the care of the disabled have been replaced by the use of hoists, but it is still important to be aware that contraction of the abdominal muscles can relieve stress on the back when lifting a patient.

In straining movements to expel contents of the pelvic organs, for example urine and faeces, the muscles of the anterior abdominal wall contract to raise the pressure inside the abdomen and pelvis.

The function of the muscles in breathing will be discussed in Section 10.4.3.

The **posterior abdominal wall** between the 12th rib and the posterior part of the iliac crest is formed by the *quadratus lumborum*. This muscle lies lateral to psoas and deep to the origin of transversus (Fig. 10.4).

Contraction of the quadratus lumborum on one side only, assists lateral flexion of the trunk. Acting in reverse, the muscle can lift the pelvic brim on the same side. When both sides contract, the lumbar vertebrae and the pelvis are stabilized for strong activity of the upper trunk and upper limb.

10.3.2 Muscles moving the head and neck

The two main functions of the muscles of the head and neck are to support the head so that it is held upright on the trunk, and to turn the head in all directions so that the eyes can focus over a wide field of vision.

Two of the muscles supporting the head on the trunk are the upper fibres of trapezius (described in Chapter 5) and the upper

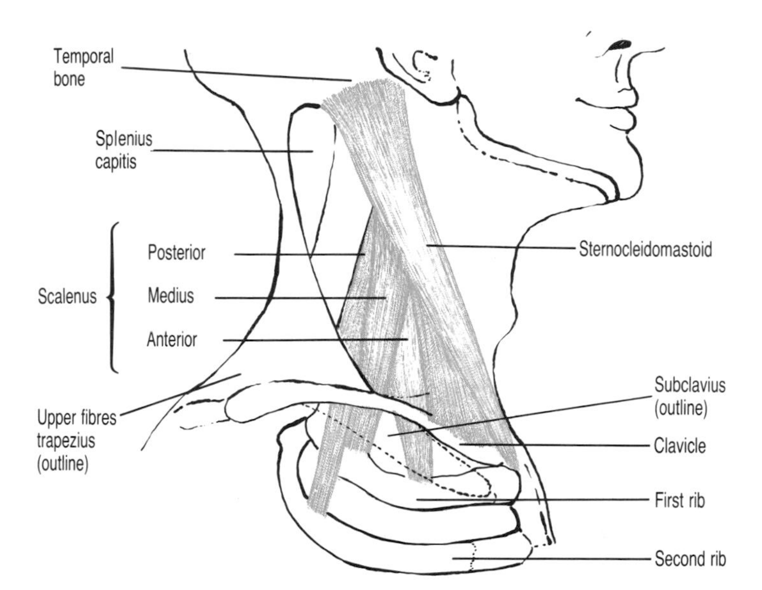

Fig. 10.8. Side view of the neck. Right sternocleidomastoid, right scalenus anterior, medius and posterior.

part of the erector spinae (Section 10.3.1). Lying in between these two muscles at the back of the neck is another pair of muscles, the **splenius capitis** and the **cervicis** (Fig. 10.8).

Holding down the deep muscles of the neck in this region, the splenius capitis has been called the 'bandage muscle'. The splenius muscles are attached to the lower part of the ligament in the midline of the neck (ligamentum nuchae), and the spines of the upper four thoracic vertebrae. Passing upwards and laterally, the capitis is inserted on the base of the skull, to the mastoid process of the temporal bone and adjacent occipital bone. The splenius cervicis inserts onto the transverse processes of the cervical vertebrae 1–4. Working statically, the splenius muscles prevent the head from falling forwards. Both sides working together pull the head backwards in extension. If one side only contracts, the head is rotated to turn the face to the same side.

The most superficial muscle on the front of the neck, clearly visible in action is the **sternocleidomastoid**, often shortened to sternomastoid. This strap like muscle crosses the neck diagonally, and combines with other muscles to perform all the movements of the head. Its name indicates the attachments of this muscle. From the upper end of the sternum and the medial end of the clavicle, the sternocleidomastoid crosses upwards and outwards to end on the mastoid process of the temporal bone of the skull, extending medially to meet the upper fibres of the trapezius (Fig. 10.8).

Both sides of the sternomastoid working together draw the head forwards and act strongly to lift the head up when lying supine. One side contracting produces lateral flexion and rotation to the opposite side. These movements are important in looking from side to side to scan the visual field. When the head is tilted backwards beyond the vertical, the sternomastoid can act as a neck extensor.

A group of three muscles in the lateral part of the neck are the **scalenes**; the scalenus anterior, medius and posterior (Fig. 10.8). Attached centrally to the transverse processes of the cervical vertebrae, the scalenes pass downwards and laterally to the 1st and 2nd ribs. These muscles are an important landmark in the location of the brachial plexus, which passes between the scalenius anterior and scalenius medius in its course towards the 1st rib.

The scalenes flex the cervical spine if both sides contract, or if one side only is active, produce lateral flexion. The muscles are also used to fix the first two ribs in deep inspiration prior to a powerful or long exhalation as when singing or playing a wind instrument.

- *LIE SUPINE and lift the head. Feel the sternomastoid and scalenes in action.*
Turn the head to the right and feel the left sternomastoid in action.

The sternomastoid and the splenius muscles combine to produce most of the turning movements of the head.

10.4 Movements of the thorax and abdomen in breathing

The action of the muscles moving the ribs, and the muscle dividing the thorax and abdomen (the diaphragm), combine to change the size of the thoracic cavity and to ventilate the lungs. The abdominal muscles are also involved in breathing, since their activity affects the position of the diaphragm.

The two lungs fill the thoracic cavity, apart from the space occupied by the heart and major blood vessels. Shaped like two cones, the base of each lung sits on the diaphragm and the apex of each lies above the clavicle. Each lung is surrounded by a narrow airtight space called the pleural cavity. The pleural membranes which form this cavity are attached to the outer surface of the lungs and the inner wall of the thorax. The cavity between the membranes is a completely enclosed space in which the pressure is lower than the pressure of the air outside the thorax. As the thorax expands due to muscle contraction, the lowered pressure in the pleural cavity causes the lungs to be expanded also. The two layers of pleura remain in contact like the sides of a new plastic bag when you try to separate them. When the lungs expand, the air pressure within the air sacs is reduced and atmospheric air is drawn in through the nose and trachea to equalize the pressure inside the lungs. Relaxation of the muscles reduces the size of the thorax to the resting volume and the pressure in the air sacs rises, therefore air passes out into the atmosphere.

The exact amount of air entering and leaving the lungs at any one time depends on the amount of movement of the thorax. Other factors that influence the volume of air breathed are the elasticity and inertia of the lung tissue, and the resistance offered by the airways in the lungs.

The increase in overall size of the thorax by muscle action is essential for the inspiration of air into the lungs. In quiet breathing, expiration is passive and relaxation of the muscles active in inspiration allows the thorax to return to the resting size. The extra air ventilated by the lungs in deep breathing is the result of additional muscle activity in inspiration, and the process of expiration becomes active. Movements of the ribs and the diaphragm occur together in breathing.

10.4.1 Movements of the ribs in quiet breathing

The twelve ribs articulate with the thoracic vertebrae posteriorly. The first seven ribs articulate directly with the sternum in front, while the 8th, 9th and 10th ribs link indirectly to the sternum by their costal cartilages. The 11th and 12th are small and free anteriorly, they play little part in breathing.

• *LOOK at the position of the ribs on an articulated skeleton. Posteriorly, identify the position of two synovial joints, one between the head of the rib and the body of the vertebra, one between the tubercle of the rib and the transverse process.*
Anteriorly, the 1st to 7th ribs have sternocostal joints.

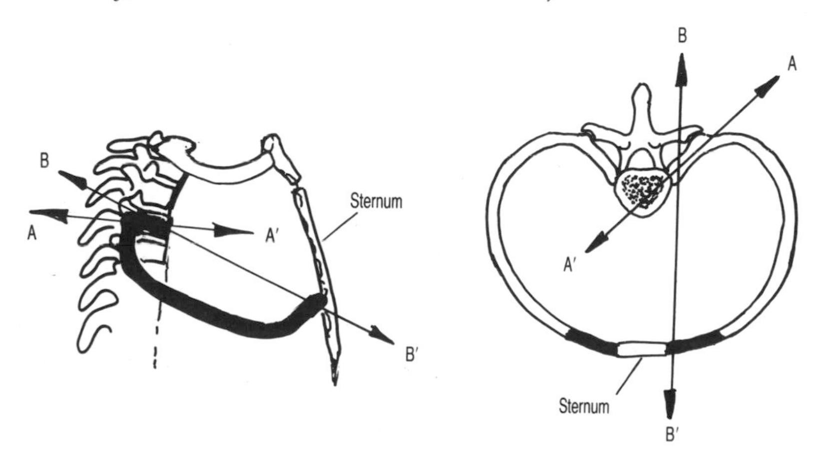

Fig. 10.9. Movement of a rib in side view and plan view. A-A′ axis through the neck of the rib, B-B axis through vertebral and sternal ends of the rib.

Note two things about the general direction of ribs 2–7, the anterior end is lower than the vertebral articulations, the central part of each rib is lower than either of the two ends, (view from the side).

Movement at the joints of the 2nd to 7th ribs occurs about the two axes simultaneously. Figure 10.9 shows the two axes — AA′ and BB′. One axis (A–A′) passes through the neck of each rib. When the rib moves about this axis, the sternum is raised upwards and forwards to increase the anterior to posterior diameter of the thorax.

The other axis (B–B′) passes through the angle of the ribs posteriorly and the sternocostal joints anteriorly. Movement about this axis lifts the middle of the rib upwards and outwards to increase the transverse diameter of the thorax.

The 8th, 9th, and 10th ribs have no sternocostal joints and therefore only move about one axis.

• *PLACE YOUR HANDS on the thorax of a partner, first at the sides over the lower rib cage. Ask your partner to breathe in deeply and watch how your hands move further apart, i.e. the thorax becomes wider. Next, stand at the side and place one hand flat on the sternum, the other hand flat on the thoracic vertebrae. Again ask your partner to breathe in deeply, and notice how the hand on the sternum moves forwards and upwards. These two movements occur together each time the ribs move.*

Muscles acting on the ribs

The muscles which move the ribs in quiet breathing form two layers in the space between adjacent ribs, known as the *external* and *internal intercostal muscles.*

The fibres of the **external intercostal muscles** pass obliquely from the lower border of one rib to the upper border of the rib below. At the anterior end of each intercostal space, the muscle is replaced by membrane. The posterior fibres pass downwards and laterally, and the more anterior fibres lie downwards and medially, i.e. in the same direction as the external oblique abdominal muscles. The first rib does not move in quiet breathing. Figure 10.10 shows the position of the external intercostal muscles in the spaces between ribs 1–6.

Contraction of the external intercostals lifts the ribs about the two axes described. The thorax increases in size by expanding in a forwards and sideways direction and air is drawn into the lungs.

The **internal intercostal muscles** lie deep to the external intercostals, and their fibres are at right angles, downwards and backwards from one rib to the one below. The muscle fibres are

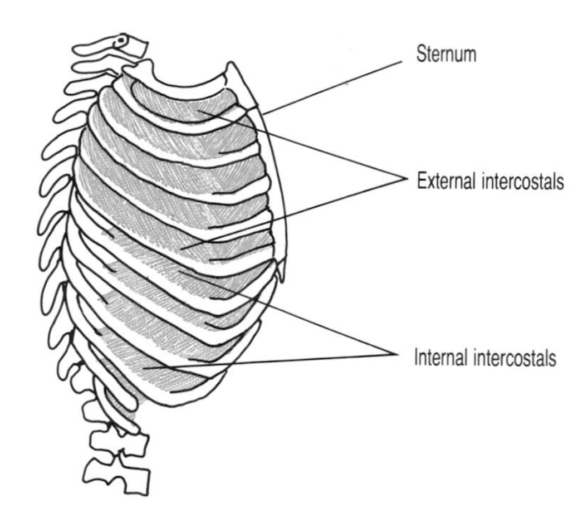

Fig. 10.10. Side view of the thorax to show intercostal muscles. External intercostals shown in upper 5 spaces. Internal intercostals shown in lower 6 spaces.

replaced by membrane at the posterior end of the intercostal space, between the angle and head of each rib. Figure 10.10 shows the position of the internal intercostal muscles in the spaces between ribs 6–10. There is conflicting evidence about the action of the internal intercostals. It has been shown that the anterior fibres between the costal cartilages are active in inspiration. Other studies have shown activity during speech, which is expiratory. The contribution of the internal intercostals to rib movements probably depends on which fibres are active, and on the level of inflation of the lungs.

Relaxation of the intercostal muscles lowers the ribs to their resting position, and air leaves the lungs. Expiration in quiet breathing is therefore passive.

10.4.2 Movements of the ribs in deep breathing

Elevation of the shoulder girdle and the upper ribs allows the thorax to expand further.

• *WATCH the neck of a person breathing deeply to see the activity in neck muscles.*

The main muscles that are recruited to increase the depth of inspiration are the *sternomastoid,* the *scalenes* and the *pectoralis minor.* Figure 10.8 (p. 223) shows the muscles of the neck. If the attachments of the sternomastoid to the head, and the scalenes to the cervical vertebrae, are fixed, these two muscles will pull the clavicle and first two ribs upwards.

The pectoralis minor is a small muscle attached to the coracoid process of the scapula, and its fibres pass downwards to the 3rd, 4th and 5th ribs. If the scapula is fixed, this muscle will also lift the upper ribs.

The action of these three muscles increases the size of the thorax further and more air is drawn into the lungs in inspiration.

The *latissimus dorsi*, which wraps round the rib cage from the lower back to the shoulder, can compress the ribs further in expiration if the humerus is fixed. In deep breathing, expiration becomes active instead of passive.

10.4.3 Movements of the diaphragm

The diaphragm is a dome shaped muscle which forms the floor of the thoracic cavity. At rest, the fibres of the peripheral part of the dome are almost vertical. Converging inwards, the muscle fibres end in a central tendon, a strong flat aponeurosis shaped like a trefoil or clover leaf. The central tendon is nearer to the front of the thorax than the back, so that the posterior fibres are longer. The heart lies immediately above the central tendon, and the pericardium, the membrane round the heart, is attached to it.

• *LOOK at an umbrella (with a very curved edge if possible). The ribs of the umbrella are in the direction of the muscle fibres of the dome of the diaphragm. Imagine the point of the umbrella compressed into a flat trefoil shape to understand the position of the central tendon.*

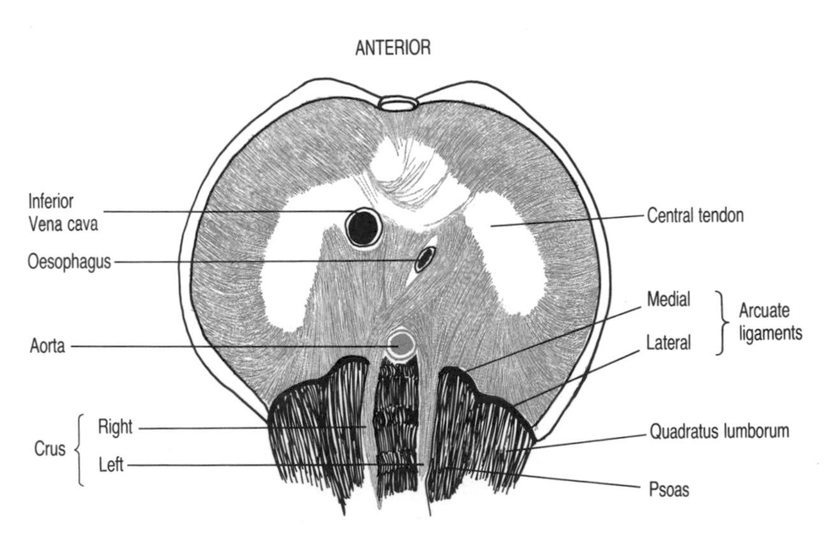

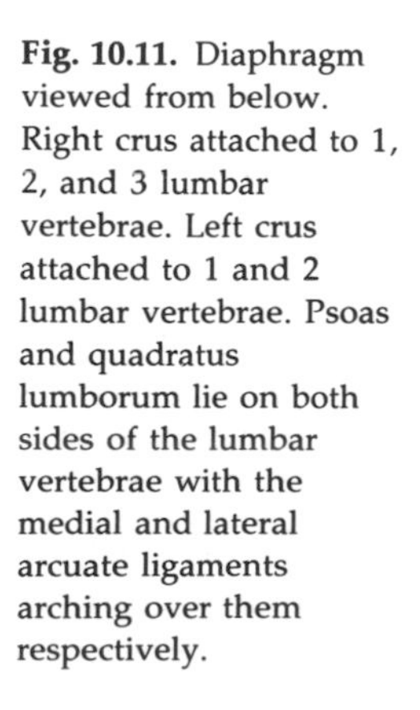
Fig. 10.11. Diaphragm viewed from below. Right crus attached to 1, 2, and 3 lumbar vertebrae. Left crus attached to 1 and 2 lumbar vertebrae. Psoas and quadratus lumborum lie on both sides of the lumbar vertebrae with the medial and lateral arcuate ligaments arching over them respectively.

Figure 10.11 is a view of the diaphragm from below (i.e. in the abdomen looking up to the under surface of the muscle). The diaphragm originates all round the lower margin of the thorax.

Beginning anteriorly, fibres originate from the xiphoid process of the sternum. Next, the 7th to 12th ribs and their costal cartilages form the largest surface for the attachment of fibres. Posteriorly, the origin from the 12th rib is interrupted by the muscles of the posterior abdominal wall, the quadratus lumborum and psoas. These two muscles are bridged by fibrous bands, known as the lateral and medial arcuate ligaments, which provide a base for the attachment of the diaphragm. The most posterior fibres originate from the sides of the lumbar vertebrae by two bands, the right crus (from L1, L2, and L3) and the left crus (from L1 and L2), which arch over the aorta in the midline. The right crus is longer to overcome the resistance of the larger liver lying below the diaphragm on the right side.

Figure 10.11 follows the complete circle that forms the origin of the diaphragm, which can be summarized:

Sternal fibres: from xiphoid process of the sternum
Costal fibres: from inner surfaces of the 7th to 12th ribs
Lumbar fibres: from the arcuate ligaments over the posterior abdominal wall muscles, and from lumbar vertebrae by two crura.

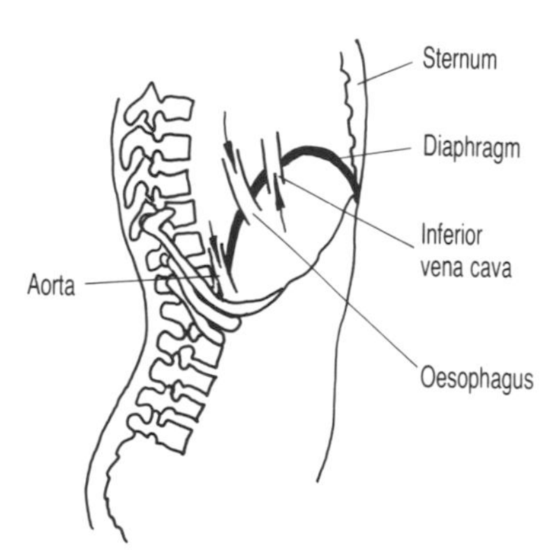

Fig. 10.12. Diaphragm: view from the side.

Figure 10.12 shows how the sternal origin is higher than the lumbar origin. The inferior vena cava passes through the central tendon, and the oesophagus passes through the muscular part just towards the left of the midline. The aorta lies posteriorly against the vertebral column.

Action of the diaphragm. When the diaphragm is *active*, the contractile muscle fibres pull the central tendon *downwards* and the dome becomes flatter.

When the diaphragm *relaxes*, the muscle fibres return to their resting length and the central tendon moves *upwards*.

10.4.4 The abdominal muscles in breathing

The muscles of the anterior abdominal wall can actively participate in breathing out. Contraction of the abdominal muscles raises the pressure inside the abdomen and the diaphragm is pushed **up**. The vertical diameter of the thorax is decreased and air is expelled from the lungs in expiration. The diaphragm and abdominal muscles cooperate in breathing movements in the following way:

inspiration: diaphragm descends, abdominals relax
expiration: abdominals contract, diaphragm moves upwards

• *PLACE YOUR HANDS on the anterior abdominal wall.*
Breathe in deeply, lifting the ribs and feel the abdominals relax as the diaphragm moves down.
Breathe out deeply, contracting the abdominals to expel more air.

During quiet breathing, the contribution of the abdominal muscles to the ventilation of the lungs varies in different individuals. In deep breathing, the use of the abdominal muscles to control and increase the depth of expiration is important in singing and in some relaxation techniques.

10.5 Movements of the pelvis

The bony pelvis is formed by the two hip bones, which articulate together anteriorly by a cartilaginous joint (the pubic symphysis), and with the sacrum posteriorly at the sacroiliac joints. The weight of the trunk above tends to tilt the upper end of the sacrum forwards and the lower end backwards. This tendency for rotation

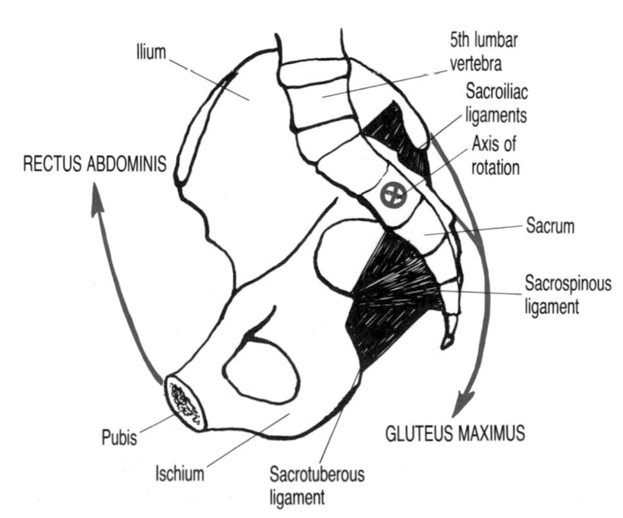

Fig. 10.13. Inner surface of the right innominate bone with the lower lumbar vertebrae and the sacrum. Arrows indicate the direction of pull of rectus abdominis and gluteus maximus in controlling pelvic tilt.

of the sacrum is prevented by two strong ligaments binding the sacrum to the hip bone (Fig. 10.13).

The pelvis is a staging post for muscles passing upwards to the trunk or downwards to the lower limbs.

Muscles acting on the trunk are the rectus abdominis, the oblique abdominals, the erector spinae and the quadratus lumborum.

Muscles acting on the lower limbs are the gluteus maximus, medius and minimus, the hamstrings, the hip adductors, the rectus femoris, the tensor fascia lata and the sartorius.

The tilt of the pelvis in relaxed standing largely depends on the opposing tension in the rectus abdominis pulling the pubis up towards the ribs, and the gluteus maximus pulling on the posterior surface of the sacrum in the opposite direction (Fig. 10.13).

Lateral tilting of the pelvis when one leg is lifted off the ground is counteracted by contraction of the gluteus medius and minimus on the supported side (see Chapter 8). When the glutei and the knee flexors are weak, the pelvis can be raised, to allow the leg to swing in walking by contraction of the quadratus lumborum and latissimus dorsi, this is known as 'hip hitching'.

The iliopsoas links the lumbar spine and pelvis with the femur and is used to raise the legs when lying supine. If the knees are extended, the iliopsoas has to develop a very large force to lift the weight of the leg acting on a long lever arm, and may pull on the lumbar spine causing back strain. Therefore, double leg raising, often used to exercise the abdominals, should be avoided.

The pelvic floor

The muscles of the pelvic floor are suspended from the bony walls and from the muscles lining the pelvis, the *obturator internus* in particular. The fibres of the obturator internus originate on the inside of the hip bone and then pass backwards to leave the pelvis below the ischial spine. The pelvic fascia, covering all the walls of the pelvis, is thickened into a fibrous arch over the obturator internus extending onto the pubis anteriorly, and the spine of the ischium posteriorly. The main muscle of the pelvic floor, the *levator ani*, is attached to this fibrous arch. The fibres of the levator ani descend and then turn inwards to meet those from the opposite side in the midline (Fig. 10.14).

The functions of the pelvic floor are to support the pelvic organs, and to withstand any increase in pressure in the abdomen and pelvis, for example in lifting, coughing and sneezing.

If the muscles become weak, particularly in women after

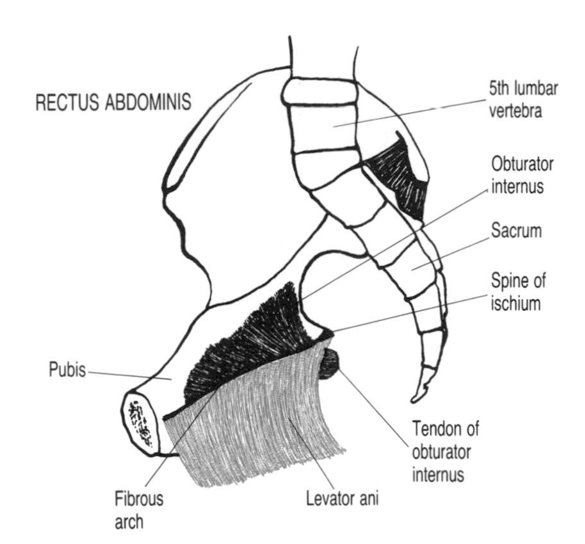

Fig. 10.14. Inner surface of the right innominate bone to show obturator internus and levator ani (cut).

childbirth, the pelvic floor sags like a loaded hammock suspended by the two bony points, the pubis and the ischium.

10.6 Nerve supply of the muscles of the trunk

The **posterior primary rami** of the spinal nerves supply all the deep muscles of the back, including the erector spinae, and in the cervical region supply the splenius capitis and cervicus.

All the other muscles of the trunk are supplied by branches of the **anterior primary rami** of spinal nerves. The spinal accessory (cranial) nerve, together with branches of C2 and C3, supplies the sternomastoid. Branches of C6, C7 and C8 supply the scalene muscles, and S3 and S4 supply the muscles of the pelvic floor.

10.6.1 Phrenic nerves

The phrenic nerves are formed from branches of the third, fourth and fifth cervical nerves in the neck, and supply the diaphragm. Each nerve passes down the neck deep to sternomastoid and enters the thorax. The right phrenic nerve lies on the pericardium covering the right atrium and pierces the central tendon of the diaphragm with the inferior vena cava. The left phrenic nerve lies on the pericardium over the left ventricle and pierces the diaphragm in front of the central tendon (Fig. 10.15).

The phrenic nerve is the motor and sensory supply to the corresponding side of the diaphragm. Injuries to the neck involving the roots of the phrenic nerves result in loss of the action of the diaphragm in breathing.

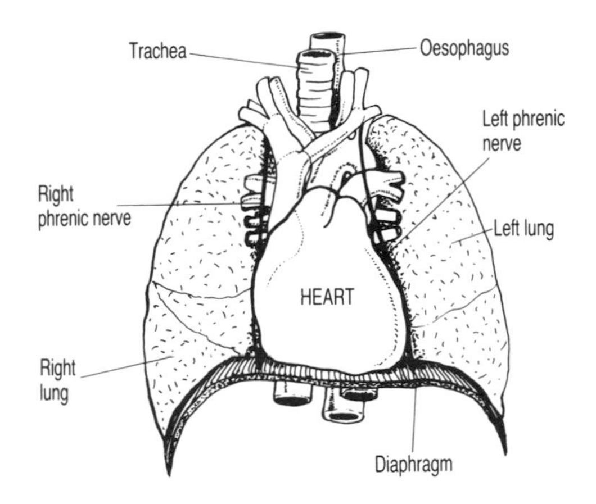

Fig. 10.15. Position and relations of the phrenic nerves in the thorax.

10.6.2 Intercostal nerves

The muscles of the thoracic and abdominal walls are supplied by the intercostal nerves, T1–T12.

Thoracic nerves 1–6 run parallel to the corresponding rib and deep to the internal intercostal muscles.

Thoracic nerves 7–12 also supply the corresponding intercostal muscles, and continue forwards from the intercostal spaces to the muscles of the anterior abdominal wall. Each layer of the abdominal wall (the rectus abdominis, the external oblique, the internal oblique and the transversus) receives branches of thoracic nerves 7–12 from above downwards.

Thoracic nerve 12 is known as the subcostal nerve and branches to supply the quadratus lumborum.

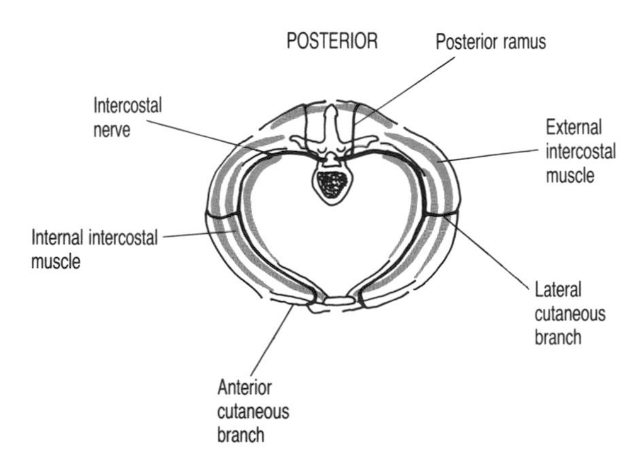

Fig. 10.16. Transverse section of an intercostal space with one pair of intercostal nerves.

10.7 Summary of the muscles of the trunk

1 *Muscles moving the head and neck.*
Sternocleidomastoid; scalenus anterior, medius and posterior; splenius capitis and cervicis.

2 *Muscles moving the thorax.* Ventilation of the lungs.
External and internal intercostals, diaphragm.

3 *Deep posterior muscles of the back.*
Erector spinae (sacrospinalis).

4 *Abdominal muscles.*

Anterior abdominal wall.	Rectus abdominis, external oblique, internal oblique, transversus.
Posterior abdominal wall.	Quadratus lumborum.

5 *Pelvic floor.*
Levator ani and coccygeus, perineal muscles.

At the end of this chapter you should be able to:
1 Outline the function of the vertebral column in maintaining the upright posture.
2 Describe the movements of the trunk and the muscles producing them.
3 Outline the position and functions of the muscles of the anterior abdominal wall.
4 Describe the muscle action involved in ventilation of the lungs in quiet and deep breathing.
5 List the muscles acting on the pelvis. Describe the position and function of the main muscle of the pelvic floor.
6 Outline the nerve supply to the muscles of the trunk, including the phrenic and intercostal nerves.

Further Reading

Caillet R (1982) *Hand Pain and Impairment.* FA Davis Co., Philadelphia.

Carlsoo S (1972) *How Man Moves. Kinesiological Methods and Studies.* Heinemann, London.

Basmajian JV (ed) (1980) *Grant's Method of Anatomy.* Williams & Wilkins, London.

Hamilton WJ (1976) *Textbook of Human Anatomy.* MacMillan, London

Kapandji JI (1970) *The Physiology of Joints.* Vols 1, 2 and 3. E & S Livingstone, Edinburgh.

Kapit W & Elson LM (1977) *The Anatomy Colouring Book.* Harper & Row, London.

Landsmeer JMF (1976) *Atlas of Anatomy of the Hand.* Churchill Livingstone, Edinburgh.

Lehmkuhl LD & Smith LK (1983) *Brunstrom's Clinical Kinesiology.* FA Davis Co., Philadelphia.

Moffat DB & Mottram RF (1987) *Anatomy and Physiology for Physiotherapists.* 2nd Edn. Blackwell Scientific Publications, Oxford.

Rasche PJ and Burke RK (1978) *Kinesiology and Applied Anatomy, The Science of Human Movement.* Lea & Febiger, Philadelphia.

Thompson CW (1978) *Manual of Structural Kinesiology.* CV Mosby Co., St Louis.

Section 3 Integration of Movement

SENSATION, ACTION AND PERFORMANCE

11 / Sensory Background to Movement

All movement starts with a background of sensory information about the space around us and about the position of the body entering the central nervous system. As movement proceeds, this sensory activity changes from moment to moment.

• *THINK about the variety of incoming signals as you walk along a rough path towards a gate:*

From the eyes scanning the visual field ahead, sensing the movement of objects on either side of the path, and anticipating obstructions that must be avoided.

From the skin of the feet detecting the roughness of the path.

From the attitude and movement of the head in relation to the body to keep the balance.

From the joints and muscles in the moving body parts.

We are aware of some of the changes, but many of the responses to the changing input are entirely automatic. We do not fall over, we do recognize obstacles in our path and avoid them by changing direction. In the absence of information from the eyes, we rely more heavily on information from the other sources of input including sound and smell. Some of the automatic responses to the changing input are basic protective reflexes, and we may find ourselves doing them even when there is no threat. For example, we may blink and 'duck' the head when a bird flies towards the windscreen as we drive in a car along the road.

If part of the sensory input is absent, due to damage or disease, there may be insufficient background of sensory input for movement to proceed normally.

11.1 Sensory subsystems in movement

The sensory system can be divided into subsystems, each providing specific information to the central nervous system. Three of these subsystems will be considered with emphasis on their role in movement. Note that the familiar idea of 'five senses' — vision, hearing, taste, smell and touch omits the important 'body sense' or proprioception originating in muscles, joints and the vestibule of the ear. The three main subsystems which provide the monitoring of changes during movement are the *somatosensory system*, the *vestibular system* and the *visual system*.

The **somatosensory** (or somaesthetic) **system** monitors a wide variety of stimuli from all over the body. The skin is part of this subsystem. The skin is not only a simple sense organ for touch, but responds to the particular pressure and temperature of surfaces, such as tools, furniture or the ground in contact with it. Pressure receptors in the skin of the soles of feet monitor the distribution of body weight over the feet. Receptors lying in muscles and tendons are also part of this subsystem. Input from the muscle spindles and from joint receptors report the changes in length and tension of muscles, and the angulation of the joints from moment to moment as the body moves. Receptors in the muscles of the upper limb also give us knowledge about the weight of objects held in the hand. If you hold someone's hand, you are feeling touch, also the temperature and weight of the hand, and you may sense whether the person is tense or relaxed. All these cues are possible due to the large variety of receptors in the skin and the muscles which collectively respond to many types of stimulation.

The **vestibular system** is concerned with sensory information from the position and movements of the head. The receptors for this subsystem are found in the vestibule and semicircular canals of the ear, lying in the cavity of the temporal bone of the skull, behind the organ for sound (the cochlea). We are largely unaware of activity in this system, except when we are in a jerky lift or at the fairground. The vestibular system gives us our sense of stability, and it plays a vital role to keep the body in balance during movement.

The **visual system** contributes to our sense of balance by providing visual clues about the orientation of objects in the space around us. We use vertical and horizontal structures, such as walls, doors and furniture to align the position of the body. Information from the eyes about the movement of objects held in the hand also allows corrections to be made to make accurate and precise movements. When the eyes are closed, all movement becomes more difficult, and highly skilled movements become almost impossible.

11.2 Input from the skin, joints and muscles

Receptors found in the skin, muscles and joints respond to changes in the external environment and the changing patterns of movement of the body parts. This information is relayed to the brain by ascending pathways in the spinal cord. The thalamus acts as an area for reception of somatosensory input at the subconscious level. The primary area for analysis at the conscious level is the postcentral gyrus of the cerebral cortex.

The main ascending pathways to the somatosensory cortex are: (i) the *anterolateral pathway* — spinothalamic tracts; and (ii) the *medial lemniscus pathway* — posterior (dorsal) column. Both systems link the receptors on one side of the body with the opposite

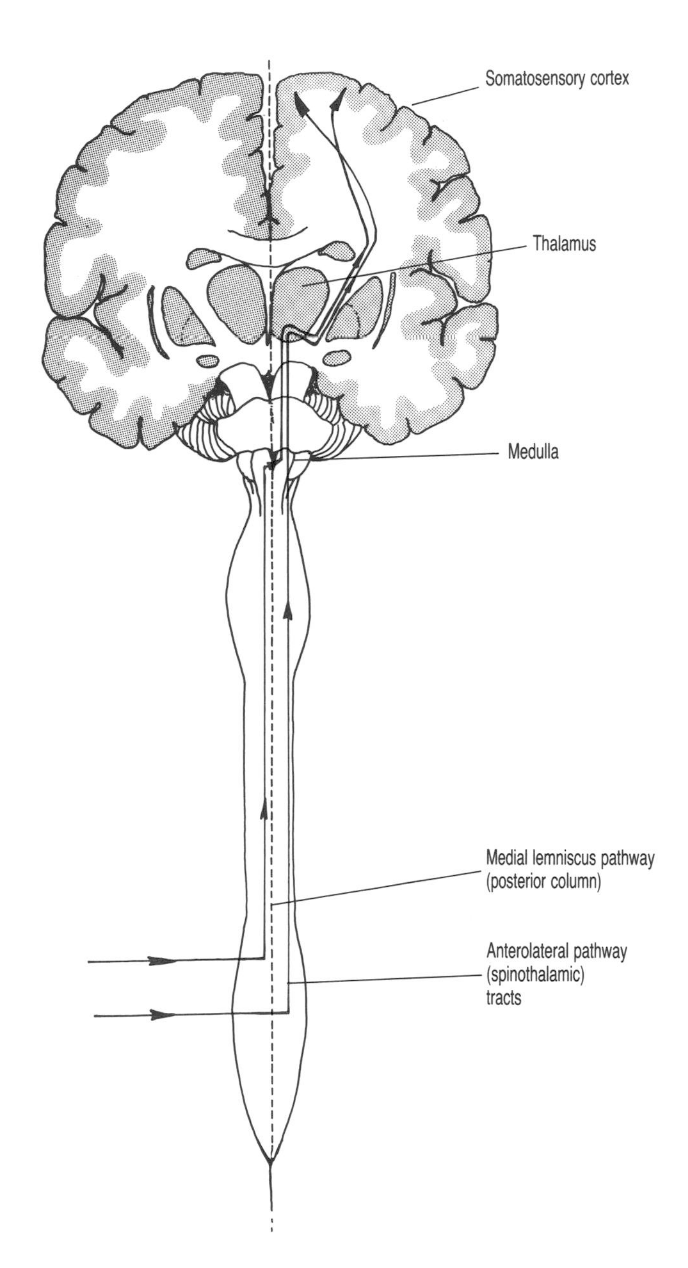

Fig. 11.1. Frontal section of the brain with spinal cord, ascending pathways to the sensory cortex — general plan.

sensory cortex. The anterolateral pathway crosses at the spinal level, while the medial lemniscus crosses in the sensory decussation in the medulla of the brain. The pathways converge in the brain stem and both are integrated in the thalamus.

• *LOOK at Fig. 11.1 to see the general pattern of these two pathways. Before proceeding to the detail of each route —*

• *REVISE the composition of a spinal nerve from Chapter 4, and the position of ascending tracts of the spinal cord described in Chapter 3.*

11.2.1 **Anterolateral pathway** (spinothalamic tracts)

Activity in the anterolateral pathway originates in sensory neurones with slowly adapting receptors in the skin and small diameter axons (Fig. 11.2). These sensory neurones enter the spinal cord and synapse in the posterior horn before crossing to the opposite side to enter the spinothalamic tracts.

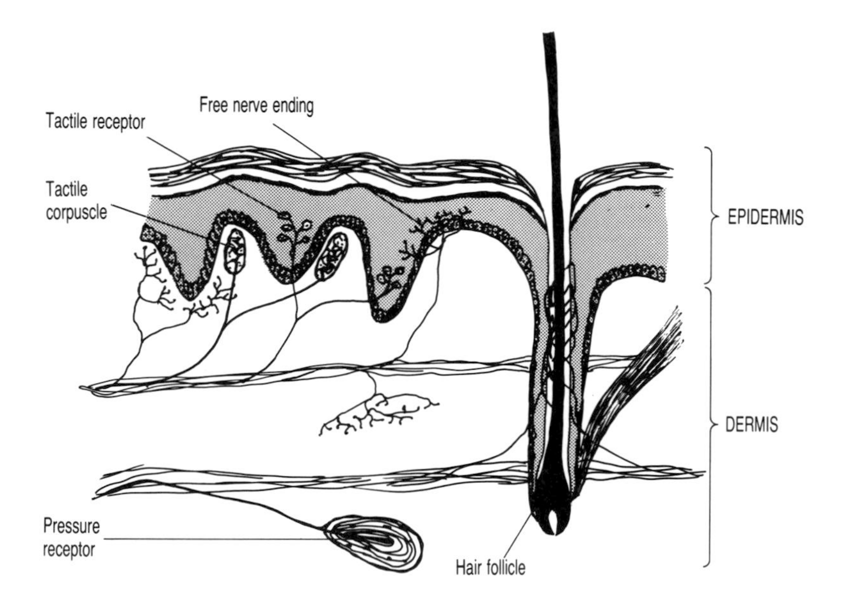

Fig. 11.2. Section of the skin showing encapsulated receptors and free nerve endings.

• *LOOK at Fig. 11.3 to trace the route followed by these impulses in the central nervous system. Notice there are three neurones in the pathway.*

Some of the first order neurones in this anterolateral pathway synapse with interneurones in the spinal cord before entering the ascending tracts, and the second order neurones branch in the brain stem to link with the reticular formation in particular.

The sensations carried by the anterolateral pathway are mainly light touch, temperature and pain. The branching of this route in the brain stem allows subconscious integration with the motor system.

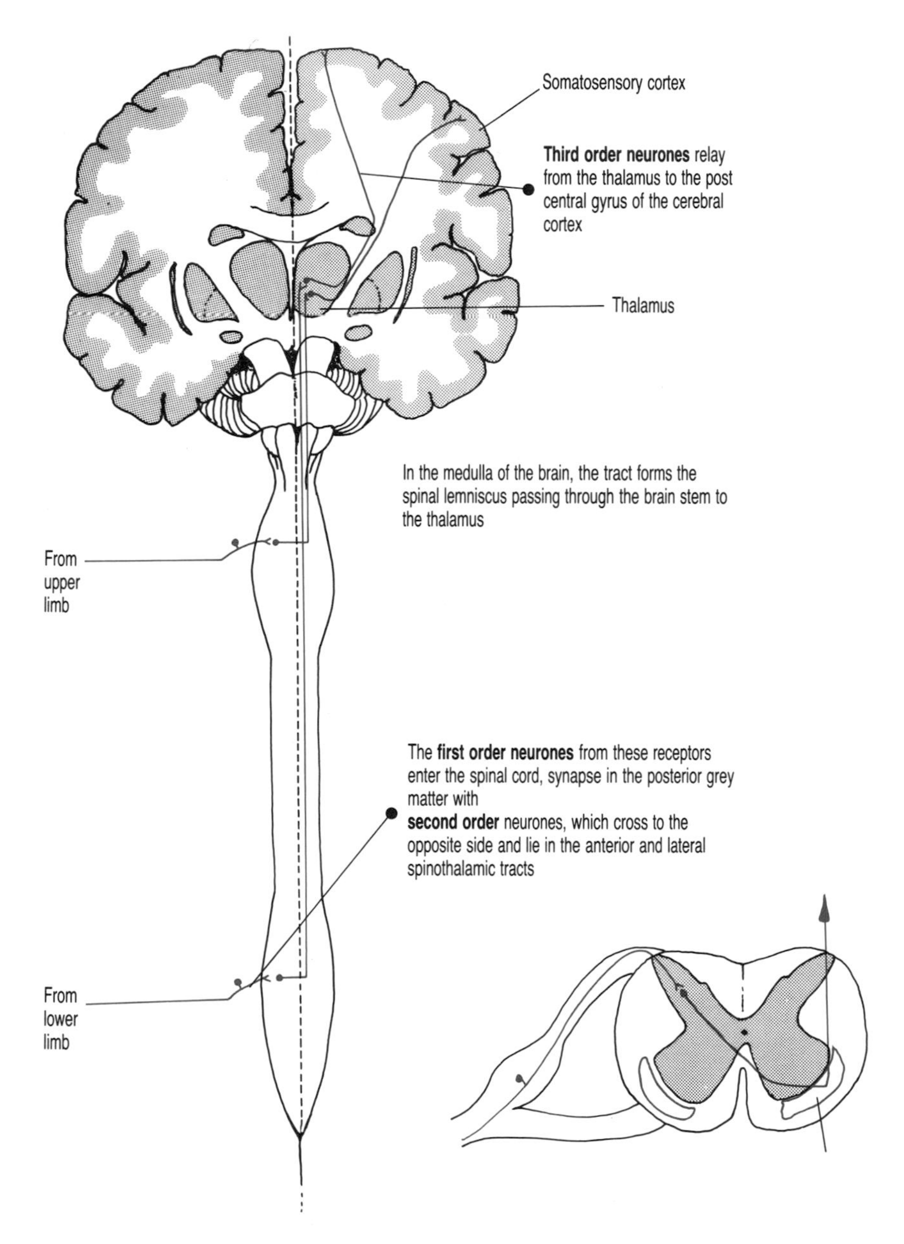

Fig. 11.3. Frontal section of the brain with spinal cord. Anterolateral pathway (spinothalamic). Transverse section of the spinal cord shows the position of the tracts.

11.2.2 Medial lemniscus pathway (posterior/dorsal column)

Sensory neurones with fast adapting receptors and large diameter axons are the origin of activity in the medial lemniscus pathway. These receptors are found in the skin and also lying in muscles, tendons and joints (proprioceptors). The sensory neurones enter the

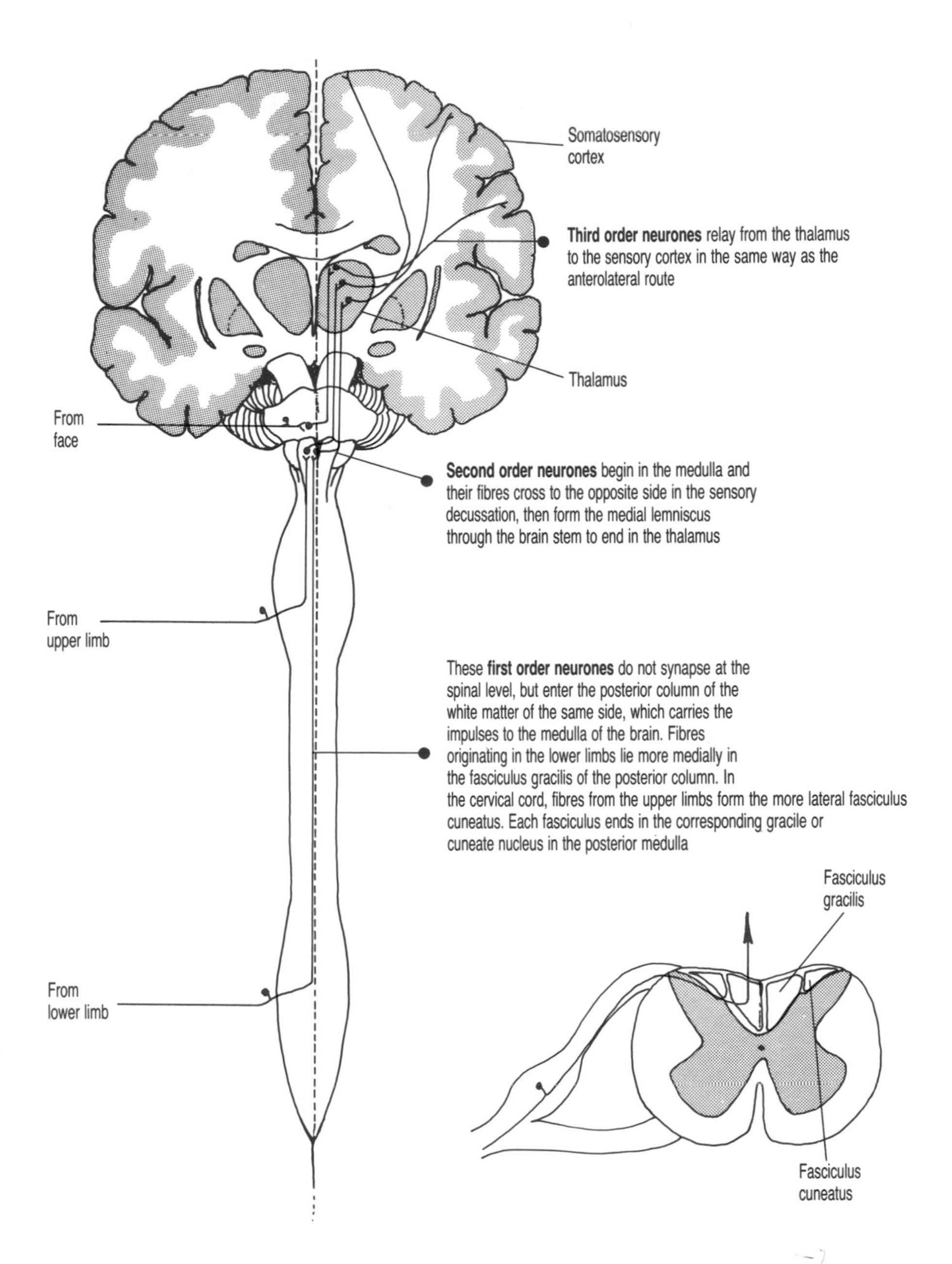

Fig. 11.4. Frontal section of the brain with spinal cord. Medial lemniscus pathway (posterior/dorsal column). Transverse section of the spinal cord shows the position of the posterior column with fasciculus cuneatus and fasciculus gracilis.

posterior horn of the spinal cord, alongside the small diameter axons, and then pass into the posterior (dorsal) column of white matter of the same side to reach the brain. (Branches from these first order neurones also synapse with interneurones in the posterior horn at the spinal level of entry.) The posterior column ends in the medulla of the brain in the gracile and cuneate nuclei. At this level, the axons of the second order neurones cross to the opposite side and pass through the brain stem in the medial lemniscus to the thalamus.

• *LOOK at Fig. 11.4 to trace the route followed by this pathway in the central nervous system. There are three neurones in this route. Return to Fig. 11.1 to find the fibres in the brain stem lying alongside those from the anterolateral pathway.*

Proprioceptive information from the lower limb is carried by fibres in the posterior spinocerebellar tract as far as the medulla, where it relays in the gracile nucleus before entering the medial lemniscus in the brain stem.

The medial lemniscus pathway provides fast and precise information to touch sensation from the skin, and proprioception from the joints and muscles.

All the fibres from *both* the anterolateral and medial lemniscus pathways pass through the internal capsule to end in a specific part of the primary sensory cortex where the body parts are represented in a particular order.

Interruption of the somatosensory fibres in the brain may result in loss of skin sensation and body awareness in the opposite side of the body.

Sensation from the face

The importance of sensory information from the skin and muscles of the face and mouth must not be forgotten. Try to imagine washing, shaving, putting on make up, and smiling, with no feeling in the muscles and skin of the face on one side. Sensory fibres from the face enter the brain stem mainly in the *trigeminal* (fifth cranial) nerve and synapse in the sensory nuclei of this nerve. Second order neurones cross to the opposite side and lie alongside the medial lemniscus to reach the thalamus. Third order fibres end in the region representing the face in the somatosensory cortex. Input from this trigeminal system is important in the movements of facial expression, swallowing and speaking.

11.2.3 The interpretation of pain

Pain is a subjective sensation and many aspects of the experience of pain cannot be explained by activity in one specific route from pain receptors to the somatosensory cortex. The Gate Theory of Pain attempts to explain how a painful stimulus may be interpreted in different ways at different times and by different individuals. There are interneurones in the substantia gelatinosa of the posterior horn of the spinal cord that act as a spinal control mechanism for the transmission of pain into the ascending pathways to the brain. Figure 11.5a shows large and small diameter sensory fibres synapsing with cells in the substantia gelatinosa (SG) and with larger transmission cells (T cells) before linking with the antero-lateral pathway to the brain. The interneurones form what is known as the *pain gate* (Fig. 11.5).

Activity in the small diameter fibres stimulates the T cells and impulses enter the anterolateral system so that pain is felt. The pain gate is **open** (Fig. 11.5b).

If there is activity in the large diameter fibres from touch receptors in the skin, and proprioceptors in muscles, the SG cells are stimulated and these in turn *inhibit* the transmission cells. This prevents activity entering the anterolateral pathway and no pain is felt. The gate is **closed** (Fig. 11.5c).

In this way, the transmission of pain impulses depends on the balance of activity in the small and the large diameter fibres entering the posterior horn of the spinal cord.

The pain gate theory explains how stimulation of the skin over a painful area by rubbing often reduces pain; the tactile stimuli increase activity in large diameter fibres and close the gate. The effectiveness of acupuncture in the relief of pain may be partly explained by the stimulation of large diameter fibres. Electrical devices applied to the skin, or implanted in the posterior column of the spinal cord can be used to relieve severe intractable pain. Patients can then control the stimulation of large diameter fibres and close the gate.

Descending pathways in the spinal cord from the *reticular formation* of the brain stem also synapse with interneurones in the facsiculus proprius (see Fig. 3.27, p. 79) at the spinal level and in turn *inhibit* the transmission cells (Fig. 11.5a). This suggests how higher centres can also influence whether or not pain is felt. A footballer or athlete may not feel the pain from an injury while the descending pathway is active during the match or race.

11.3 Input from the position of the head

The vestibular system monitors the position of the head in relation to the body and the movement of the head in relation to the space around the body. The brain processes this information to control the muscles of the trunk (axial muscles) and keep the body in balance. The vestibular system also links with the nerves supplying the eye muscles, via the brain stem, so that the eyes can be kept 'on target'. This will be discussed in more detail in Section 11.4.

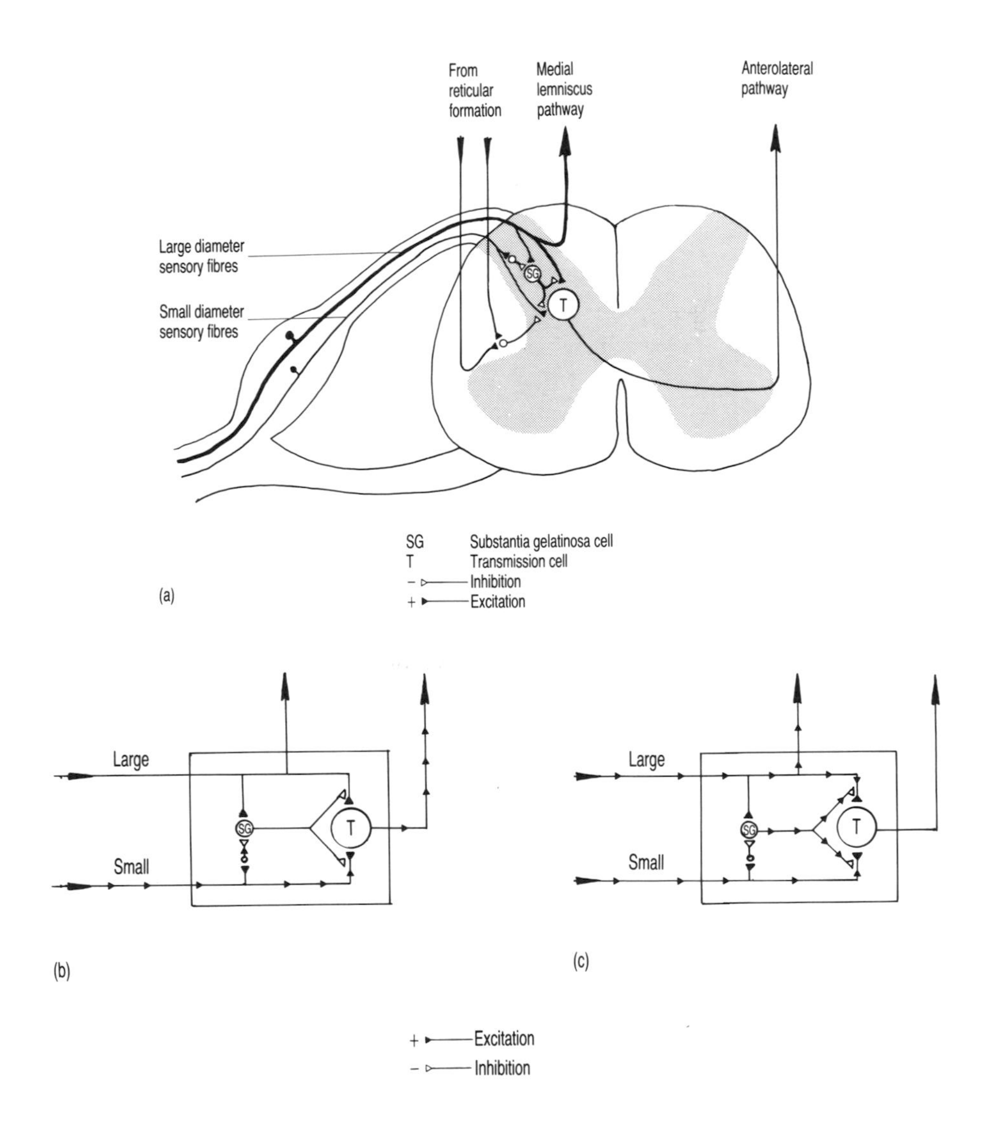

Fig. 11.5. Pathway of the 'pain gate' in the spinal cord: (a) spinal cord to show sensory neurones, interneurones and route to the anterolateral pathway; (b) pain gate (diagrammatic): gate open; less inhibition of T cells by SG cells; and (c) pain gate (diagrammatic): gate closed; more inhibition of T cells by SG cells. (SG — cells in substantia gelatinosa, and T — transmission cells.)

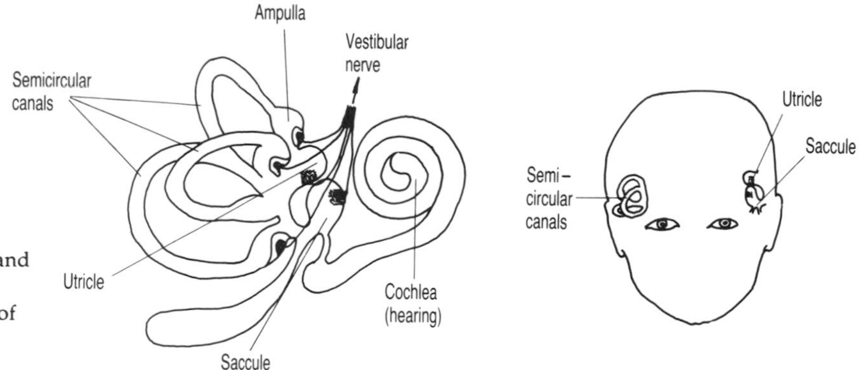

Fig. 11.6. Vestibule (utricle and saccule) and semicircular canals. Position in the head of the semicircular canals.

The receptors which activate the vestibular pathway are in the *vestibule* and *semicircular canals* of the inner ear which consist of five fluid filled sacs communicating with each other, arranged in the form of: (i) two oval bulbs, about 5 mm in diameter, known as the utricle and saccule, that together form the **vestibule**; and (ii) three **semicircular canals**, about 1 mm in diameter, lying above and behind the utricle and saccule. One canal lies in each of the three planes of the head — superior, posterior and lateral (Fig. 11.6).

The proprioceptors found in the walls of these sacs respond to movement of the fluid in the sacs as the head moves in space and in relation to gravity. Each receptor responds to a particular direction and velocity of head movement. Together, the receptors are important for maintaining the balance of the body, and for monitoring head movement. We are not generally aware of vestibular activity, so that it may be difficult at first to appreciate its importance in everyday movement.

Receptor areas in the walls of the utricle and saccule are called otoliths or maculae. The receptor cells have projecting cilia embedded in a jelly like mass, which contains particles of calcium called otoconia (Fig. 11.7a). If the head tilts, the fluid in the sacs lags behind the movement of the utricle and saccule walls, the cilia are bent and the sensory cells are stimulated (Fig. 11.7b). Sideways tilting of the head results in increased firing of impulses from one saccule, and less from the opposite saccule. The otoliths on the base of the utricle signal when the head is bent forwards and backwards. In horizontal movement of the head and body, for example when sitting in a car or a train moving forwards, the otoconia lag behind the movement of the wall of the sac, the cilia are again bent and the sensory cells stimulated (Fig. 11.7c).

A simple way to try to understand the mechanism of the utricle or saccule is to imagine a football filled with fluid. If the football is

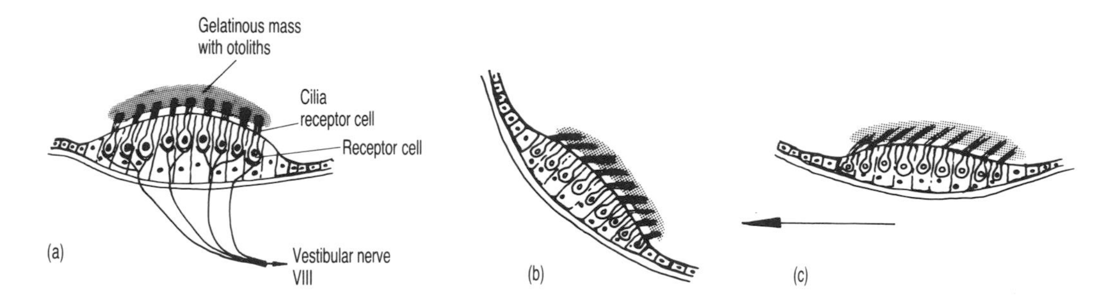

Fig. 11.7. Otolith receptors in the utricle and saccule; (a) head upright; (b) head tilted; and (c) head movement.

tilted or moved steadily in a horizontal direction, there is always a delay in the movement of the fluid inside the football. This moment of delay would be signalled by flexible pins projecting from the inner side.

The **semicircular canals** are thin tubes with relatively less fluid than the utricle and saccule. Receptor areas in the semicircular canals are found in the ampulla, a swelling at the base of each canal. Sensory cells in the ampulla also have cilia embedded in a jelly like structure called the *cupula*, but there are no calcium particles (Fig. 11.8). The cupula forms a flap like a swing door, moving backwards and forwards in response to movement of the fluid along the canal as the head moves. As the cupula bends, the cilia move and the hair cells are stimulated. Rotation of the head affects the canal lying in the same plane of movement. Figure 11.8 shows the direction of head movement that stimulates each of the canals on the left side of

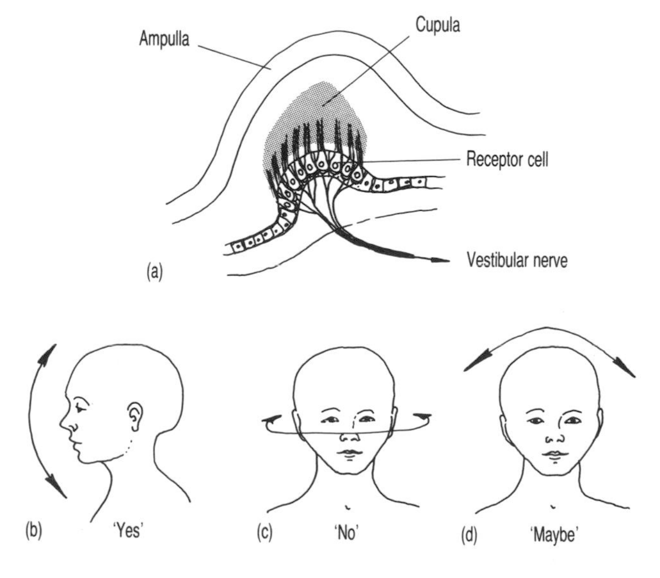

Fig. 11.8. (a) Cupula in the ampulla of a semicircular canal: (b), (c) and (d) directions of head movement for stimulation of each canal on the left side of the head, (b) 'yes'; (c) 'no'; and (d) 'maybe'.

the head. The five fluid filled sacs of the vestibule and semicircular canals form the anatomical structure known as the membranous *labyrinth*. Although individual receptors may respond to a greater extent in particular movements of the head, it is the combined effect of fluid movement in all the cavities that is integrated in the vestibular nucleus. The automatic responses to stimulation of these receptors are known as vestibular or labyrinthine reflexes.

Return to Chapter 4, Section 4.4.2, and Fig. 4.13 (pp. 93–96) to revise the pathways in and out of the vestibular nucleus. The vestibulo-ocular reflex, which maintains a constant image on the retina as the head moves, has already been described. The vestibulospinal reflex keeps the body in balance if you start to fall to one side. The vestibule signals the change in head position, and the extensor muscles on the same side of the body increase their activity via the vestibulospinal tracts. The neck muscles on the opposite side also contract to keep the head upright.

11.4 Input from the eyes

Light entering the eyes passes through several transparent layers of cells and blood vessels to reach the rods and cones, the primary receptor cells. The retina itself is like a 'mini brain' and some processing occurs in its layers of cells before transmission along the fibres of the optic nerve to the brain.

One function of the retina is to act as a system for signalling movement in the environment around the body. When we move in a particular direction, the eyes are fixed on the centre of the visual field ahead. Images of objects moving on either side of us are signalled by the receptors in the periphery of the retina. When the eyes follow a moving object, such as a tool held in the hand, images from the tool remain stationary on the retina and those from the background sweep across the retina. In both cases, the retina is transmitting information about the changes in the environment during movement.

The visual pathway, from the eyes to the striate cortex in the occipital lobes of the cerebral hemispheres (Fig. 11.9), branches in the midbrain just before the lateral geniculate nucleus. These fibres, which synapse in the superior colliculus, allow the visual system to cooperate with the vestibular system in keeping the body balanced during movement.

It is easy to demonstrate that the eyes contribute to our sense of balance.

- *STAND on one leg with your eyes open, and then with your eyes*

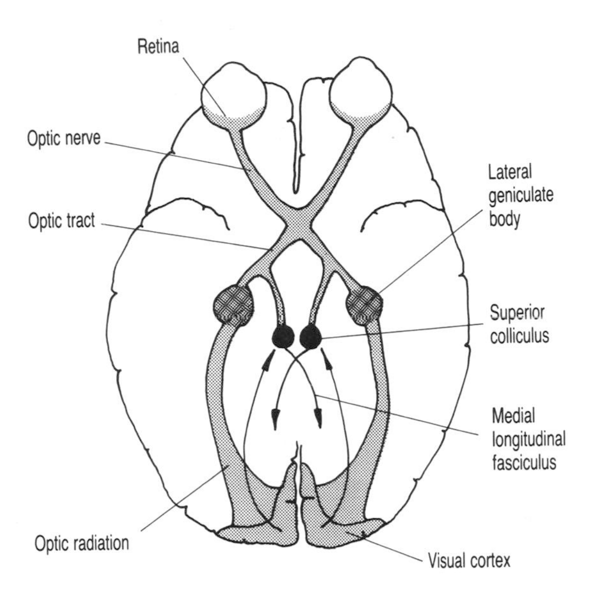

Fig. 11.9. The visual system.

closed. Notice how much you sway, and may fall over when the eyes are closed.

- *TRY running over rough ground at night. The fine adjustments needed to place the feet accurately and with the correct force are more difficult without the visual input.*

The visual input becomes more important during movement as a target is approached. In reaching out to grasp a glass of water, or running to jump on a bus, it is the final stage of gripping the glass or negotiating the step that requires more visual information.

In this chapter we have outlined the variety of sensory information entering the central nervous system during movement. Within the central nervous system the overall patterns of signals at any one moment are synthesized and processed at all levels. Like a radio receiver with an infinite number of channels, the sensory input is tuned to 'listen' to particular combinations of signals so that they can be recognized and acted on.

Some of the pathways in the sensory system have been described to explain the possible results of injury to the main ascending routes, which include loss of skin sensation and loss of body awareness. In damage to the brain, the integration of many parts of the sensory system may be interrupted, and the outcome is always complex. A comprehensive assessment of sensory loss is essential for each individual patient. The sensory system has a remarkable capacity to adapt to loss, particularly in the young. The reorganization of neurones to compensate may involve axons growing new branches to make new connections, and diverting activity along unused routes.

At the end of this chapter, you should be able to:

1 Describe the two main ascending pathways in the central nervous system for information from the skin, muscles and joints to the somatosensory cortex.

2 Outline the role of the vestibule of the ear in monitoring the position and movement of the head.

3 Summarize the variety of sensory information received by the central nervous system that provides the background to movement.

12 / Motor Control

All the movements we make in daily activities, such as dressing, speaking, eating, writing and walking, are controlled by motor centres in the brain. Activity passes down from these motor centres in descending pathways to the motor neurones of the cranial nerves in the movements of speaking, eating and facial expression, and to the motor neurones of spinal nerves in the movements of the limbs and trunk. The motor system moves the arms in skillful activity, the legs in walking, and controls the background posture of the whole body. The same system is involved in movements of the tongue, lips and larynx needed for speech. The motor cortex directs the conscious control of movements, while subcortical areas, such as the basal ganglia and cerebellum, control the sequence and timing of all the muscles involved in the performance of smooth coordinated movements. Motor nuclei in the brain stem are particularly concerned with the control of the balance of the body as the movement proceeds. The brain stem nuclei also relay into descending pathways to the spinal cord. Motor control in the spinal cord is the result of the variety of influences from the descending pathways, as well as from the proprioceptors in the muscles and tendons themselves.

The activity of the spinal motor neurones will be considered first and then the influence from descending pathways from the brain.

12.1 Lower motor neurones. Motor activity at the spinal level

The lower motor neurones form the final route to the muscles in all movement, both voluntary and reflex. The cell bodies of the lower motor neurones lie in the anterior horn of the spinal cord and in the nuclei of the cranial nerves. The axons of the lower motor neurones lie in the peripheral nerves supplying the muscles (see Fig. 1.15, p. 23).

Lower motor neurones are stimulated by activity in: (i) the descending pathways which terminate in the anterior horn of the spinal cord at all levels (see Section 12.2); and (ii) local spinal reflexes. Two of the local spinal reflexes are the *myotatic (stretch) reflex* and the *golgi tendon reflex*.

12.1.1 The myotatic reflex. Static and dynamic activity

Activity in the myotatic unit, which maintains a muscle at constant length to hold a position, has been described in Chapter 1 (Section 1.6). During movement, muscles **do change in length**, and the level of reflex activity is modified by changing the 'setting' of the

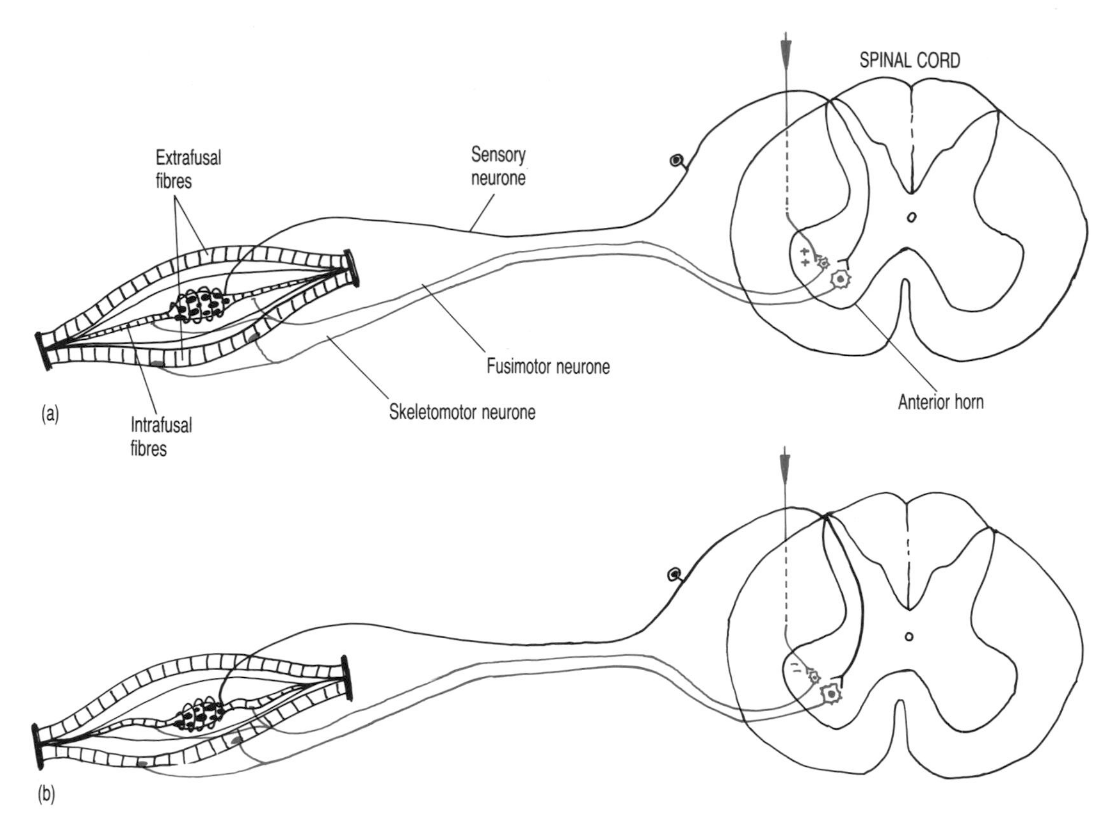

Fig. 12.1. Muscle spindle sensitivity: (a) high. Descending pathways stimulate fusimotor neurones, the intrafusal fibres contract and the muscles spindle is very sensitive to distortion; (b) low. Descending pathways inhibit fusimotor neurones, intrafusal fibres are slack and the muscle spindle is less sensitive to distortion.

spindles. The intrafusal fibres of the muscle spindles are supplied by the smaller fusimotor (gamma) neurones (see Fig. 12.1a). If these neurones are excited by impulses from descending pathways in the spinal cord, the intrafusal fibres of the spindle contract. The spindle then becomes taut and more sensitive to length changes in the muscle (Fig. 12.1a). On the other hand, if the impulses from descending tracts inhibit fusimotor neurones, the spindle becomes slack and only responds to marked changes in length of the muscle (Fig 12.1b). In this way, the higher levels of the central nervous system modify spinal stretch activity during movement.

Consider the hand performing fine manipulative movements, such as doing up buttons and tieing shoe laces. Stretch reflex activity must be damped in the muscles of the hand to allow rapid length changes to occur. At the same time, muscles of the shoulder and arm perform background activity to hold the position of the limb and allow the fingers to move accurately. The spindles in the supporting muscles are set at a high level, so that any change in length is resisted. During movements of the whole limb, the setting of the spindles in all the active muscle groups is continually monitored by the cerebellum as the movement proceeds.

Static and dynamic intrafusal fibres

If we look in more detail at the structure of the muscle spindle shown in Fig 1.18 (p. 27), two different types of intrafusal fibres can be identified. Both types have a primary sensory ending wound round the central area (the annulospiral endings described in Chapter 1). In addition, some of the intrafusal fibres have secondary sensory endings towards the periphery of the fibre, known as flower spray endings, which respond to the rate of change in length of the muscle during movement. The two types of intrafusal fibre are: (i) *nuclear bag fibres* with a bulge in the middle where the nuclei are found, and secondary sensory endings are present; and (ii) *nuclear chain fibres* which are thinner and their nuclei are lined up in a row.

The **nuclear bag** (dynamic) **fibres** respond to rapid changes in length of the muscle, while the **nuclear chain** (static) **fibres** respond to prolonged slow stretch. The muscles spindles, therefore, relay detailed information about both the length, and the rate of change in length, of a muscle during movement to the spinal cord. Stretch reflex activity can then be adjusted to the appropriate level during the progress of a movement.

12.1.2 Golgi tendon reflex

Receptors are also found in the tendons of muscles. They lie embedded in the collagen fibres of the tendon and in series with the muscle fibres. Remember that the muscle spindles lie in parallel with the muscle fibres. The Golgi tendon organs are not activated by the tension present in muscles at rest, since they have a higher threshold of stimulation than muscle spindles. When the tension in a muscle rises rapidly, the muscle pulls on the tendon and the Golgi tendon organs are stimulated. Figure 12.2 shows the Golgi tendon reflex pathway. The sensory neurones synapse with small interneurones which are inhibitory to the lower motor neurones. The result is that *less* activity reaches the muscle, and the tension is reduced by *relaxation* of the muscle. The tendon organs therefore act as a protective mechanism to prevent sudden rise in tension which might tear a muscle or tendon, and also cooperate with the muscle spindles by providing a control system for muscle tension.

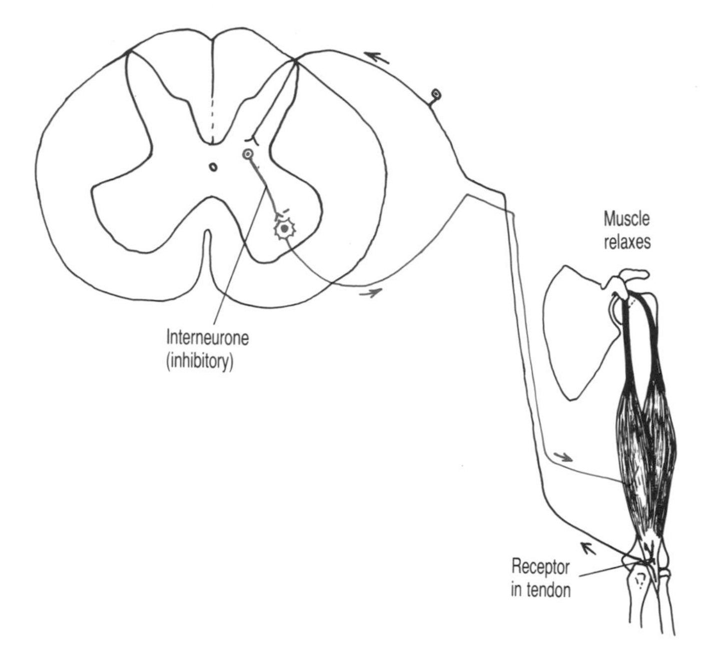

Fig. 12.2. Golgi tendon reflex. Receptor in the tendon of the muscle is stimulated by a rapid rise in tension. The sensory neurones synapse with inhibitory interneurones in the spinal cord. The lower motor neurones are inhibited and the muscle relaxes.

12.1.3 Interruption

Interruption of lower motor neurones may be due to damage to the cells in the anterior horn of the spinal cord, or to the axons in peripheral nerves. The result is: (i) flaccid paralysis of muscles which cannot respond to voluntary or reflex stimulation; and (ii) loss of muscle tone (*atonia*). The muscles feel soft and have no 'life' in the absence of stretch reflex activity.

12.2 Upper motor neurones

The neurones which link motor areas in the brain with the spinal cord and with the cranial nerve nuclei in the brain stem are known as upper motor neurones. They lie in descending tracts in the brain stem and the spinal cord. The collective influence of these upper motor neurones on the lower motor neurones of the spinal cord determines the level of activity in muscles during movement.

Upper motor neurones synapse at all levels in the spinal cord with alpha motor neurones supplying Type I and Type II muscle fibres (see Section 1.4.1, p. 16), and also with gamma motor neurones innervating the intrafusal fibres of muscle spindles. In this way, upper motor neurones determine the recruitment of motor units and also the level of stretch reflex activity in the muscles.

The upper motor neurones in the central nervous system have been divided into the following.

1 *Direct descending pathways* from the motor cortex to the lower motor neurones, are known as the *pyramidal system.*

2 *Descending pathways* which include all the upper motor neurones not in the pyramidal system, are known as the *extrapyramidal system*.

The division into pyramidal and extrapyramidal was originally based on the assumption that the motor cortex, relaying into the pyramidal system is concerned with voluntary movement, while the extrapyramidal system is involved in background postural activity during movement. Recent evidence suggests that the only special function of the pyramidal system is in the control of precision movements of the hands and feet. The extrapyramidal system is involved in all movements, both voluntary and reflex.

12.2.1 Descending pathways from the motor cortex

Upper motor neurones originating in the motor cortex (described in Chapter 3, Section 3.4.1) form a fast direct route via the brain stem and the spinal cord to the lower motor neurones. The axons of these neurones form the following tracts.

1 Lateral and anterior *corticospinal* tracts.

2 *Corticobulbar* (or corticonuclear) tracts.

The **corticospinal tracts** have their cells of origin in the primary motor cortex. Some cells of the pre motor cortex and the primary sensory area also contribute to the tracts.

The corticospinal fibres converge as they enter the posterior limb of the internal capsule (see Chapter 3, Fig. 3.15, p. 66). Passing into the brain stem, the fibres lie anteriorly in the cerebral peduncles of the midbrain, and continue down through the pons. At the level of the medulla, 85% of the fibres cross to the opposite side and enter the lateral white matter of the spinal cord to become the *lateral corticospinal tract*. The other fibres continue anteriorly in the white matter of the spinal cord as the *anterior corticospinal tract*. The area where the fibres cross in the medulla is known as the decussation of the pyramids (see Fig. 3.17c, p. 68). Fibres of both tracts terminate in the spinal cord, where they synapse with lower motor neurones. The anterior corticospinal fibres cross at the level of the segment they supply. Some corticospinal fibres relay via interneurones at the spinal level.

The **corticobulbar tract** fibres originate in the same cortical areas. In the midbrain, pons and medulla of the brain stem the fibres terminate in the motor nuclei of cranial nerves (III–XII).

Voluntary movement executed by the motor cortex on one side of the body is relayed via the corticospinal and corticobulbar tracts to the muscles of the opposite side. The corticobulbar fibres link via cranial nerves with the muscles of the face, and the corticospinal fibres link via spinal nerves with the muscles of the limbs and trunk. The most important function of the corticospinal fibres is the voluntary control of skilled precision movements of the distal muscle groups of the limbs. The muscles involved in speech, facial expression and eye movements are controlled via the corticobulbar tracts.

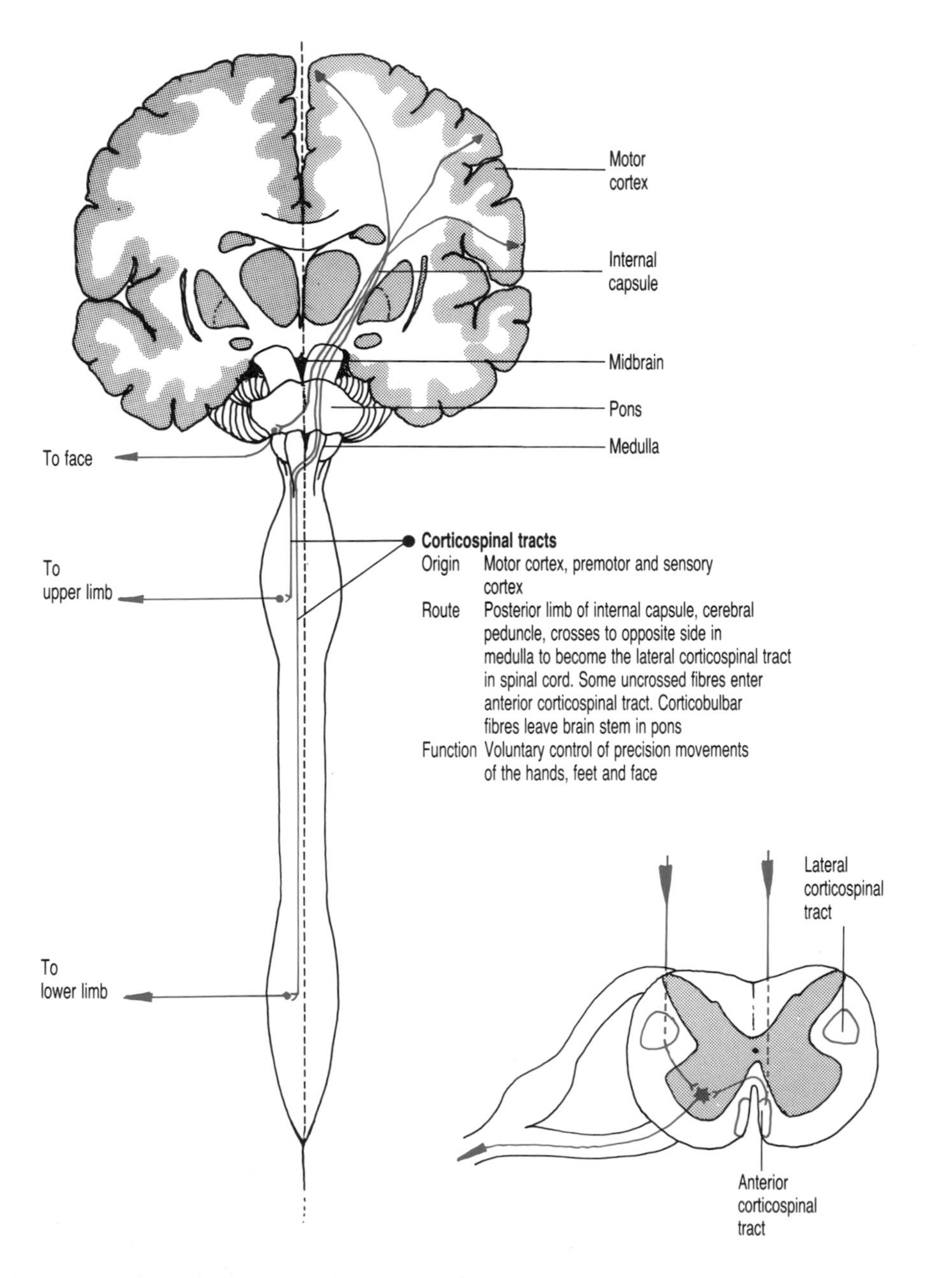

Fig. 12.3. Frontal section of the brain with the spinal cord. Pathway of the corticospinal and corticobulbar tracts. Transverse section of the spinal cord shows the position of corticospinal tracts.

• *LOOK at Fig. 12.3 to follow the route of the first order neurones in the cortibulbar and corticospinal tracts.*

Interruption

Interruption of the corticospinal and corticobulbar pathways in the brain affects the muscles of the opposite side of the body, and there is loss of voluntary movement, particularly fine movements of the hands and feet.

12.2.2 Descending pathways from other motor centres in the brain

The upper motor neurones that originate in motor areas below consciousness relay at various motor nuclei in the brain on their way to the lower motor neurones, which means that the routes are polysynaptic, unlike the direct link of the pyramidal pathways.

Figure 12.4 shows diagrammatically the brain areas included in the motor system. Input to these centres includes the eyes, ears (sound and balance), the basal ganglia, the cerebellum and the motor cortex. The descending tracts to the spinal cord are also indicated. The location of these tracts is found in Chapter 3, Fig. 3.27 (p. 79).

• *REVISE the position and function of the basal ganglia, vestibular nuclei, reticular formation and cerebellum by returning to Chapter 3.*

The axons of upper motor neurones originating in the motor areas shown in Fig. 12.4 form the following descending tracts.

1 Rubrospinal tract.
2 Reticulospinal tracts.
3 Vestibulospinal tract.
4 Tectospinal tract.

Figure 12.5 shows the position of three of the centres of origin in the brain stem and indicates their descending tracts. Figure 12.6 shows the position and termination of three of these descending tracts in the spinal cord. The tectospinal tract, concerned with the control of head movements, is only present in cervical segments.

The descending tracts synapse with skeletomotor (alpha) and fusimotor (gamma) motor neurones at the spinal level. Some fibres are excitatory and others are inhibitory. The overall action of these pathways on the fusimotor neurones is inhibition, which eliminates unwanted tone, so allowing skillful movement to take place. Positioning and support, the function of the proximal muscles of

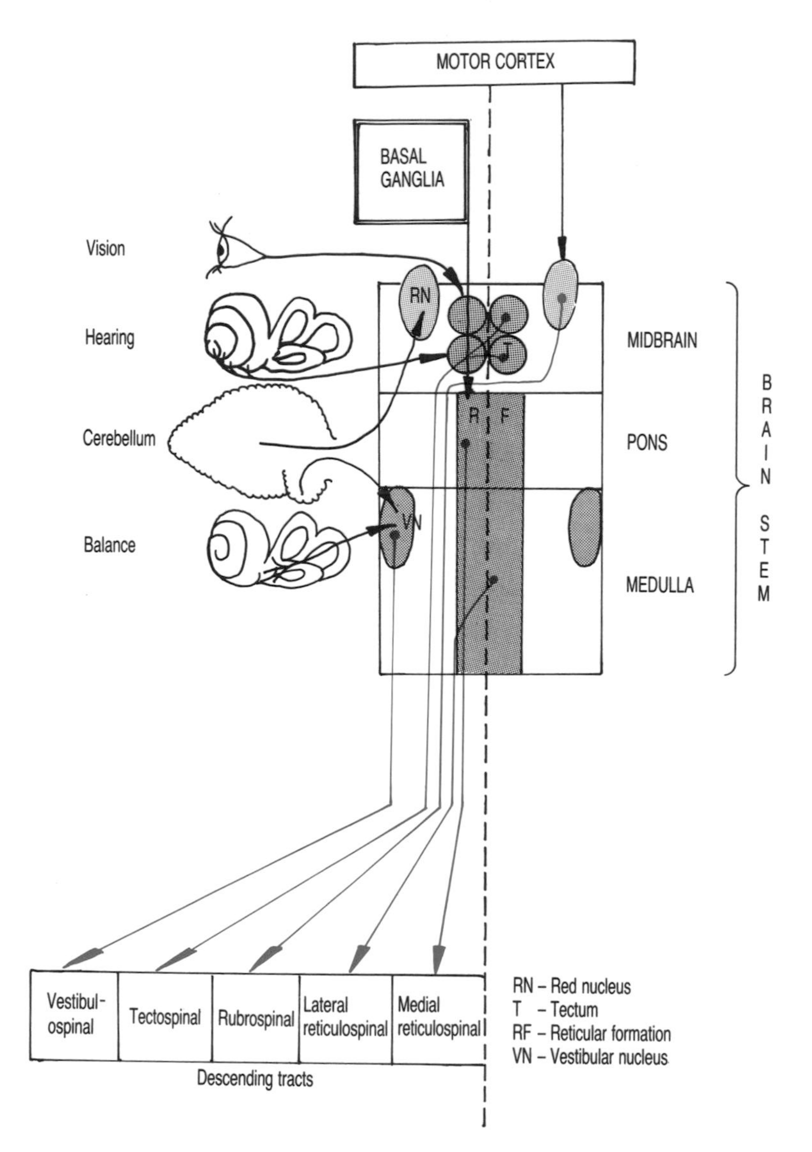

Fig. 12.4. Motor centres in the brain stem (diagrammatic). Origin of the descending tracts from the brain stem centres is shown. RN: red nucleus. T: tectum (superior and inferior colliculli). RF: reticular formation. VN: vestibular nucleus.

the limbs, are controlled by the rubrospinal and lateral reticulospinal tracts. For example the muscles around the shoulder region, the elbow and the wrist, which support the upper limb during fine manipulation movements of the hand, are mainly activated via these routes. The muscles of the neck and trunk that maintain posture and balance receive most activity from the vestibulospinal and medial reticulospinal tracts.

All the tracts that are involved in the regulation and coordination of movement also receive input from the basal ganglia and from the cerebellum via the brain stem.

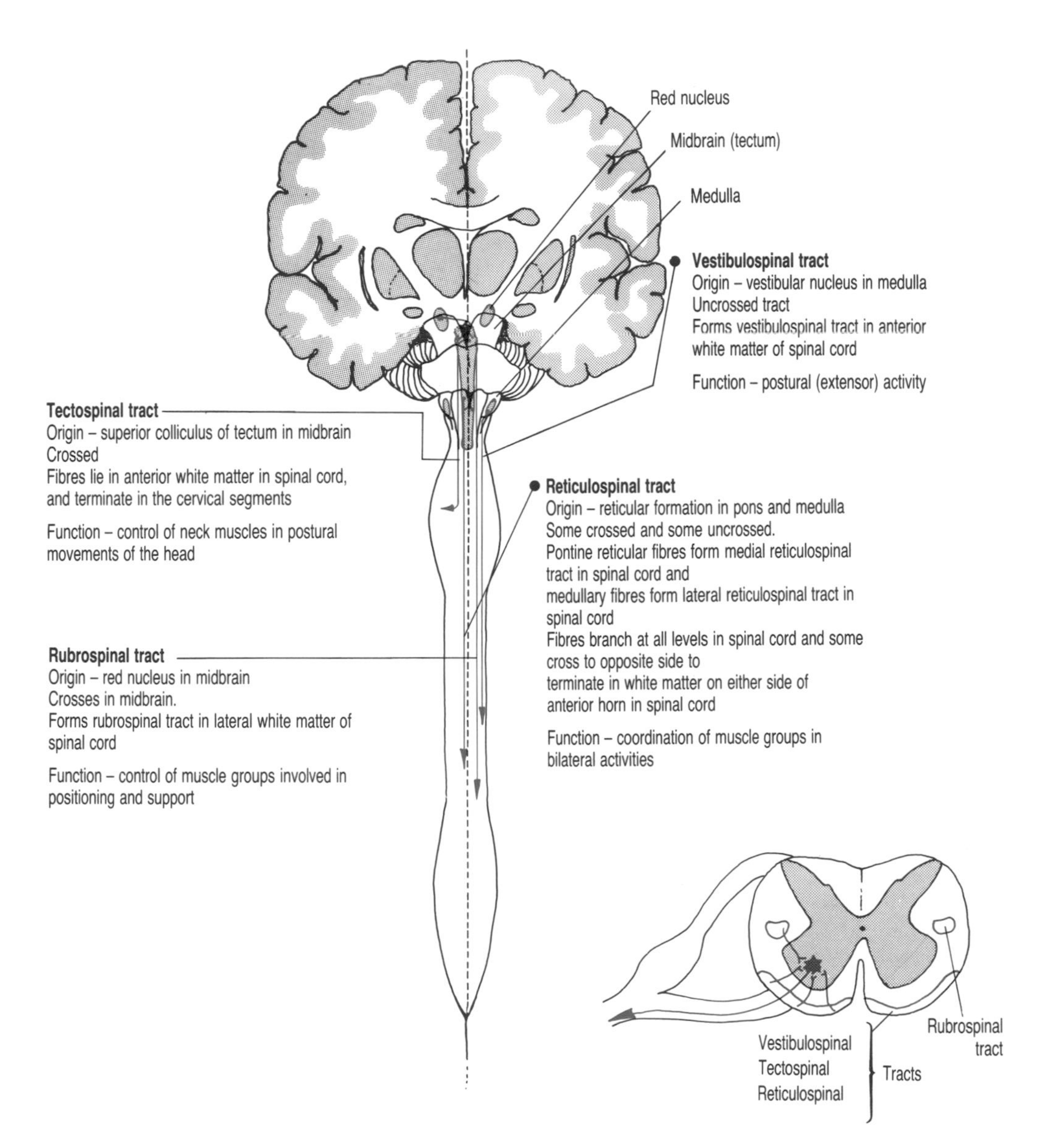

Fig. 12.5. Frontal section of the brain with spinal cord. Pathways of descending tracts — extrapyramidal. Rubrospinal tract, vestibulospinal, reticulospinal tracts, tectospinal tract. Transverse section of spinal cord shows convergence of descending tracts on the lower motor neurone.

Interruption

Interruption of these pathways in the brain results in the following.
1 Abnormal movement patterns, with spasticity in particular muscle groups of a limb. For example, the upper limb may show flexor synergy, and the lower limb may demonstrate an extensor synergy, so that normal movements are difficult to perform.

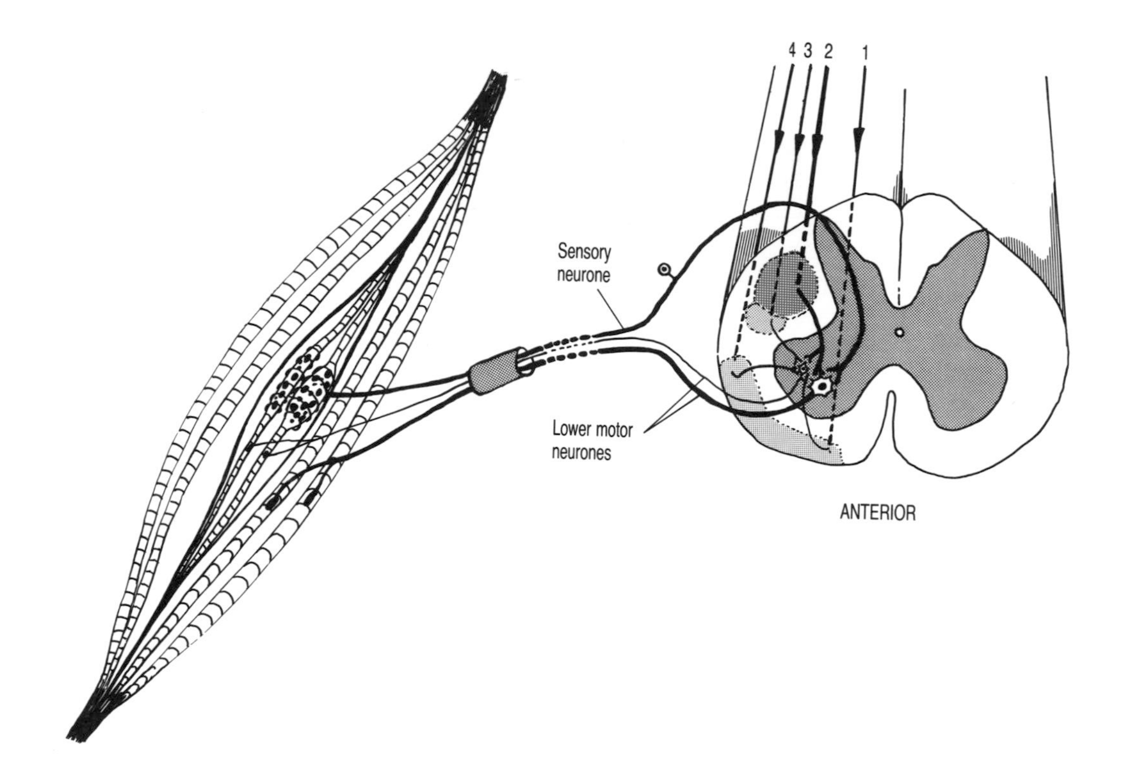

Fig.12.6. Termination of upper motor neurones on the skeletomotor and fusimotor neurones in the spinal cord: 1. vestibulospinal tract; 2. lateral corticospinal tract; 3. rubrospinal tract; and 4. tectospinal tract.

2 Reflex activity, for example tendon jerks, can be elicited but is exaggerated.

3 Abnormal muscle tone occurs, which may result in either of the following:

(a) *Hypotonia* if the excitatory pathways are affected; the hypotonic muscles are less sensitive to slow stretch.

(b) *Hypertonia* if the inhibitory pathways are involved allowing the fusimotor fibres to become overactive.

12.2.3 Summary of upper motor neurone function

The direct pathways from the motor cortex (corticospinal and corticobulbar) are primarily concerned with the execution and control of voluntary movement, particularly the fine movements of the hands and feet. At the same time, the postural muscles of the trunk, and the supporting muscles arranged around the proximal joints of the limbs, are activated via the polysynaptic pathways, involving motor areas below consciousness, working in cooperation with the motor cortex.

Upper motor neurone lesions may affect both direct and polysynaptic routes, so that the outcome is variable. The initial shock usually produces flaccid muscles and poor movement, but this may change to spasticity and abnormal movement patterns. Rehabilitation is aimed at the restoration of normal movement patterns and normal background muscle tone.

12.3 Higher centres in motor control

The distinction between conscious and unconscious motor control is difficult to define, since many movements are under conscious control at the start and end of the movement, but once started, the sequence of muscle activity proceeds automatically. In operating a typewriter or playing the piano, we first practise the movements and each finger requires conscious control. After many repetitions of the same motor commands, control passes to the subconscious level and the movements can be performed automatically. Any conscious effort can then be directed to other activities like talking to a colleague in the office, or listening to the sounds from other musical instruments in an orchestra. However, any interruption in the flow of motor commands, for example turning a page, still requires conscious control. The motor cortex is concerned with the control of the performance of voluntary movement. Large areas of the motor cortex are devoted to those areas of the body that perform the most complex movements (see Chapter 3, Section 3.4.1, pp. 57–59).

The higher centres involved in motor activity are the *motor cortex*, the *basal ganglia*, the *cerebellum* and the *motor nuclei* of the *brain stem*. We have seen how these brain areas link to the spinal level by the descending motor pathways. These motor centres also link with each other within the brain, and the basal ganglia link to almost all areas of the cerebral cortex. The options for modification of the output are numerous. To give further understanding of the particular roles of the cerebellum, the basal ganglia and the nuclei of the brain stem, some of the linking pathways will be considered in more detail.

12.3.1 Motor cortex and cerebellum

The cerebellum links with the motor cortex of the opposite side by a loop which crosses the midline (Fig. 12.7). Corticospinal fibres originating in the motor cortex branch in the brain stem to synapse in

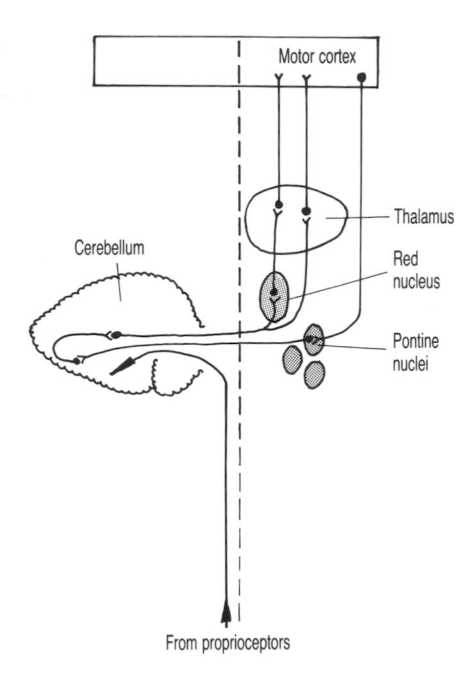

Fig. 12.7. Motor control loop between the motor cortex and cerebellum crossing the midline.

the *pontine nuclei* and then link with the opposite cerebellar cortex. Also, branches from small pyramidal cells in the motor cortex relay in the inferior *olivary nucleus* before entering the opposite cerebellum. In these two ways, there is a fast input from the motor cortex to the cerebellum which crosses the midline. A return pathway links the cerebellum across the midline to the thalamus and back to the motor cortex. The pattern of activity for the intended movement leaving the motor cortex is compared with the sensory information reaching the cerebellum from the proprioceptors in the muscles and joints of the body. The cerebellum then modifies the activity of the motor cortex via the return loop, to correct the progress and timing of the muscle activity and to bring the movement to a halt.

Interruption

Interruption of this loop between the motor cortex and the cerebellum at any point results in movements becoming jerky and uncoordinated. There is a tendency to overshoot, that is, a limb swings too far in each direction before it finally reaches the intended position — known as 'intention tremor'. If the cerebellum is involved, the effects are on the same side of the body (*ipsilateral*), whilst damage to the cortical part of the loop affects the opposite side of the body (*contralateral*).

12.3.2 Basal ganglia and motor cortex

The individual nuclei of the basal ganglia link together as a functional unit. The output from the system is from the globus pallidus and putamen, which link to the motor cortex via the thalamus. The motor cortex projects back to the globus pallidus, so that another loop is formed for the control of motor activity (Fig. 12.8). The link is with the motor cortex of the same side, unlike the loop to the cerebellum which crosses the midline.

The functions of the basal ganglia, based on evidence from the results of diseases of these motor nuclei, have been discussed in Chapter 3, Section 3.5. The basal ganglia appear to be important in the initiation of a movement and the forward planning of the programme for its performance by the motor cortex. In the absence of the motor commands from the basal ganglia, the motor cortex cannot execute the movement appropriately.

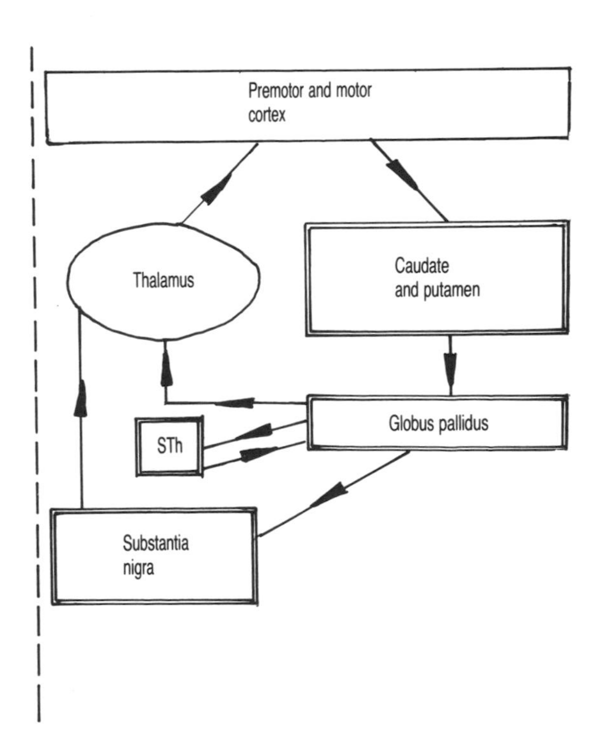

Fig. 12.8. Motor control loop between the motor cortex and the basal ganglia (STh: subthalamic nuclei).

12.3.3 Motor centres in the brain stem

The main function of the motor nuclei in the brain stem is in postural control during movement.

The **vestibular nucleus** stimulates lower motor neurones of extensor antigravity muscles which keep the body upright. Any change in the position of the head is monitored by the vestibule of the ear (see Chapter 11, Section 11.3.2 and Fig. 4.13, p. 96). Extensor activity is then increased in muscles on the side to which the head is turned and balance is maintained.

The **red nucleus** in the midbrain (sometimes included with the basal ganglia) provides the link between the cerebellum and the spinal cord. Descending pathways from the red nucleus via the rubrospinal tract are important in the coordination of movement of the limbs of the opposite side of the body.

The **tectum** of the midbrain contains two pairs of nuclei, the superior and inferior colliculli. These are stimulated by sound from the ears and visual stimuli from the eyes (see Section 3.9.1, p. 67). Activity in the tectospinal tract stimulates the muscles of the neck which then change the position of the head. Examples of this response can be recognized during such activities as team games, driving a car and keeping the body balanced on an unstable surface.

Output from the **reticular formation** in the brain stem influences activity in the lower motor neurones supplying muscles of the trunk and the proximal muscles of the limbs.

All these centres are part of the **brain stem reflexes** which function automatically to keep the body stable during movement. Fig. 12.4 shows how the output from the brain stem centres passes down the descending tracts to the lower motor neurones.

12.3.4 Summary of the three levels of motor control

The motor system functions at three levels of control.

The **spinal level** executes movement patterns based on spinal reflexes and on the activity received from higher levels of the nervous system. The muscle spindles play a major part in the spinal control of muscles during movement by adjusting activity in the lower motor neurones.

The **brain stem** contains the nuclei with the cells of origin of many of the descending extrapyramidal tracts, which modify activity at the spinal level to maintain the posture and balance of the body. Input to the brain stem nuclei is mainly from the eyes, the vestibule of the ear and from proprioceptors via the cerebellum.

The **higher centres** (motor cortex, basal ganglia and cerebellum) initiate and programme motor commands, which are passed down to the brain stem and spinal neurones in voluntary movement. Movements that are repeated can be stored as central commands, and can be performed without continuous reference to consciousness.

At the end of this chapter you should be able to:

1 Describe the position and function of lower motor neurones in the spinal cord and peripheral nerves.

2 Outline the activity of muscle spindles and Golgi tendon organs during movement.

3 Describe the position and function of upper motor neurones in the central nervous system.

4 Describe the main descending pathways in the central nervous system.

5 Summarize the functions of: (i) corticospinal and corticobulbar pathways; and (ii) descending tracts from the motor centres of the brain stem.

6 Give a brief account of the role of the motor cortex, basal ganglia and cerebellum in movement.

13 / Integration and Performance

In the performance of movement, motor commands are formed in the central nervous system and carried by the peripheral nerves to all the active muscles. The integration of the input from the sensory system with activity in the motor system is part of the process of developing the motor commands for a particular movement. Some aspects of the local integration in networks of neurones, and the links between centres in the brain and the spinal cord, will be considered in this chapter.

13.1 Integration in neurone pools

We have described how impulses pass from one neurone to another at a synapse in Chapter 1, Section 1.5.1. Collections of neurones in the central nervous system that have a particular function are called neurone pools (see Fig. 1.14, p. 23). The motor neurones in the anterior horn of the spinal cord which supply a group of muscles form a neurone pool, and the nuclei of cranial nerves are other examples of neurone pools. We will now consider some of the ways in which groups of neurones can regulate their own input and output.

Within a network of neurones, the axons of each neurone branch to synapse with many other neurones (divergence), and each cell body receives branches from many other neurones (convergence) (Fig. 13.1). An example of convergence is in the motor neurone pools of the spinal cord where neurones receive input from all the descending pathways (see Fig. 12.6, p. 262).

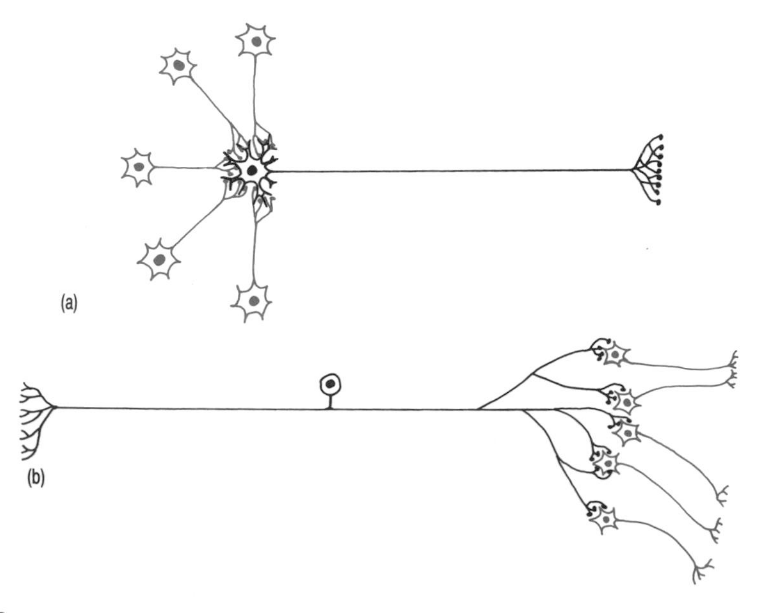

Fig. 13.1. Neurone circuits: (a) convergence — several presynaptic neurones synapse with one postsynaptic neurone; and (b) divergence — one presynaptic neurone synapses with several postsynaptic neurones.

Divergence allows the input from one source to be relayed in different directions at the same time. It has been estimated that each of the neurones in the brain has 100 inputs converging onto it, and each neurone diverges to 100 other neurones. The number of possibilities for the route of impulses through a neurone network in the brain is therefore enormous.

The balance of excitatory and inhibitory influences within a group of neurones affects its output, and there are several ways in which this balance may be changed.

Neurotransmitters

The most common neurotransmitter substance in the nervous system is acetylcholine. In the brain, several other substances have been identified such as dopamine, GABA, serotonin and enkephalin. Dopamine is an inhibitory transmitter released by neurones in the substantia nigra (one of the basal ganglia). In addition, the sensitivity of neurones to transmitter substances can be changed by the presence of other chemicals known as neuropeptides. The effect of a change in sensitivity is to regulate the amount of information entering or leaving a particular group of neurones. In diseases of the central nervous sytem, it may be the balance of excitatory and inhibitory transmitter substances produced by groups of neurones that is disturbed with the result that abnormal movements are produced. Parkinson's disease is an example of this.

Presynaptic inhibition

Inhibition of neurones also occurs in the central nervous system by a 'blocking' mechanism, whereby the excitatory transmitter substance is inhibited before it can be released. This is known as presynaptic inhibition. A presynaptic neurone releases inhibitory transmitter substances on to the synaptic knobs of the excitatory neurone as shown in Fig. 13.2. An example of presynaptic inhibition is the activity of the cells of the substantia gelatinosa in the 'pain gate' mechanism which regulates the activity of the pain

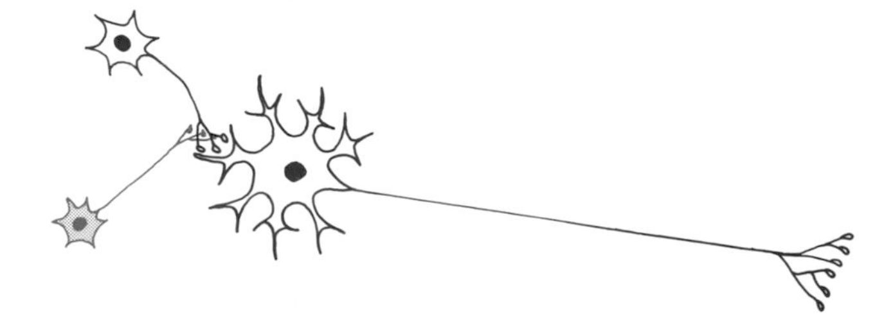

Fig. 13.2. Presynaptic inhibition. An inhibitory synapse is formed by one presynaptic neurone (red) on the synaptic knobs of a second presynaptic neurone, preventing the release of its transmitter substance.

transmission cells (see Fig. 11.6, p. 248). Another example is the suppression of stretch reflex activity by descending tracts from the medulla of the brain. Descending fibres from the medulla exert presynaptic inhibition on the sensory neurones entering the spinal cord from muscle spindles. This provides a mechanism for changing the level of stretch reflex activity.

Feedback inhibition

Neurones may inhibit themselves by collateral branches of their own axons. The lower motor neurones in the spinal cord branch in the anterior horn to synapse with small interneurones known as **Renshaw cells** (Fig. 13.3). The effect of activity in the Renshaw cells is to inhibit the lower motor neurones. In this way, motor neurone activity can be modified at the spinal level by a local feedback circuit. Similar recurrent branches of axons are found in neurones in the cerebral cortex and the limbic system.

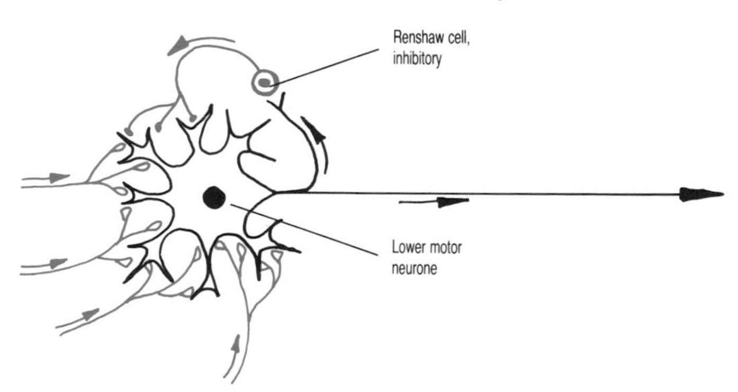

Fig. 13.3. Renshaw cell. Inhibition of a lower motor neurone via a collateral which synapses with an inhibitory interneurone (Renshaw cell).

Synergy

The performance of a particular pattern of movement, or synergy, is the combined result of different levels of activity in all the active muscle groups. Synergy is achieved by the correct modulation of activity in all the motor neurone pools in the spinal cord.

13.2 Spinal integration

13.2.1 Interneurones

The integration of spinal reflex activity is the function of the large number of interneurones which connect both sides, and different levels, of the spinal cord. The interneurones that spread activity up and down the cord lie in the intersegmental (fasciculus proprius) tract (see Figs 3.26c and 3.27, pp. 78 and 79).

The flexor withdrawal reflex, a protective reflex which draws a limb away from a harmful stimulus, has already been described in Chapter 3, Section 3.14.3. If the stimulus is strong, the activity spreads across the spinal cord to stimulate the extensors of the opposite limb. The other pair of limbs may also be affected via interneurones connecting spinal segments up or down the cord. This is an example of integration at the spinal level. The pattern of movement performed in flexor withdrawal and crossed extensor reflex is basic to many natural movements, such as walking. Consequently, if we try to oppose this pattern by swinging the same arm forwards as the swinging leg, it feels unnatural and awkward.

The spread of activity across the spinal cord by interneurones is the basis of *associated reactions.*

• *ASK a partner to remove an elastic band placed round the fingers and thumb of one hand without using the other hand and watch how the complex movements attempting to release the fingers from the band are mirrored in the untied hand.*

The associated reaction movements may become exaggerated when the interruption of descending pathways releases spinal reflexes from the control by higher centres.

13.2.2 Reciprocal innervation

All movements require the integration of activity in opposing muscle groups, which is the result of reciprocal innervation of the lower motor neurones of the particular muscle groups involved. Opposing muscles groups, for example flexors and extensors, abductors and adductors, acting round a joint cooperate during movement. Excitation of one group is accompanied by inhibition of the antagonist group which then relaxes and allows the agonist to contract. In spinal reflex movements, the sensory neurones entering the spinal cord branch in the posterior horn of the grey matter. A branch of each neurone excites the alpha motor neurones of one group of muscles, while the other branch relays to interneurones that form an inhibitory synapse with the motor neurones supplying the opposing muscle group (Fig. 13.4). This is known as reciprocal innervation or reciprocal inhibition, whereby the activity in opposing muscle groups is balanced and graded during movement.

The *development* of integrative activity in opposing muscle groups is seen in young babies. The dominance of flexor activity is evident at birth when the baby lies with bent arms and legs, and

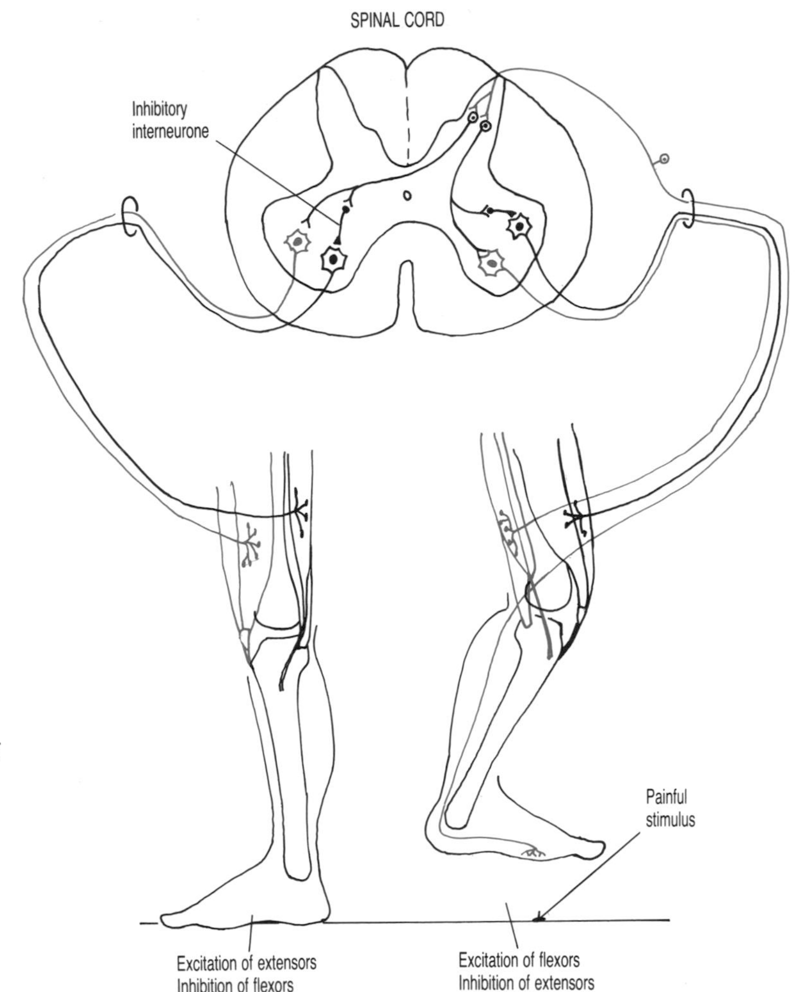

Fig. 13.4. Reciprocal innervation. Activity in opposing muscle groups balanced by excitation of the lower motor neurones of the agonist and inhibition of the lower motor neurones of the antagonist. Pathway is shown for reciprocal innervation involved in flexion of one limb and extension of the opposite limb.

curled fingers (Fig. 13.5a). By 4 months, the extensors of the arm can be used to lift the head and chest from the floor in prone lying (Fig. 13.5b). By 6 months, the legs extend to take the body weight if the baby is held upright. The arms also reach out in extension to grasp objects in the hand. The early movements are flexor/extensor patterns. Later the ability to integrate activity in muscles acting as synergists and fixators allow more complex movement patterns to develop.

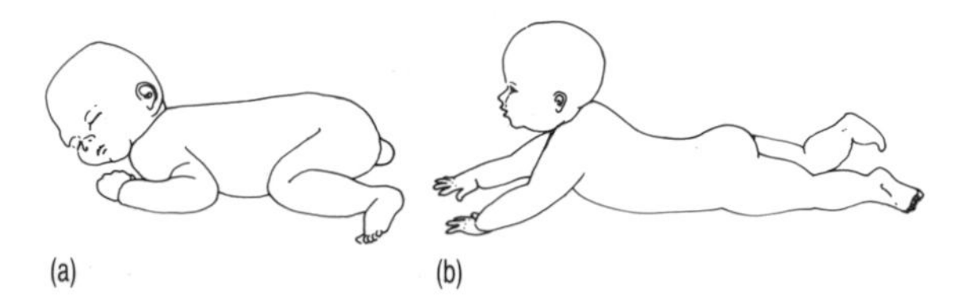

Fig. 13.5. Early development of extensor activity: (a) flexed position of the newborn in prone lying; and (b) by 3 months, the neck extends to lift the head, the arms and the legs are extended.

13.3 Integration in the brain stem

The brain stem is largely concerned with the integration of activity related to postural support during movement.

The development of brain stem reflexes can be seen in the young child. The first stage in control of posture is head control. At birth, there is no head control, but by 1 to 2 months, the *tonic labyrinthine reflex* develops and the vestibular nucleus in the brain stem can integrate the position of the head with the body, so that the head moves with the body (Fig. 13.6a). Movements of the head stimulate the *tonic neck reflexes* in the young baby. Turning the head to one side results in extension of the limbs on the same side and flexion of the opposite limb (Fig. 13.6b).

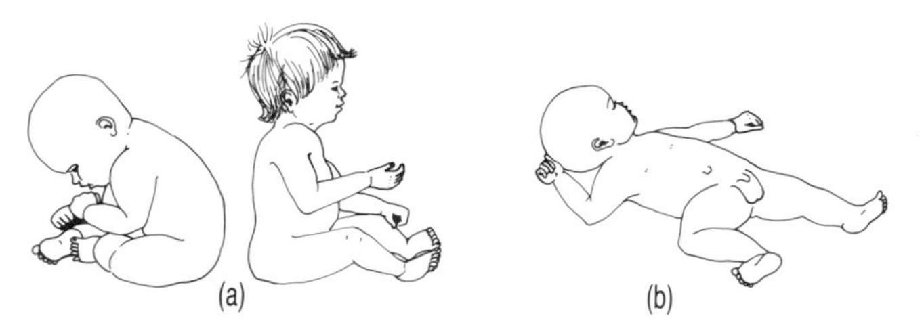

Fig. 13.6. Development of brain stem reflexes: (a) labyrinthine reflex. Newborn held in sitting, head falls forwards. By 6 months, the baby sits independently and the head is held upright on the trunk; and (b) asymmetric tonic neck reflex. Turning the head to one side produces extension of the limbs on the same side and flexion in the opposite limbs.

Turning the head upwards increases activity in the flexors of the lower limbs, turning the head downwards increases flexion in the upper limbs. (These are known as *symmetrical tonic neck reflexes*).

By 7 months, the baby has voluntary control of the head and eye movements, and the righting reflexes are established. The neck and body *righting reflexes* allow the baby to rotate about the vertical axis and roll over from prone to supine, and supine to prone lying. The rotation of the head stimulates stretch receptors in the neck muscles, and the response is contraction of the rotators of the trunk so that the body starts to turn. Pressure on the side of the body then stimulates the body righting reflex which rotates the pelvis to complete the turn.

The final stage is the development of the more complex *equilibrium reactions* which occurs between 1 and 2 years of age. These automatic adjustments of the head and body position allow the body segments to align over the feet in standing and walking. The tonic neck and righting reflexes become modified with the development of equilibrium reactions, but they remain as basic movement patterns. For example, the position of the head and arms in the asymmetrical tonic neck reflex is often adopted in dance movements.

In the adult, highly skilled movement patterns are developed by the higher centres, and brain stem reflexes are modified.

Interruption of the higher centres by disease or injury in the adult may result in the reappearance of brain stem reflexes as abnormal patterns of movement.

Fig. 13.7. Equilibrium reactions. Girl standing on a balance board. As the body is thrown off balance, the body segments realign to keep the balance and to keep the eyes looking straight ahead.

• *OBSERVE the equilibrium reactions in the adult by asking a partner to try to balance on a board resting on a cylinder or pipe as shown in Fig. 13.7. As each foot moves, the body is thrown off balance. Notice the simultaneous changes in the hips, the trunk and the head as the body tries to maintain balance over the feet. Also notice how the eyes remain looking ahead.*

13.4 Motor planning and coordination

The basal ganglia and cerebellum together plan and coordinate movement.

The **basal ganglia** do not have direct input from the sensory system. The complex system of the basal ganglia has: (i) *input* from all areas of the cortex and from the thalamus; and (ii) *output* to the muscles via the red nucleus and reticular formation, and to the motor cortex for execution of movement. Some of the links between the nuclei of the basal ganglia, the thalamus and the cerebral cortex have been shown in Fig. 3.12, p. 62. The output to the brain stem is mainly from the globus pallidus to the recticular formation (Fig. 13.8).

The basal ganglia select and generate the motor programme for a movement.

The **cerebellum** monitors and modifies the progress of a sequence of movements, acting as a correction device. When the cerebellum is damaged, tremor, ataxia and overshoot of movements occur. This can be seen when the feet try to walk on a straight line, or the hands try to make accurate placement. The cerebellum has been compared with the control system of a guided missile which ensures its arrival on target.

Figure 13.9 shows how the cerebellum receives and sends the following.

1 It receives **input**

(a) from all the sensory pathways of the somatosensory and vestibular system, and

(b) from the motor cortex which is concerned with the execution of voluntary movement.

2 It sends **output**:

(a) to all the muscles of the body (including axial, proximal and distal muscles).

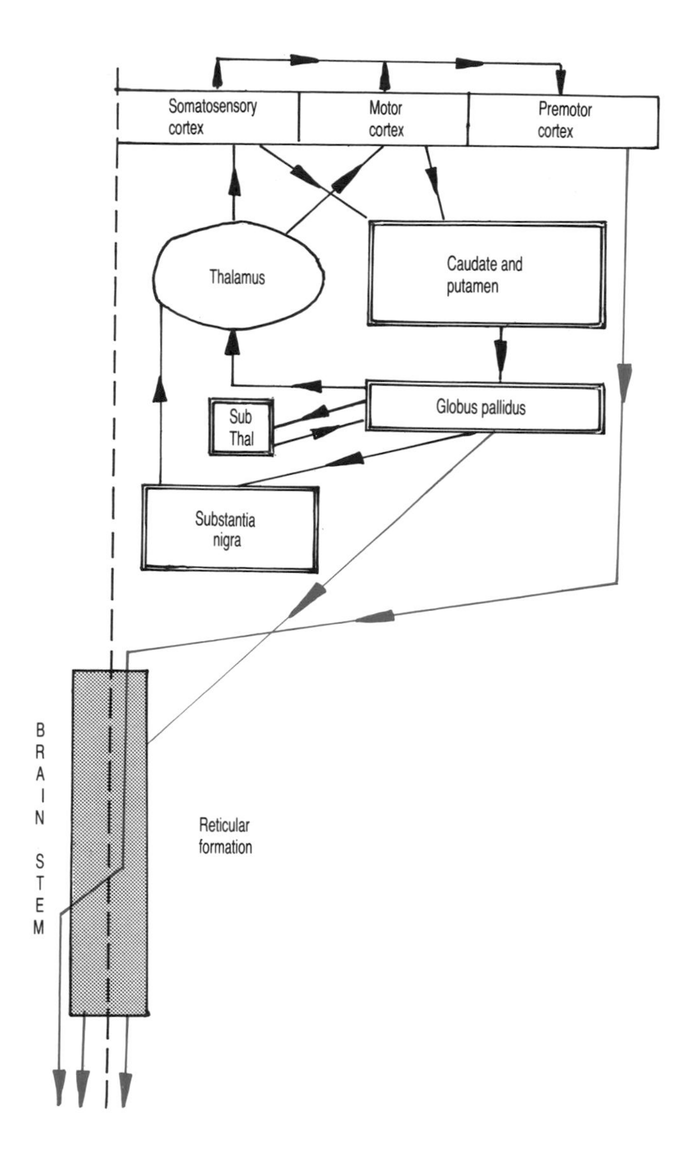

Fig. 13.8. Links between the basal ganglia, thalamus and cerebral cortex. Output of the system to the brain stem and spinal cord.

(b) back to the sensorimotor cortex.

Motor commands are sent out by the motor cortex via the corticospinal tracts to the muscles. The same commands are also sent to the cerebellum through the pontine nuclei and by a second pathway through the olivary nuclei. These commands are then compared with the sensory information reaching the cerebellum from the muscles. The modification of muscle activity required to correct any error in the movement is achieved by cerebellar output to the red nuclei in the brain stem and via the rubrospinal tracts back to the motor neurones of the spinal cord.

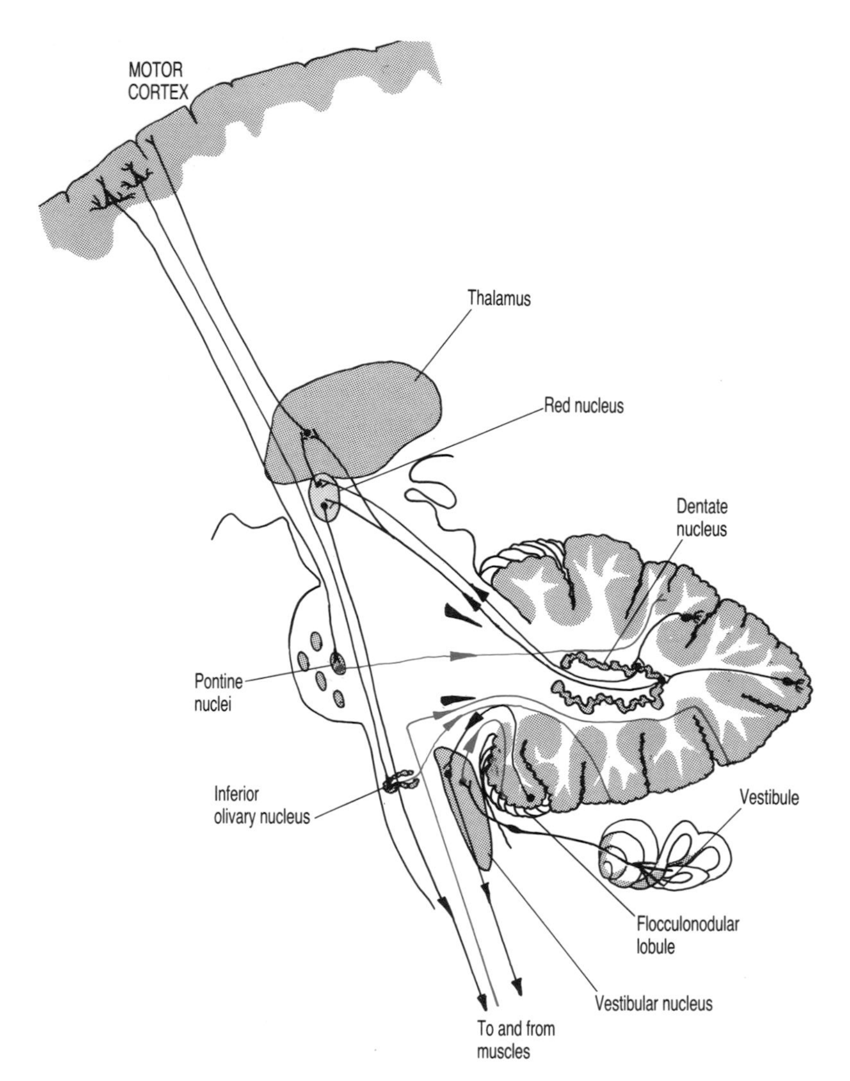

Fig. 13.9. Pathways in and out of the cerebellum. **Incoming** from the pontine nuclei, olivary nuclei, the vestibule of the inner ear and the muscles. **Outgoing** to the thalamus, red nucleus and vestibular nucleus.

- *LOOK at Fig. 13.9 to follow the routes* ***in*** *and* ***out*** *of the cerebellum.*

The cerebellum regulates the magnitude and timing of activity in all the muscles involved in a movement as it proceeds, and in bringing the movement to a halt when the goal has been reached.

The output from the motor cortex at any one moment is preprogrammed by the basal ganglia and modified by the cerebellum.

13.5 Behavioural aspects of movement

The way we perform daily tasks such as dressing and cooking depends on our mood and on how alert we feel at the time. In this section, a brief outline of the brain areas concerned with behavioural aspects of movement will be given.

13.5.1 Arousal. Reticular formation (Fig. 13.10)

By its position in the central core of the brain stem, and its numerous links with the sensory and motor systems, the reticular formation has an effect on the way we perform movement. If there is a high level of activity in the reticular formation, the level of arousal is high, and the muscles seem to 'spring into action'. A low arousal state (which we may feel on waking from a night's sleep for example), reduces our capacity to move with speed and accuracy. This reflects the absence of overall stimulation of the cerebral cortex from the reticular activating system (ARAS) seen in Fig. 3.18 (see p. 70). The addition of an extra stimulus, such as an alarm clock or door bell ringing, increases the input to the system and we feel more able to move quickly.

Neurones in the reticular formation are non specific, and they are stimulated by different types of sensation. Collaterals from the anterolateral ascending tracts are given off to the reticular formation as they pass through the brain stem. Reticular neurones project to most areas of the cerebral cortex either directly, or indirectly via the thalamus. Activity in these ascending fibres makes the cerebral cortex more receptive to stimulation, and general alertness is increased.

Another way in which the reticular formation is involved in movement is in the modification of breathing and circulation of the

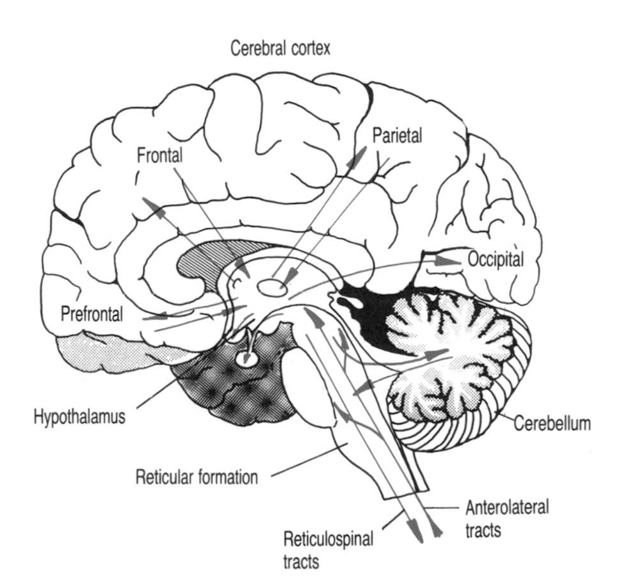

Fig. 13.10. Sagittal section of the brain to show reticular formation. Collaterals of the anterolateral system, link with the cerebellum and the arousal system to the cerebral cortex are shown.

blood to meet the oxygen demands of the active muscles. Collections of neurones in the reticular formation in the pons and medulla control the heart rate, blood pressure and breathing and are known as the 'vital centres'.

The reticular formation receives information from the various motor centres in the brain, particularly the basal ganglia and the vestibular nuclei, and relays into the reticulospinal tracts to both the alpha and gamma motor neurones in the spinal cord. In this way, the integration of activity in the reticular neurones exerts an influence on the performance of movement, especially the postural adjustments involved.

13.5.2 Mood and emotion

The brain areas concerned with our emotional feelings are the *limbic system*, the *prefrontal cortex* and the *hypothalamus* (see Section 3.8, p. 65). Some of the connections between these brain areas are shown in Fig. 13.11.

The structures which form the **limbic system** can be divided into those which connect with the forebrain, and those linking to the midbrain: these two areas are linked by the fornix, which is a large tract of white matter. The hypothalamus lies inbetween the limbic forebrain and the midbrain.

The limbic forebrain links with the **prefrontal cortex** which is

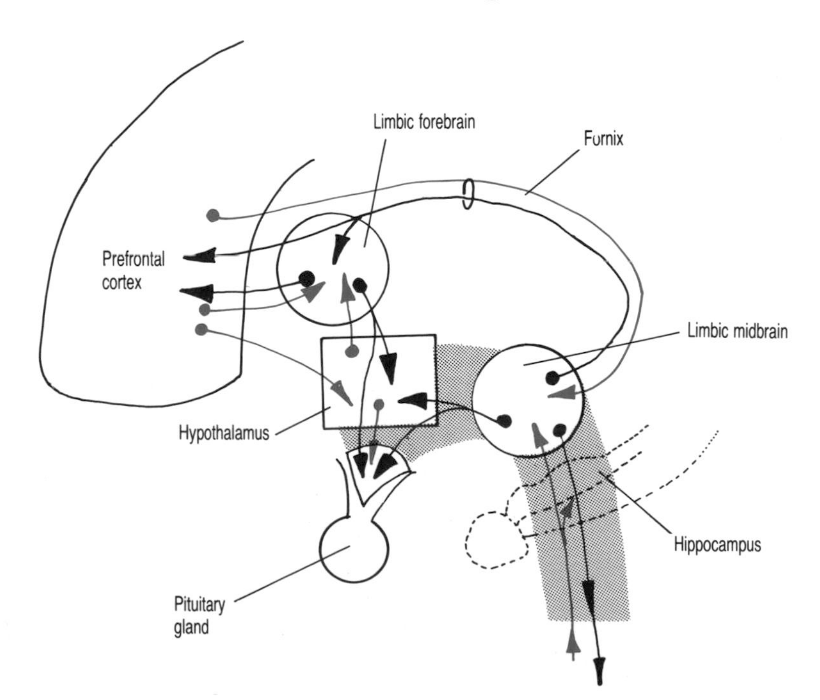

Fig. 13.11. Connections between the limbic system, hypothalamus and prefrontal cortex.

concerned with subjective feelings of emotion. Sensory information filters back down from the cerebral cortex to the limbic forebrain and affects its overall activity. The limbic midbrain links with the reticular formation. Pathways from the hippocampus to the basal ganglia suggest that the limbic system plays a role in the long-term memory of motor skills.

The **hypothalamus** is the area of the brain that responds to stress in the body. The stress stimulus (or stressor) may originate outside the body, such as pain to the skin, a loud noise, or an object rapidly approaching the eyes; or the stressor may be internal, such as feelings of anxiety or depression. The result of these stressors is activity in the sympathetic nervous system which prepares the body for action by speeding up the heart, opening up the airways of the lungs and so on. The same responses occur in the athlete preparing to start a race, so that the muscles spring into action when the starting pistol fires. A stress response of short duration is helpful in preparing the body for action, but if prolonged, then stress diseases may develop.

13.6 Performance

13.6.1 Open and closed loop movements

Voluntary movement can be divided into two types: (i) *open loop* or *ballastic* movement, which once initiated must follow its course; and (ii) *closed loop*, which is guided movement, always variable and subject to fine adjustment.

Open loop is preprogrammed, based on input from the sensory system, and once started, it proceeds without reference to any changes that may occur in the environment. An example of this type of movement is throwing a ball: once the ball has been released, control over its flight has ended, and the accuracy of this movement therefore depends on the ability to assemble the correct motor commands before the movement begins. Repetition and memory play a part in the central nervous system preprogramming, and the motor system carries out the commands. Other everyday examples of this type of movement are brushing the hair, chopping vegetables, pressing a typewriter or a piano key.

Closed loop involves a changing input to the central nervous system which is fed back to the brain as the movement proceeds. An 'error correction' is then applied to the output in the motor system to achieve the desired performance. Examples of closed loop type of movement are hand sewing, writing and drawing with care.

Most of our daily activities are a combination of **both** of these types of movement. For example, brushing the hair starts as a ballistic movement, but this becomes modified as the brush encounters different friction with each stroke, and as the hair is brushed into ones personal style.

13.6.2 Performance of skilled movements

A motor skill is a sequence of movements that are performed to achieve a particular goal with appropriate speed and accuracy. The acquisition of a motor skill involves a learning period when the correct motor commands are developed in the central nervous system, and the activity in all the groups of muscles is gradually modified to produce the correct force and timing. In this section we will only consider the role of the cerebellum in the performance of skilled movement to give an introduction to the neurological mechanisms involved.

When we learn to perform a complex motor skill, such as swinging a golf club, driving a car, or manipulating a paint brush on canvas, we make the movements slowly at first, then repeat them many times under conscious voluntary control. Eventually the movements can be performed without the constant reference to consciousness. Another example of this process is learning to operate a typewriter or piano keyboard. At first, each key is pressed with conscious effort, but after repeating the movements many times a prolonged sequence of finger movements can be performed with accuracy. During the repetition in training and practice of a motor skill, the same patterns of activity are generated from the cortex and the cerebellum to the spinal cord. It has been suggested that the cerebellum can store these motor patterns so that the same sequence of movements can be performed without reference to consciousness. In this way, the cerebellum acts as a *'skills bank'*. If there is any change in the pattern of input to the cerebellum (for example the driver may change to a new car), then the stored motor pattern can be updated, and after a short period of practice, the movements become automatic again. Various theories have been developed to explain the role of the cerebellum in the acquisition of motor skills: the theories are based on detailed knowledge of the cellular structure of the cerebellar cortex and on the results of damage to the cerebellum.

The **cerebellar cortex** has three layers which are distributed uniformly over all the surface of the cerebellum (Fig. 13.12). This is different from the cerebral cortex where the arrangement and type of neurones varies in different areas.

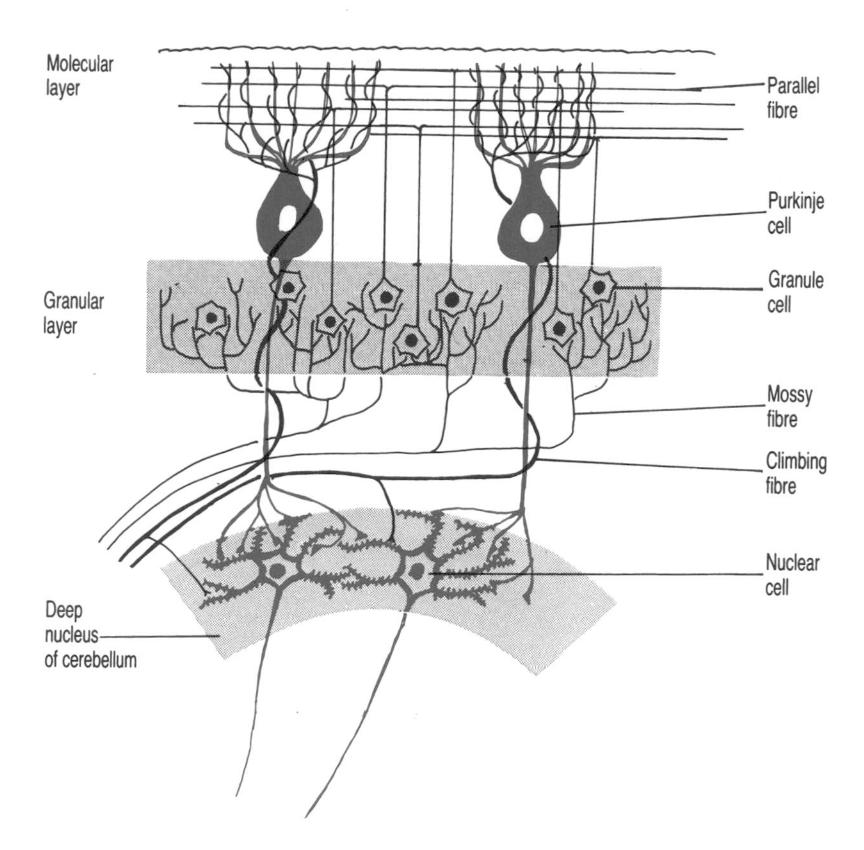

Fig. 13.12. Cellular structure of the cerebellar cortex (simplified), showing Purkinje cells, granule cells and deep nuclear cells.

The middle layer of the cerebellar cortex has a layer of large **Purkinje cells**, which have extensive and complex systems of dendrites reaching to the superficial (molecular) layer. The axons of the Purkinje cells form the only *output* from the cerebellar cortex. These axons end in the deep nuclei of the cerebellum and then relay to the brain stem.

Two systems of *incoming fibres* affect the activity of the Purkinje cells.

1 Climbing fibres which originate only in the inferior nucleus of the olive in the medulla. Each fibre winds round the dendrites of one Purkinje cell. This forms a fast route for the excitation of the Purkinje cells.

2 Mossy fibres which originate in all the other areas of input to the cerebellum (Fig. 13.9, p. 276). The mossy fibres synapse with granule cells in the deeper cell layer before sending ascending axons to the molecular layer. Each axon divides into two to form the *parallel fibres* which traverse the molecular layer of the cortex like telephone wires, and each branches to make contact with a large number of Purkinje cells. The activity in the mossy fibres provides input from a large number of sources in the body. The activity is then spread by the parallel fibres to the output from a large number of Purkinje cells.

During the early stages of motor learning, the Purkinje cells are excited by the climbing fibres, which are stimulated by the motor cortex via the olivary nucleus. The performance of each movement produces a particular pattern of sensory feedback from the muscles, skin, eyes and ears, which is carried via the mossy fibres and the parallel fibres to the Purkinje cells. After many repetitions of the same sequence, motor commands of successful movements are stored by the Purkinje cells, and excitation by the climbing fibres from the cortex is no longer required. If any updating of the sequence is required, the olivary route via the climbing fibres is involved again until a new motor pattern is generated.

13.7 Summary of some basic concepts in movement performance

Muscle components

The groups of muscles involved in a movement can be divided into three functional components.

1 **Posture** and **balance** — the function of the trunk muscles controlled by the vestibular and medial reticulospinal tracts from the brain stem, and by spinal reflex activity.

2 **Positioning** and **support** — the function of the proximal limb muscles (shoulder and elbow, hip and knee) controlled by the cerebellum and red nucleus, and the lateral reticulospinal tract from the midbrain.

3 **Skilled** and **precision movement** — the function of the distal muscle groups of the limbs (the hand and foot), the muscles involved in speech, facial expression and eye movements controlled by the primary motor and premotor cortex via the corticobulbar and corticospinal tracts.

These three components are adjusted throughout the progress of the movement.

Neural components

The **neural** components of a movement (shown diagrammatically in Fig. 13.13), can be divided into the following.

1 **Planning** and **programming** — the function of the cerebral cortex (particularly the association areas), the basal ganglia, the cerebellum, the thalamus. Rapid movement depends particularly on preprogramming by the basal ganglia, since no correction can be

applied once the movements have started. A sequence of more prolonged movements is programmed by the cerebellum which plans the progress from one movement to the next. The premotor and the sensory areas of the cerebral cortex provide essential input for planning to the basal ganglia and the cerebellum.

2 Performance — the function of the motor cortex and its output to the spinal cord. The cerebellum provides an error correction to the movements, based on sensory feedback from the muscles as movement proceeds.

The accuracy and precision of the performance of skilled movements depends on the accumulation of stored motor patterns in the cerebellum as the result of repetition of the same sequence of movements.

The acquisition of a complex motor skill, such as learning to ride a bicycle, is an event that most of us remember. The movements involved in propelling the bicycle require strength and bilateral coordination of the extensor muscles of the legs to overcome the resistance of the pedals. The leg movements must be accompanied by appropriate movements of the head and trunk to balance over the seat of the bicycle. Repetition of the combined movements of the

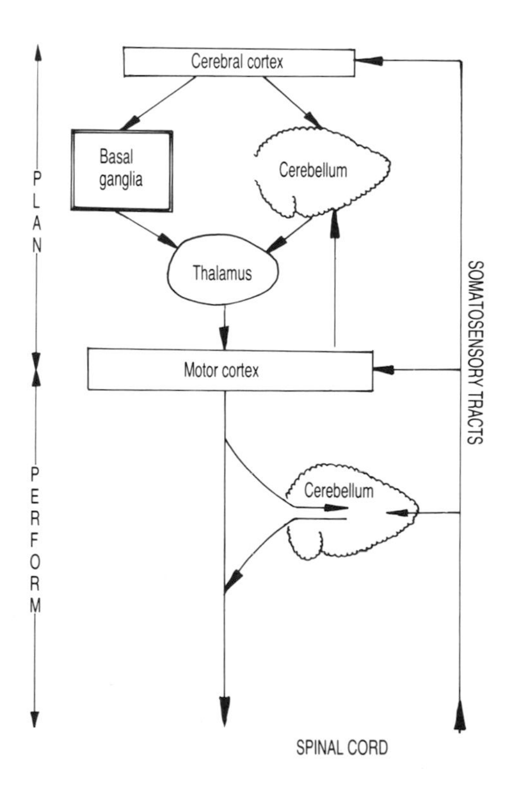

Fig. 13.13. Plan of the brain areas involved in planning and performance of movement.

legs, the trunk and the head acting together eventually achieves the balance required. Next, the responses to various external changes, such as the road surface, turning corners, and traffic, must be learnt, so updating the early stages of learning. Finally, the performance of the movements becomes automatic. All our effort can then be directed towards enjoying the scenery, and (depending on our mood at the time), the pleasure that we gain from the fresh air and physical exercise.

At the end of this chapter, you should be able to:

1 Outline the regulation of input and output of activity in groups of neurones in the central nervous system.

2 Appreciate the importance of interneurones and reciprocal innervation in spinal reflex movements.

3 Outline the brain stem reflexes and explain their role in the postural control of movement.

4 Describe the role of the basal ganglia and cerebellum in the planning and coordination of movement.

5 Summarize the function of the cerebellum in the development of skilled movements.

Further Reading

Barr ML & Kiernan JA (1983) *The Human Nervous System. An Anatomical Viewpoint.* Harper & Row, London.

Carpenter RHS (1984) *Neurophysiology.* Edward Arnold, London.

Coen CW (ed) (1985) *Functions of the brain. Edited papers from lectures at Wolfson College.* Oxford University Press, Oxford.

Eccles JC (1977) *The Understanding of the Brain.* McGraw–Hill Book Co. Ltd., Maidenhead.

Ganong WF (1977) *Review of Medical Physiology.* Lange Medical Publications, Los Altos, California.

Guyton AG (1987) *Basic Neuroscience.* W.B. Saunders, New York.

Higgins JR (1977) *Human Movement. An Integrated Approach.* CV Mosby Co., St Louis.

Illingworth RS (1980) *The Development of the Infant and Young Child.* Churchill Livingstone, Edinburgh.

Kidd GL (1986) The myotatic reflex. In: *Cash's Textbook of Neurology for Physiotherapists* (ed by Downie PA). Faber & Faber, London.

Melzack R & Wall P (1982) *The Challenge of Pain.* Penguin, London.

Noback CR & Demarest RJ (1977) *The Nervous System, Introduction and Review.* McGraw-Hill Book Co. Ltd., Maidenhead.

Schmidt RF (ed) (1978) *Fundamentals of Neurophysiology.* Springer Verlag, New York.

Schmidt RF (ed) (1978) *Fundamentals of Sensory Physiology.* Springer Verlag, New York.

Sheridan MD (1973) *Children's Developmental Progress from Birth to Five Years.* NFER Publishing Co. Ltd, Windsor, Berks.

Smyth, MM & Wing A (1984) *The Psychology of Human Movement.* Academic Press, London.

Stein JF (1982) *An Introduction to Neurophysiology.* Blackwell Scientific Publications, Oxford.

Young LZ (1978) *Programs of the Brain.* Oxford University Press, Oxford.

Appendix 1 Bones

Right clavicle — superior aspect

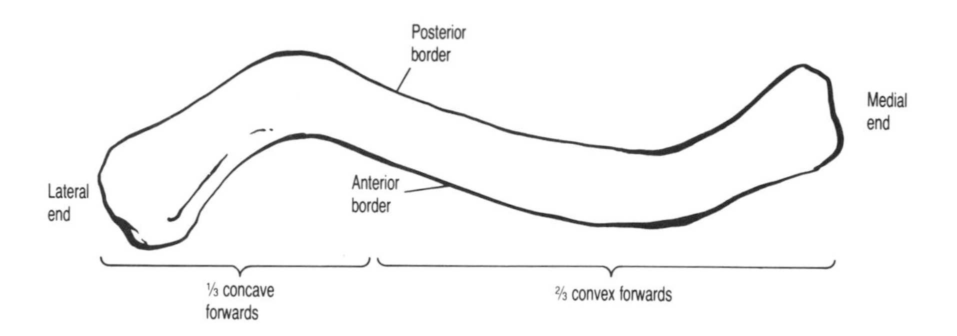

Right scapula — anterior aspect

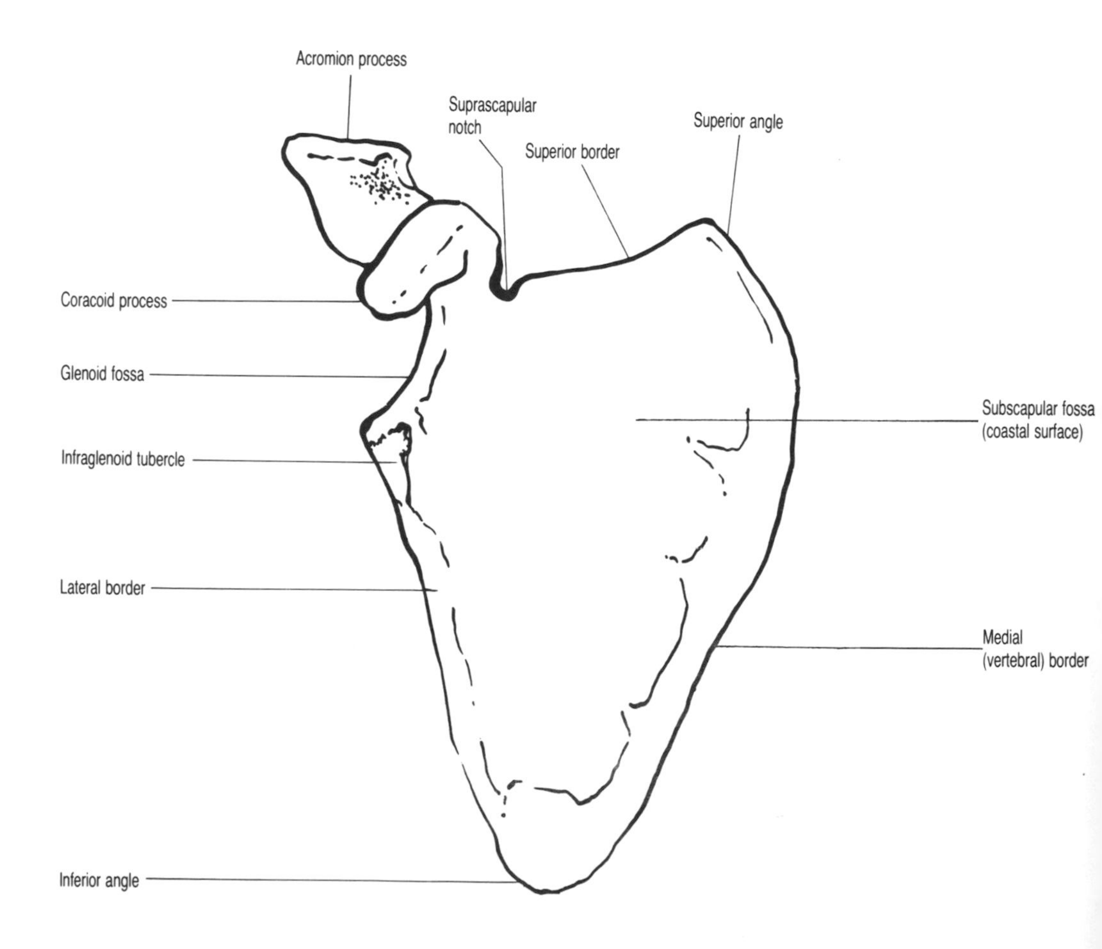

Right clavicle — inferior aspect

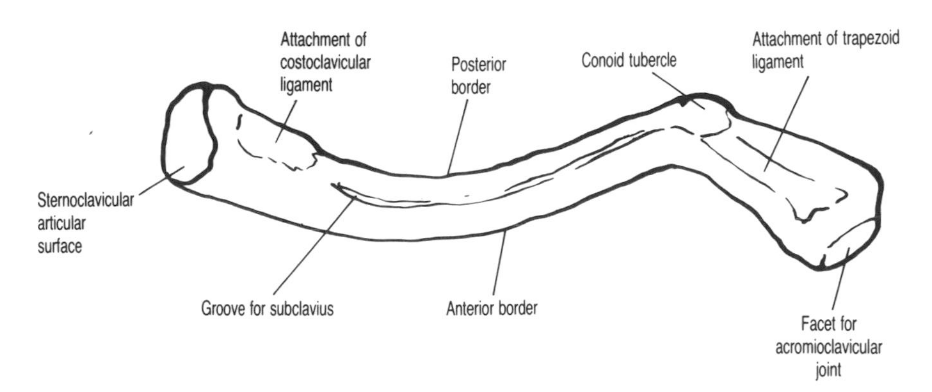

Right scapula — posterior aspect

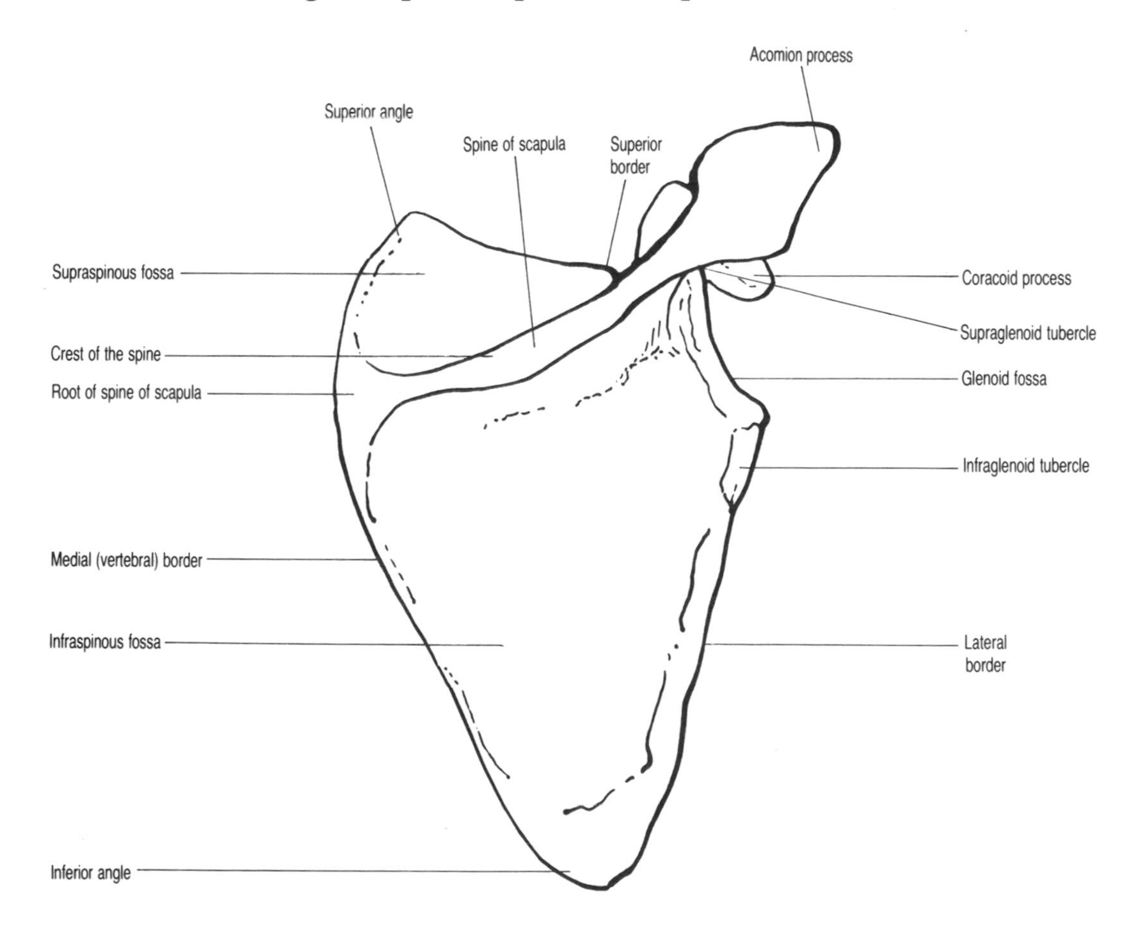

Right humerus

Anterior aspect

Posterior aspect

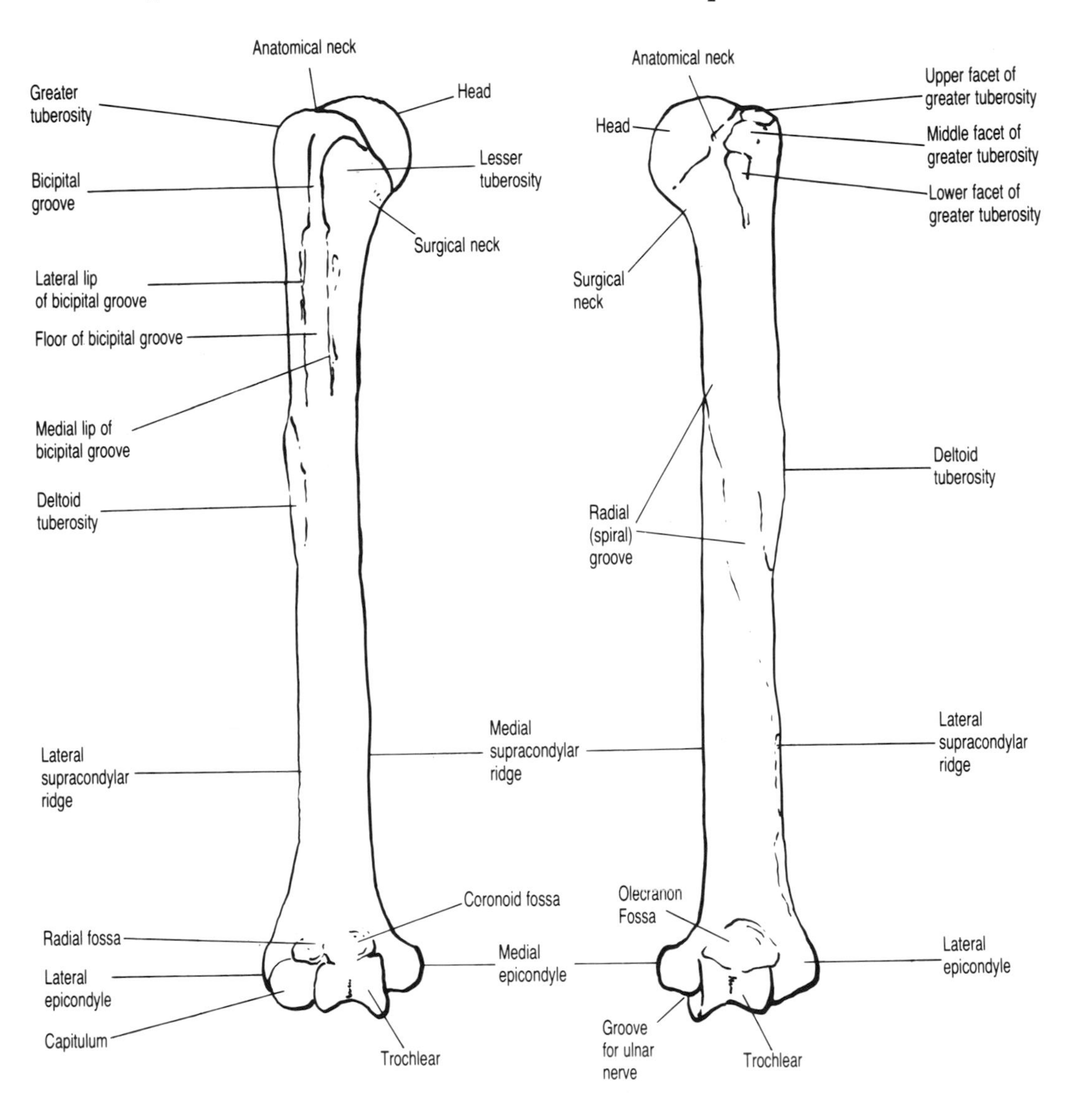

Right radius and ulna

Anterior aspect

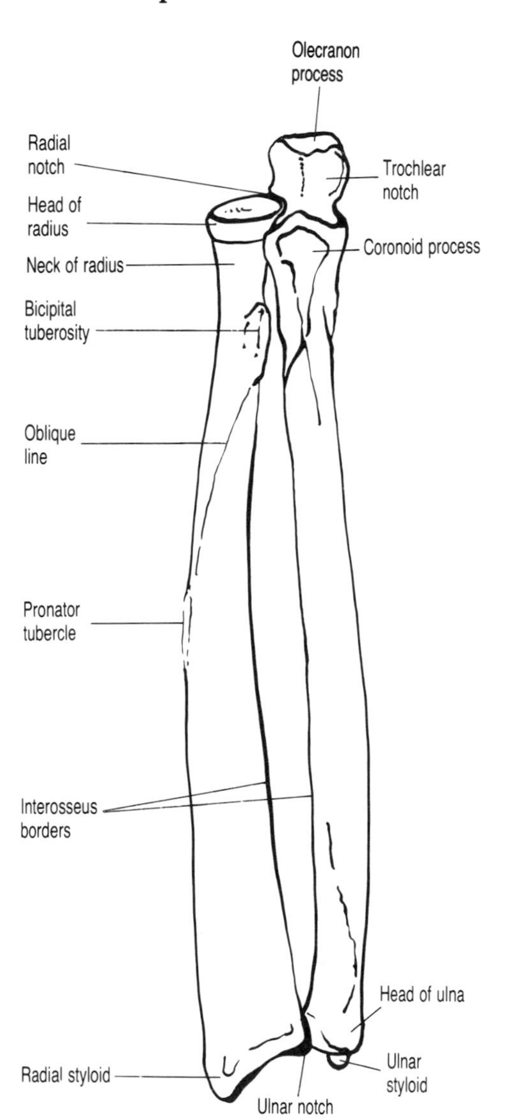

Posterior aspect

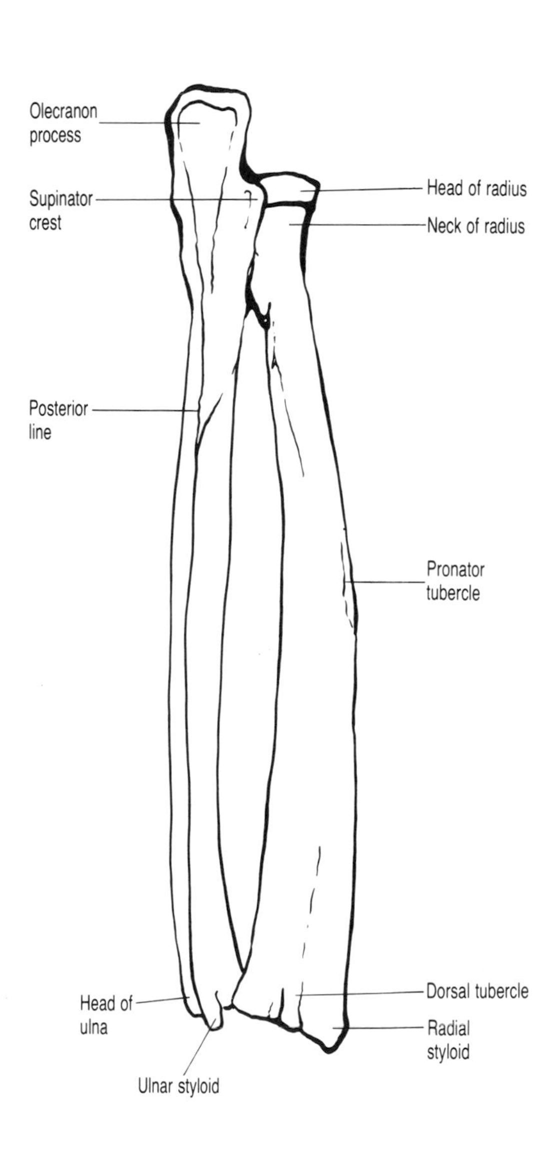

Right hand — palmar aspect

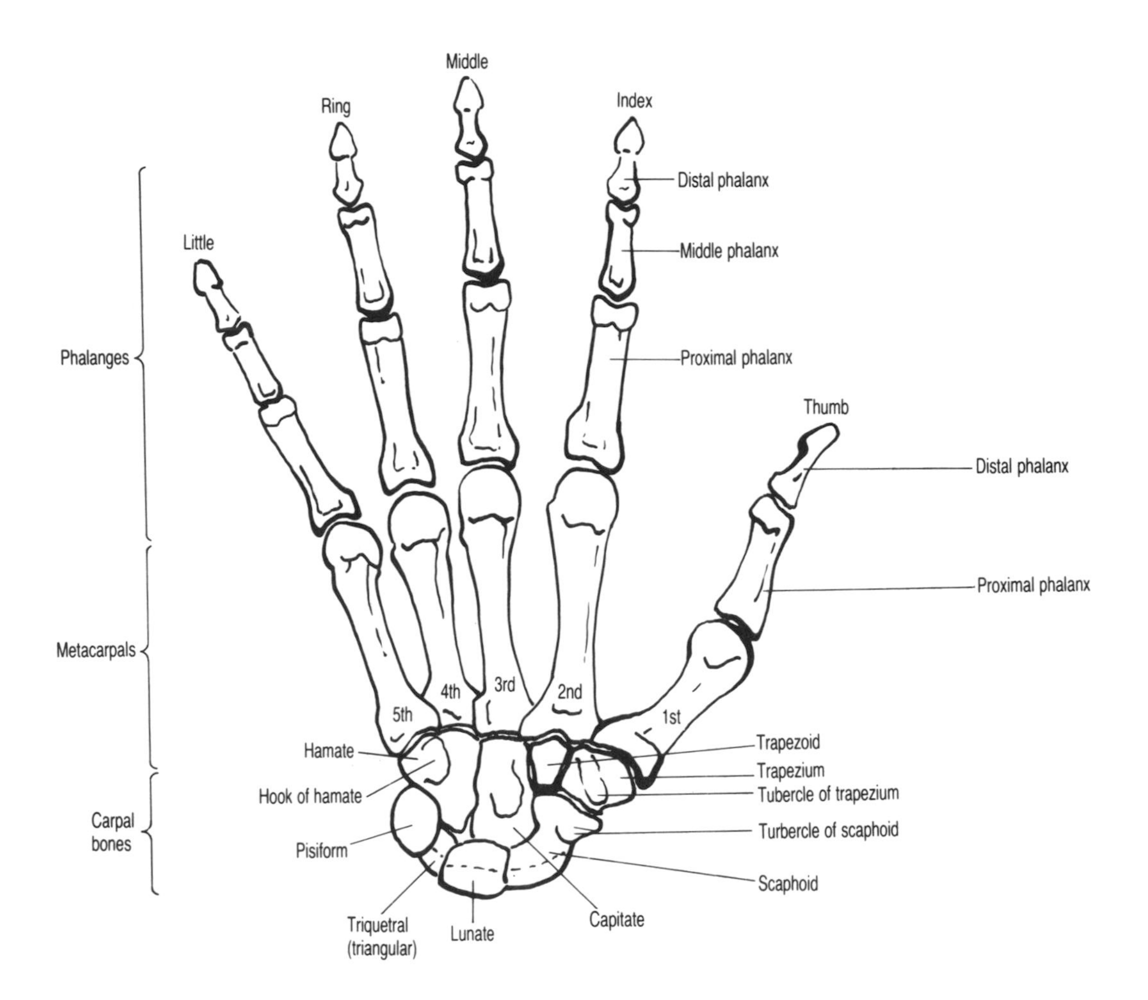

Right hand — dorsal aspect

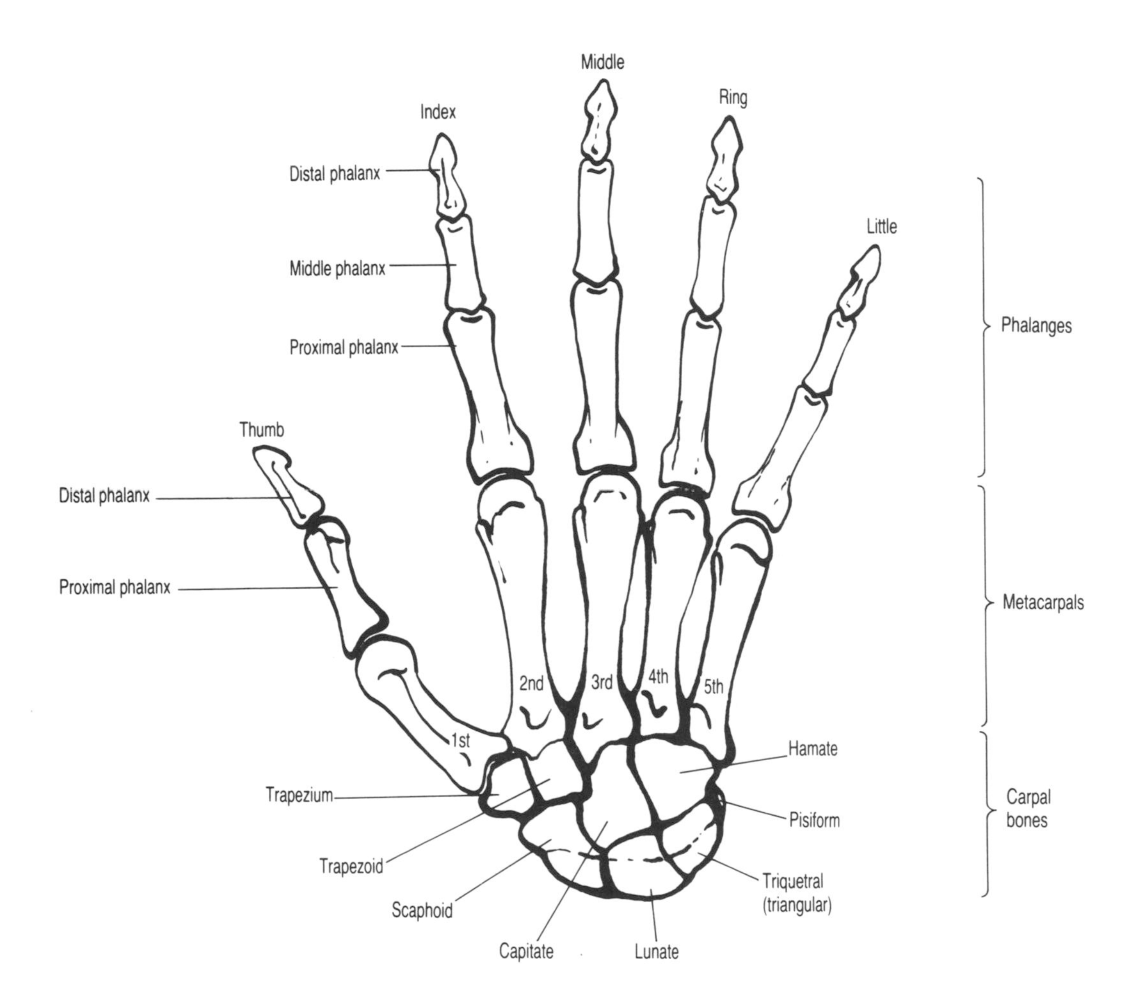

Pelvis — anterior aspect

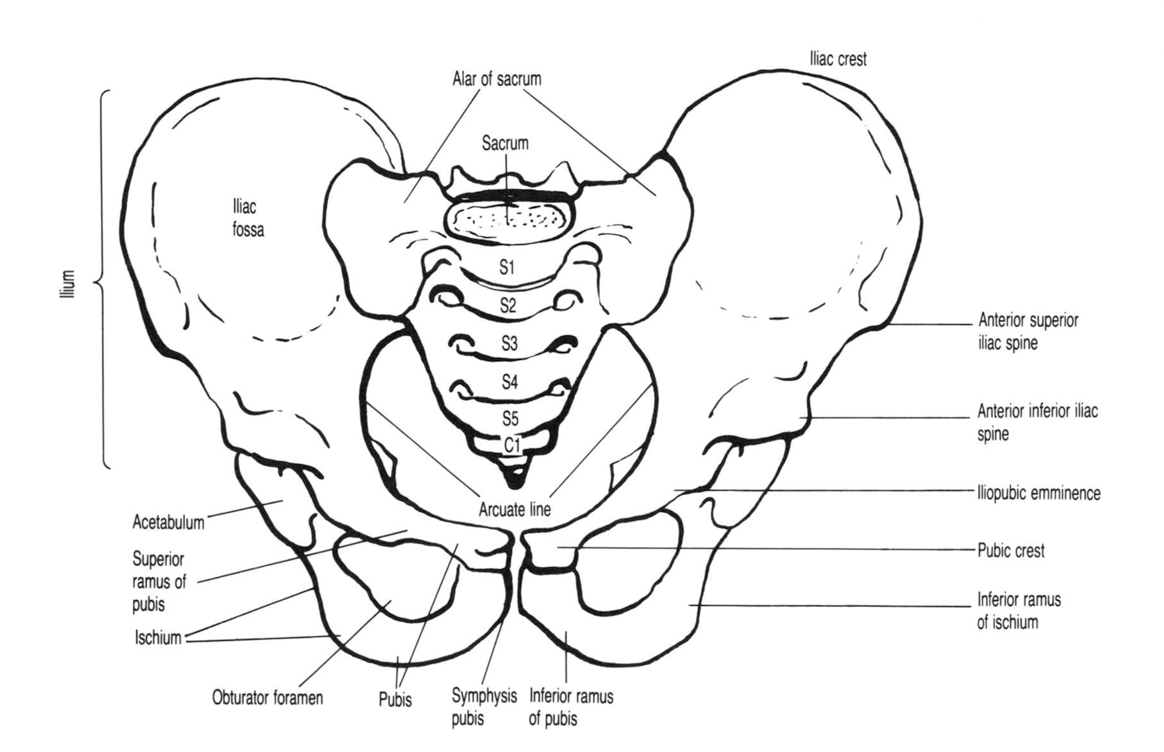

Pelvis — posterior aspect

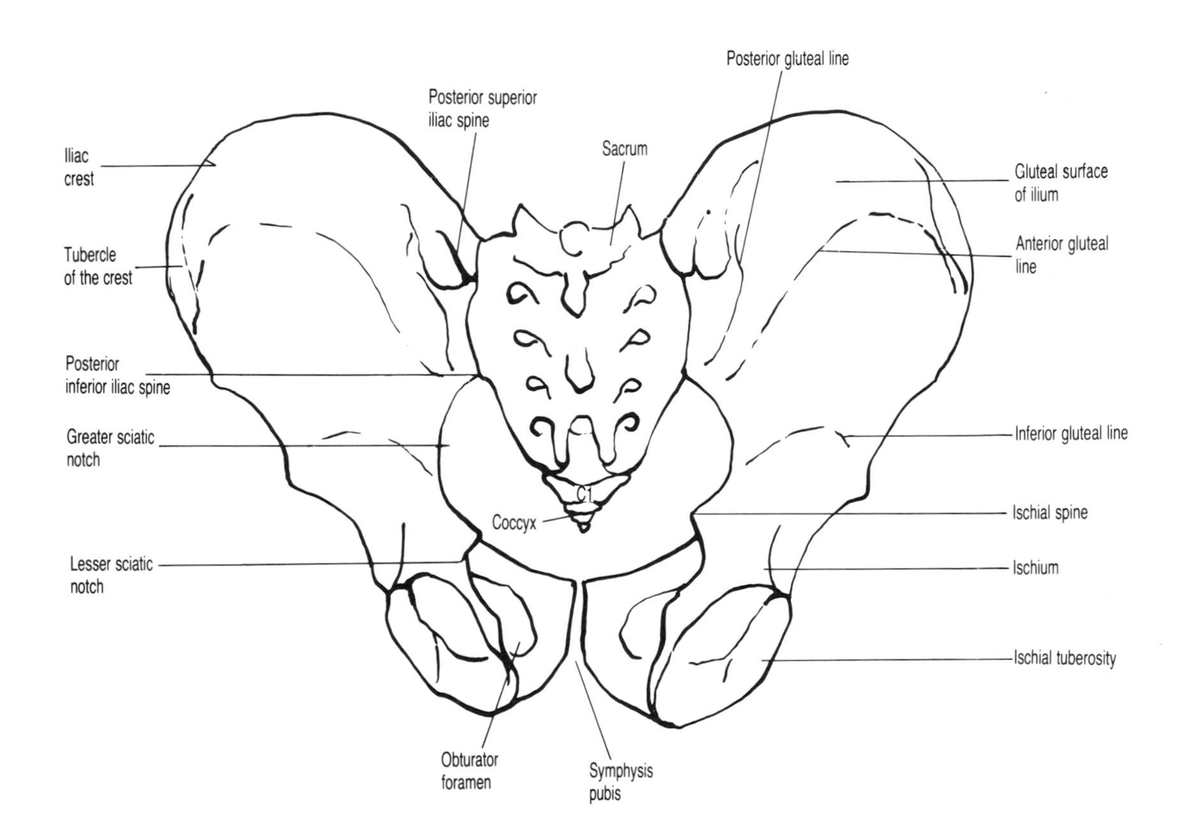

Right femur

Anterior aspect

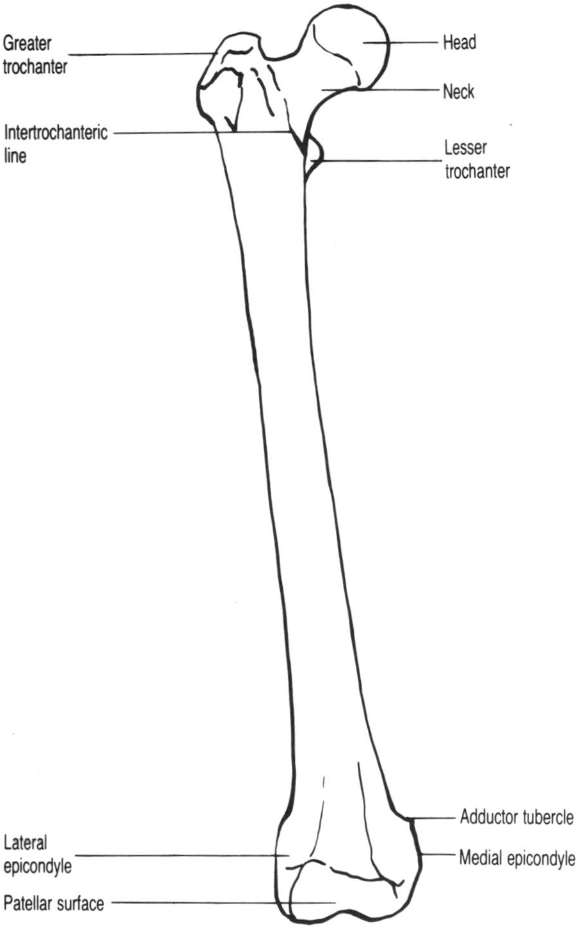

Greater
trochanter
Head
Neck
Intertrochanteric
line
Lesser
trochanter
Adductor tubercle
Lateral
epicondyle
Medial epicondyle
Patellar surface

Posterior aspect

Head
Neck
Greater
trochanter
Fossa of the
head
Quadrate tubercle
on the introchanteric
crest
Fossa of the
greater trochanter
Lesser trochanter
Gluteal tuberosity
Spiral line
Linear aspera
Medial supracondylar
line
Lateral supracondylar
line
Adductor tubercle
Popliteal surface
Lateral epicondyle
Medial
epicondyle
Lateral condyle
Medial condyle
Intercondylar
notch (fossa)

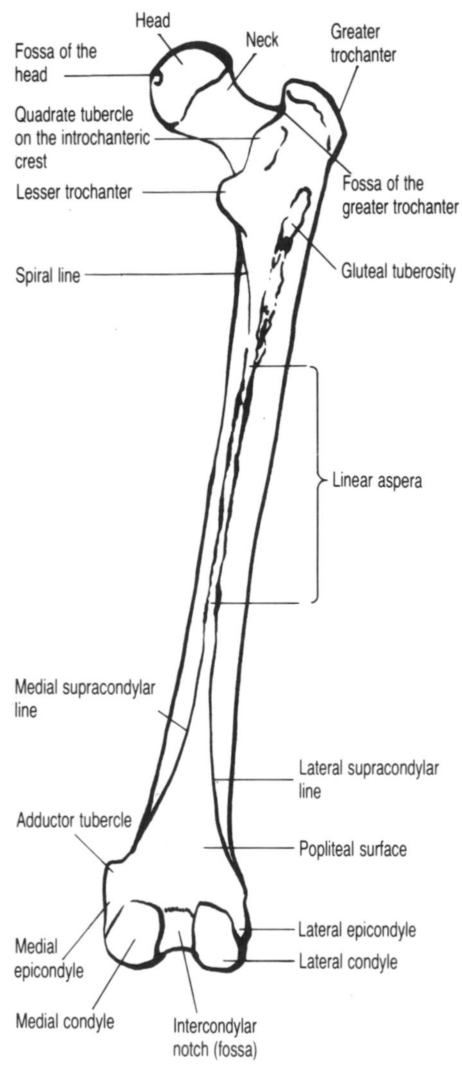

Right tibia and fibula

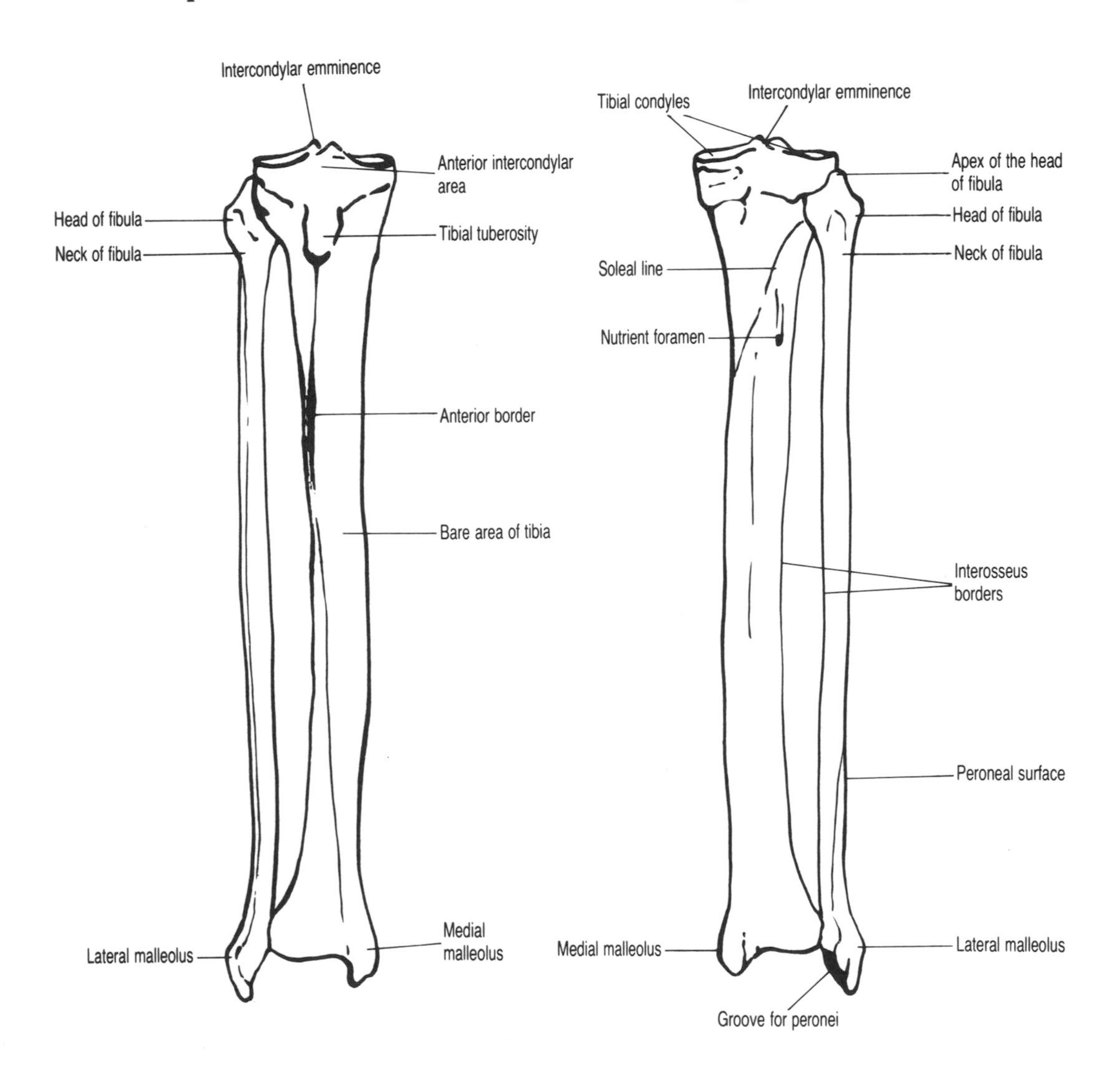

Right foot

Lateral aspect

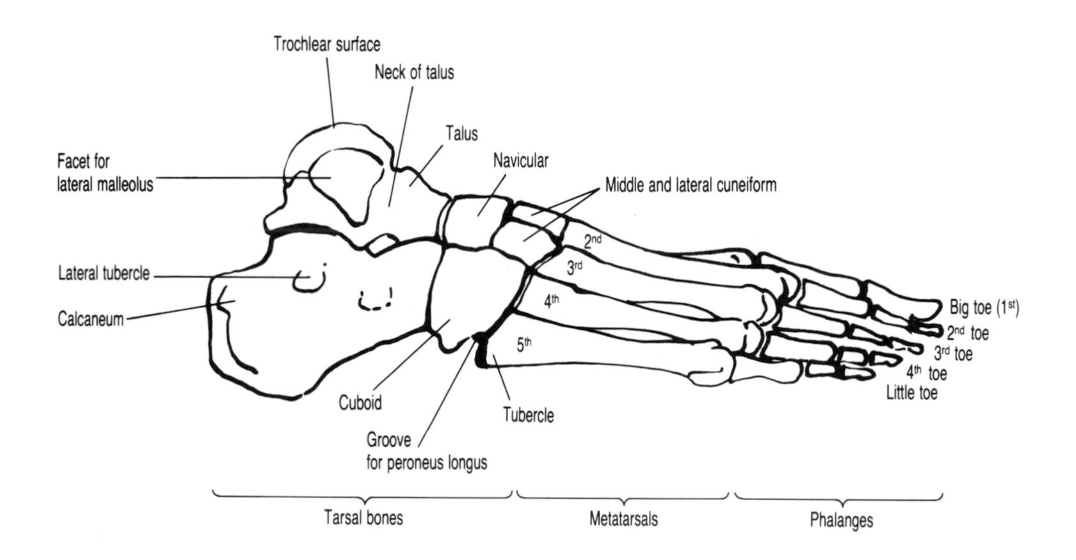

Medial aspect

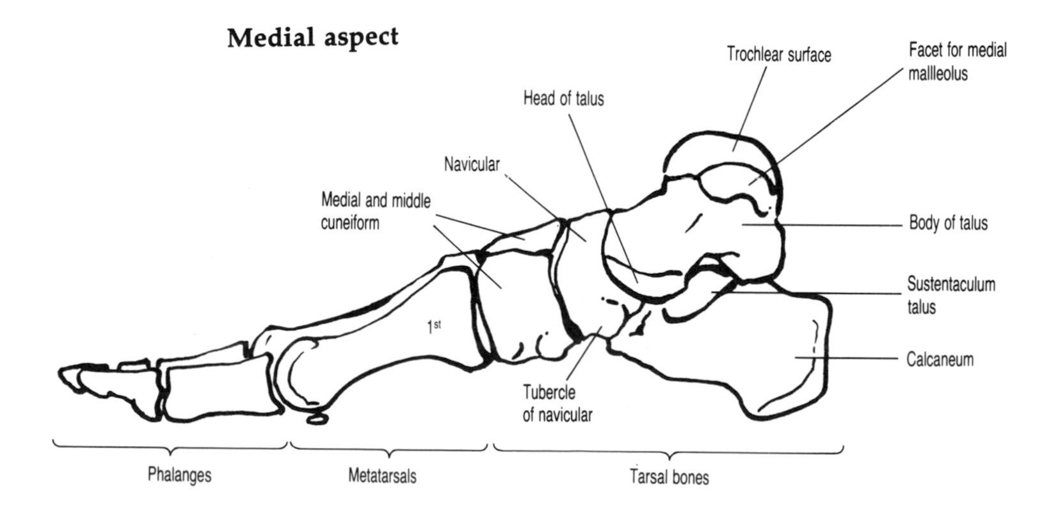

A typical (thoracic) vertebra

Superior aspect

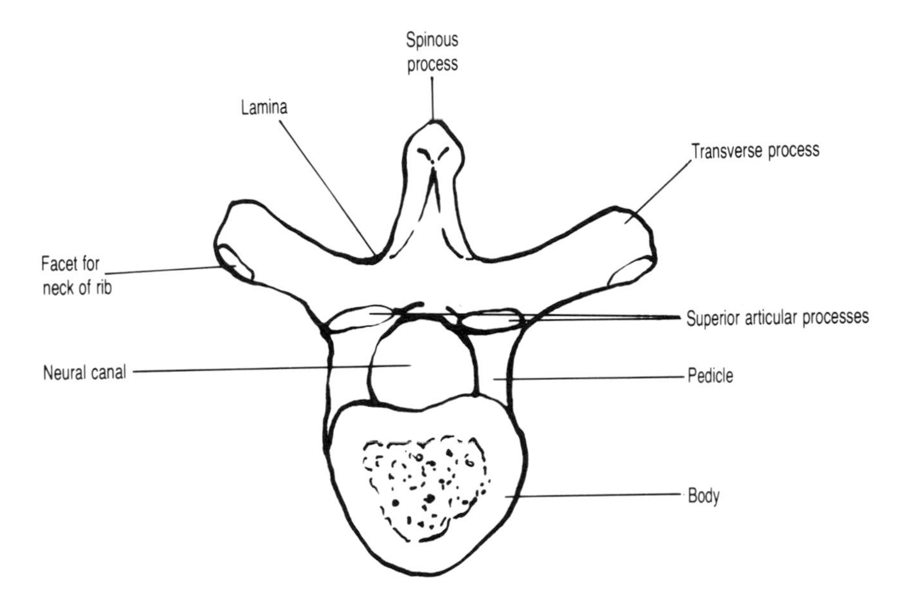

Lateral aspect

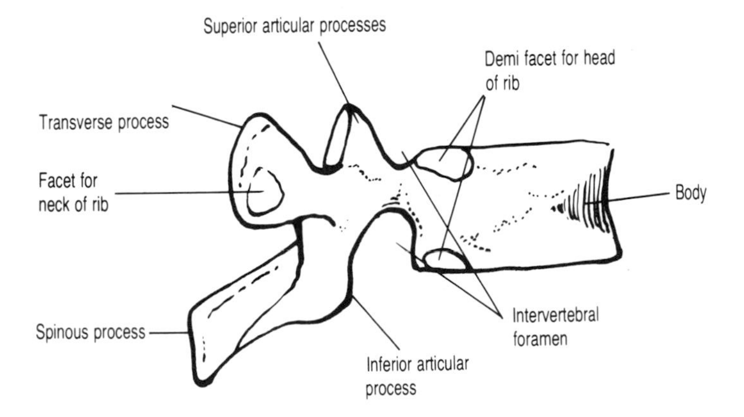

Appendix 2 Joints

Right acromioclavicular joint — superior aspect

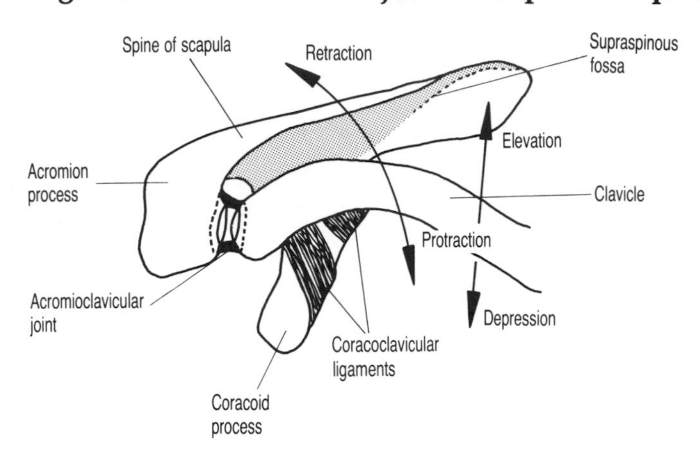

Sternoclavicular joints — anterior aspect (left joint with capsule removed)

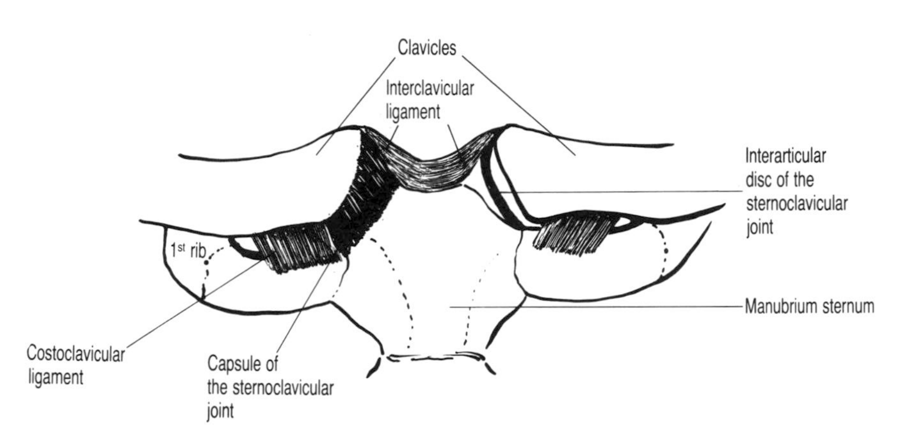

Right shoulder (glenohumeral) and acromioclavicular joints — anterior aspect

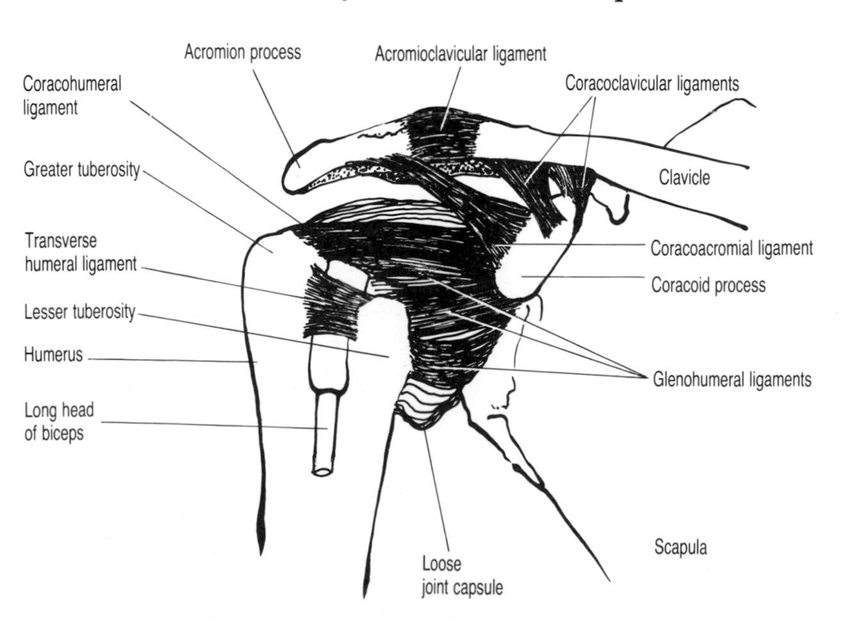

Right elbow joint

Lateral aspect

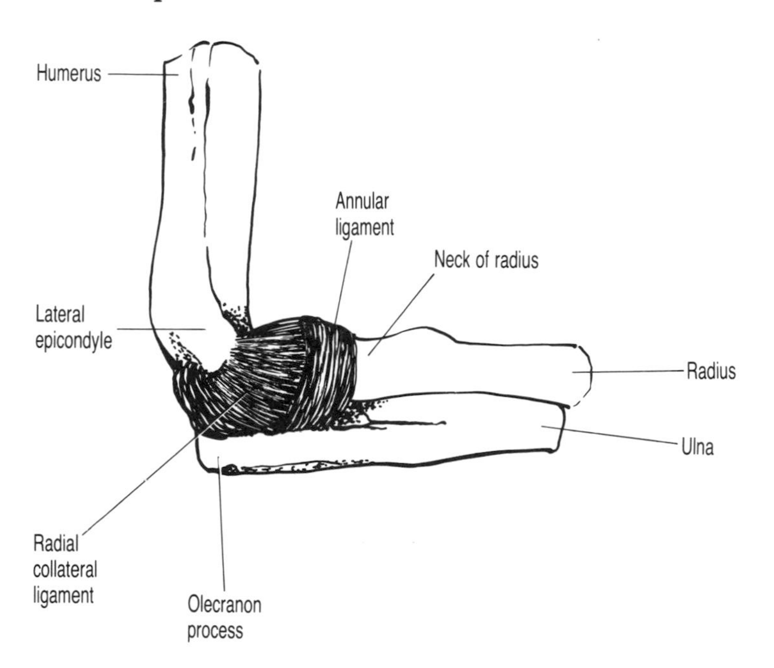

Medial aspect

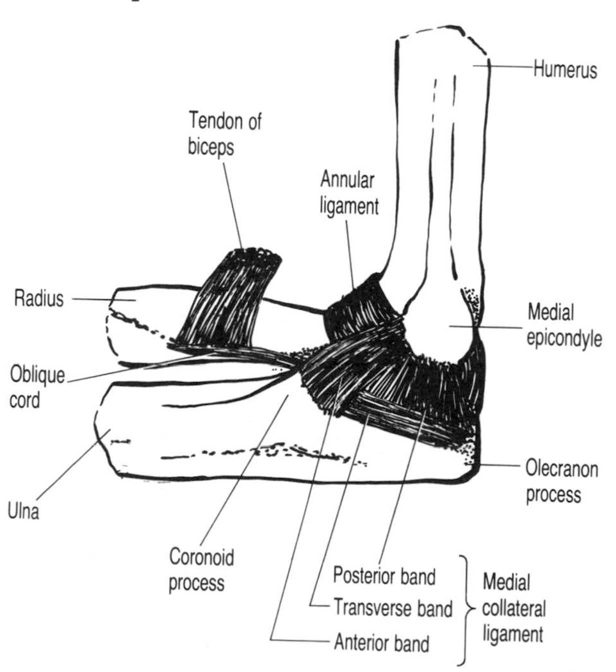

Right superior radioulnar joint (seen from above)

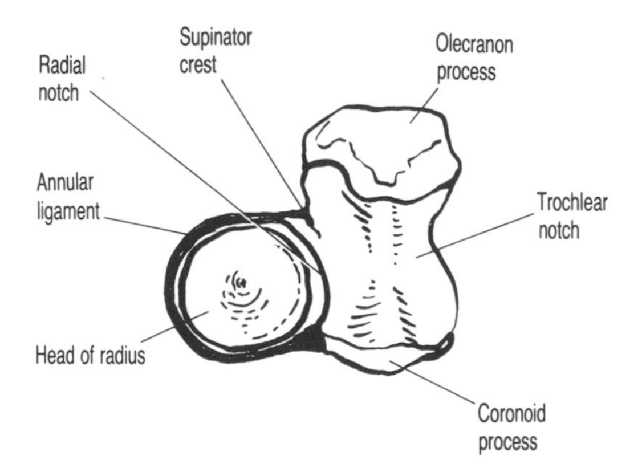

Middle radioulnar joint

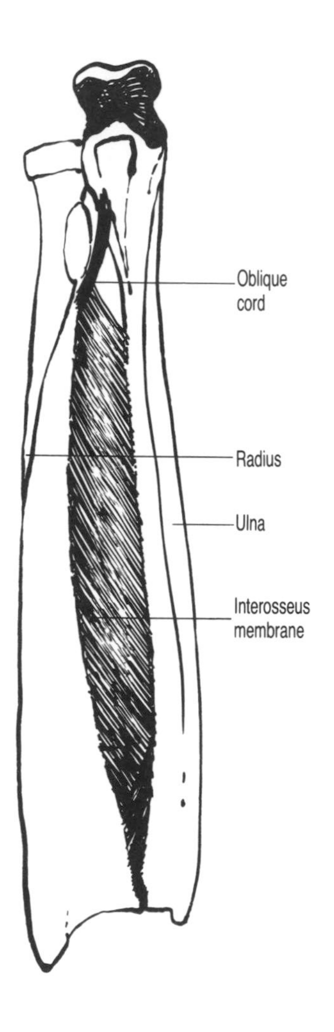

Right inferior radioulnar joint (seen from below)

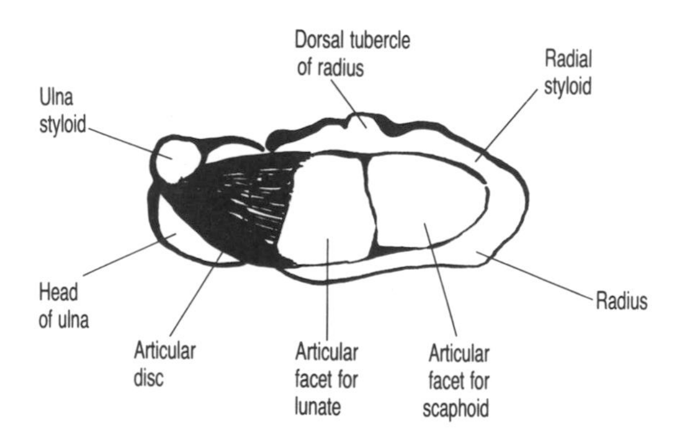

Right wrist (radiocarpal) joint — anterior aspect

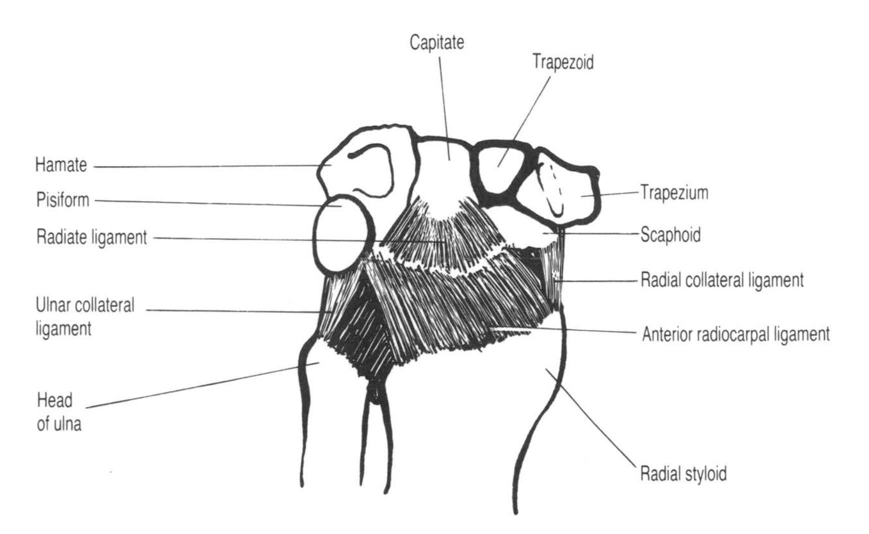

Joints of the finger — lateral aspect

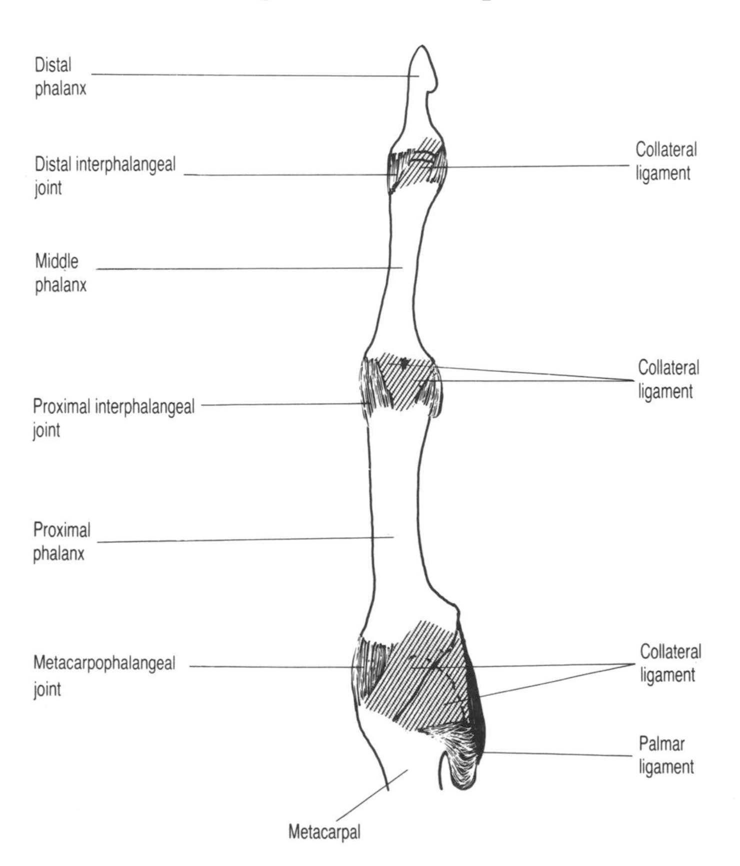

Right hip joint — anterior aspect

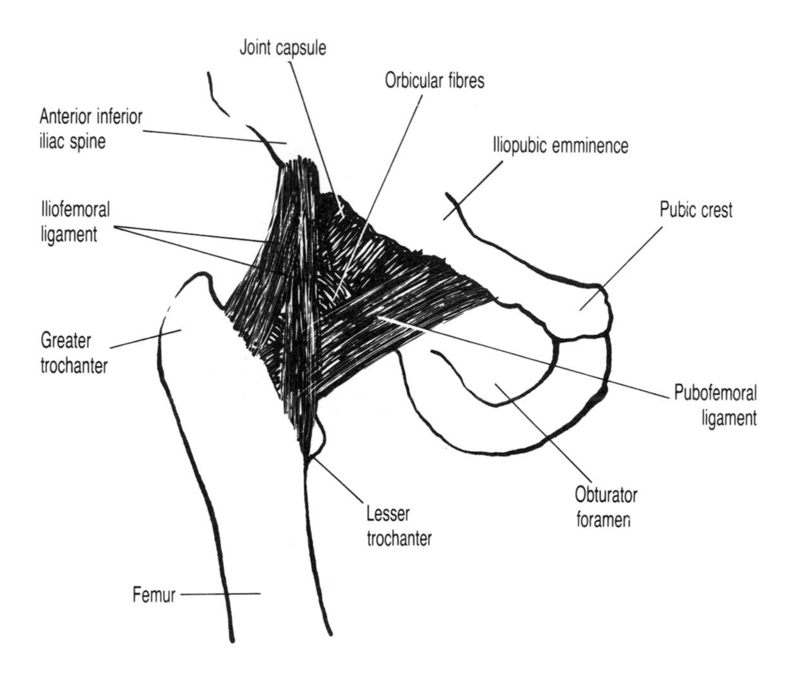

Right knee joint — anterior aspect

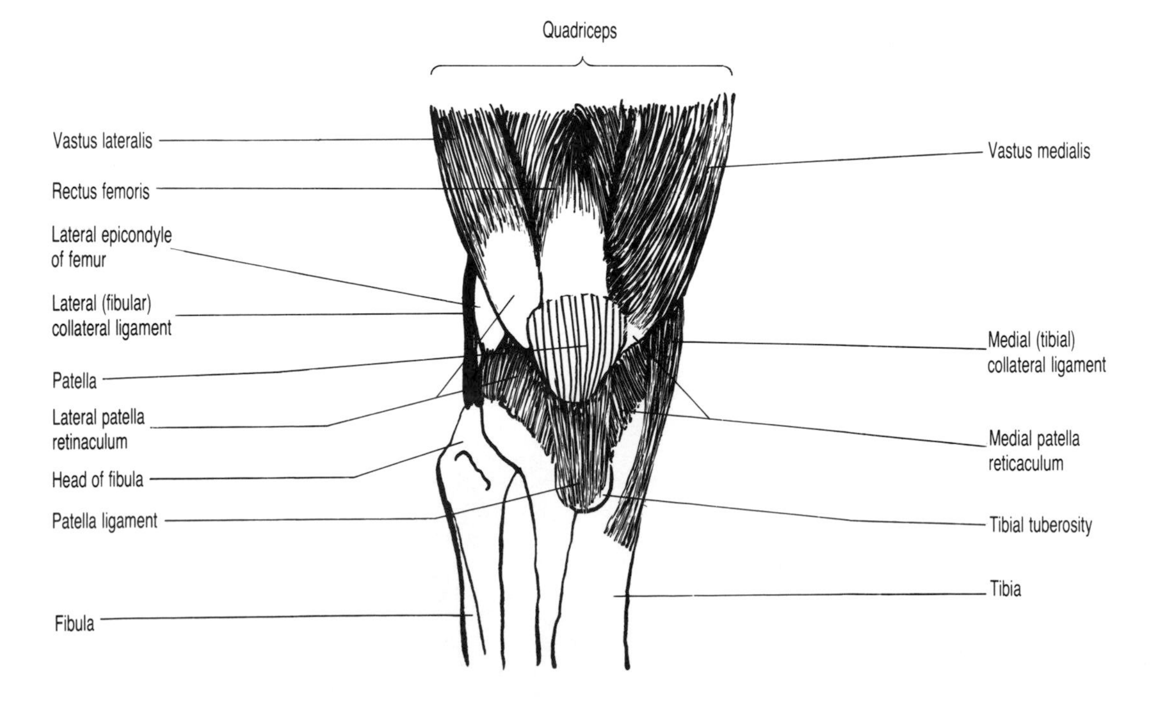

Right knee joint — anterior aspect
(Capsule and muscle removed)

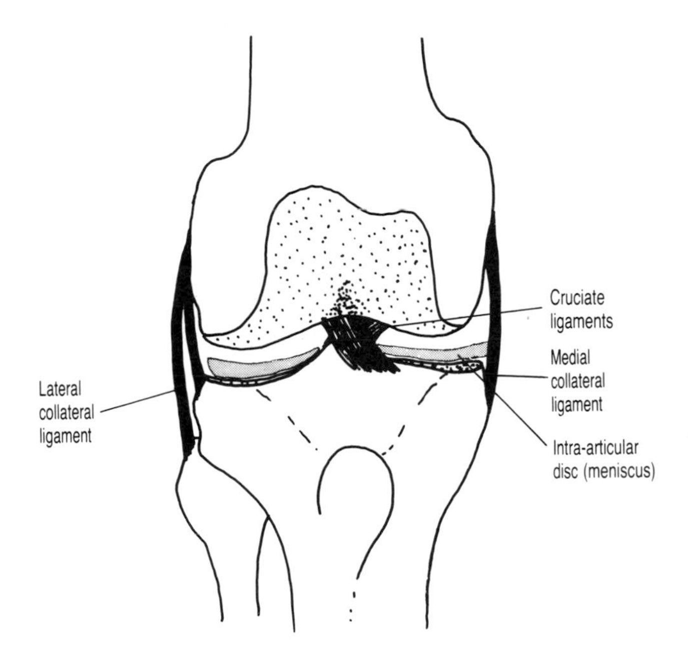

Right ankle joint

Medial aspect

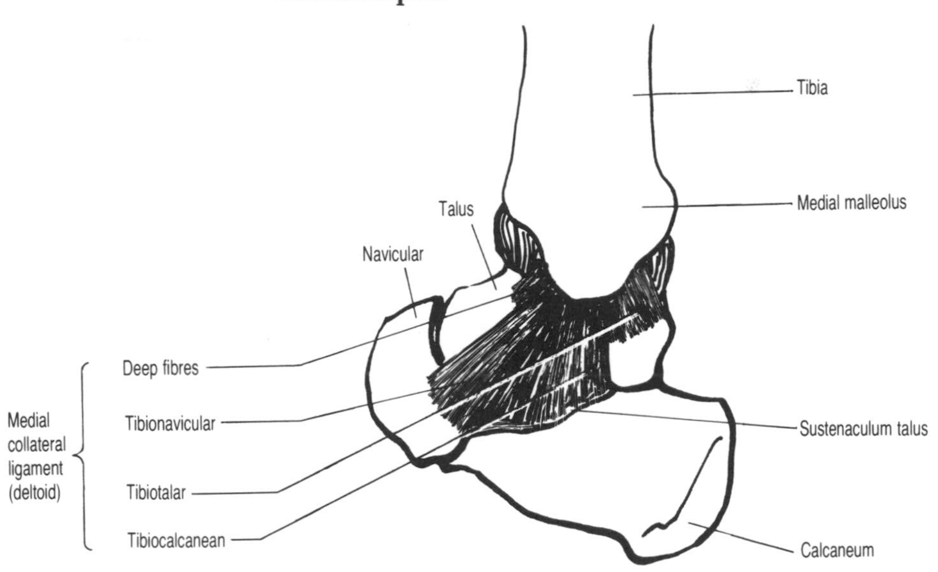

Lateral aspect

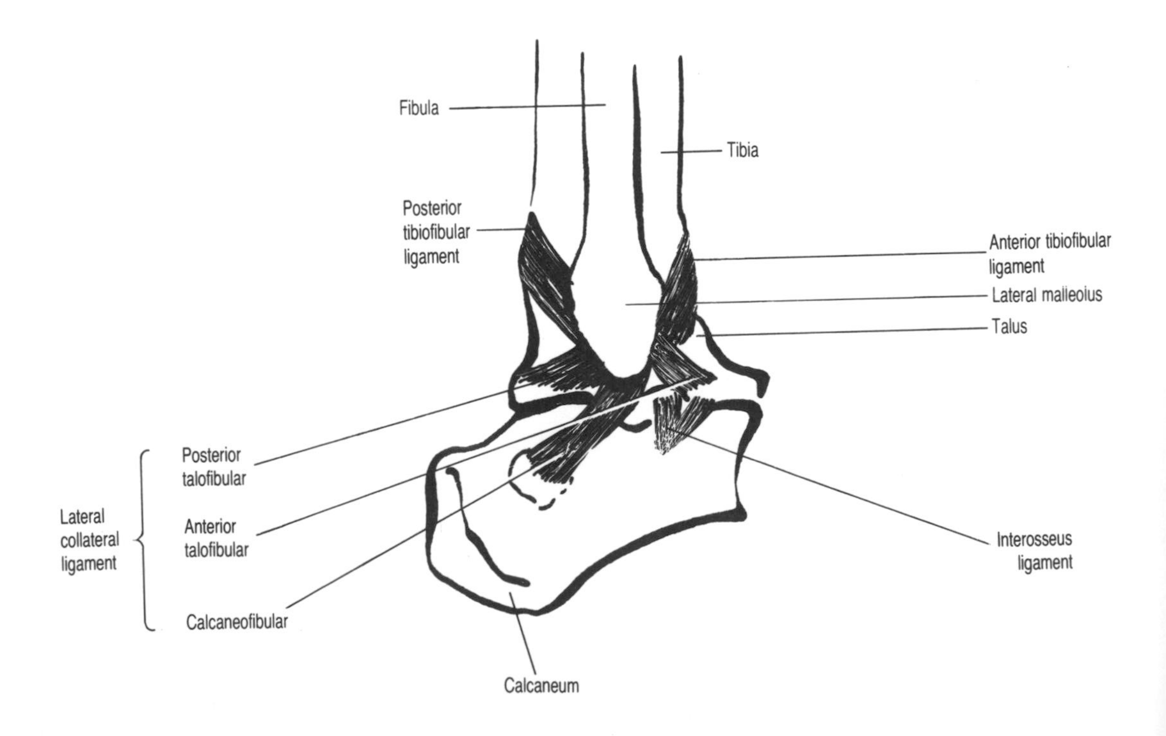

Appendix 3 Segmental Nerve Supply of Muscles

Table A3.1. Cranial nerves

I	Olfactory	Sensory from roof of the nose. Smell
II	Optic	Sensory from retina of the eye. Vision
III	Oculomotor	Motor to four of the muscles of the eye (superior, inferior and medial rectus, inferior oblique), motor to the sphincter muscle of the iris and the ciliary muscle of the lens
IV	Trochlear	Motor to the superior oblique eye muscle
V	Trigeminal	Sensory to skin of the face and anterior tongue
		Motor to salivary glands and muscles of mastication (temporalis and masseter)
VI	Abducens	Motor to the lateral rectus eye muscle
VII	Facial	Sensory to anterior tongue. Taste Motor to muscles of the face and salivary glands
VIII	Vestibulocochlear	Sensory from vestibule. Balance
		Sensory from cochlear of ear. Sound
IX	Glossopharyngeal	Sensory from posterior tongue. Taste
		Motor to salivary glands and pharynx
X	Vagus	Sensory and motor to pharynx, larynx, thoracic and abdominal organs
XI	Spinal Accessory	
	Cranial root	Motor to the muscles of the pharynx and larynx
	Spinal root (C1—C5)	Motor to sternomastoid and trapezius
XII	Hypoglossal	Motor to muscles of the tongue

Note: Cranial nerves supplying muscles contain sensory proprioceptor fibres, except the facial nerve. Proprioception from facial muscles is carried in the Trigeminal nerve.

Table A3.2. Spinal nerves. Segmental origin in the spinal cord of the nerves supplying the muscle groups moving the limbs

C5, C6	**Shoulder**	Abductors and lateral rotators
C5, C6, C7, C8		Flexors, extensors, adductors and medial rotators
C5, C6	**Elbow**	Flexors
C7, C8		Extensors
C5, C6	**Forearm**	Supinators
C6, C7, C8,		Pronators
C6, C7, C8	**Wrist**	Flexors, extensors and deviators
C7, C8, T1	**Digits**	Long flexors and extensors
C8, T1	**Hand**	Intrinsic muscles
L2, L3	**Hip**	Flexors
L2, L3, L4		Adductors
L4, L5, S1		Extensors, medial and lateral rotators and abductors
L2, L3, L4	**Knee**	Extensors
L4, L5, S1, S2		Flexors
L4, L5, S1,	**Ankle**	Dorsiflexors
L4, L5, S1, S2		Plantarflexors
L4, L5, S1	**Foot**	Invertors
L5, S1		Evertors
L5, S1, S2		Intrinsic muscles

Table A3.3. Segmental innervation of the muscles of the upper limb (after Basmajian, J. (ed) (1980) *Grant's Method of Anatomy*, 10th edn, published by Williams & Wilkins).

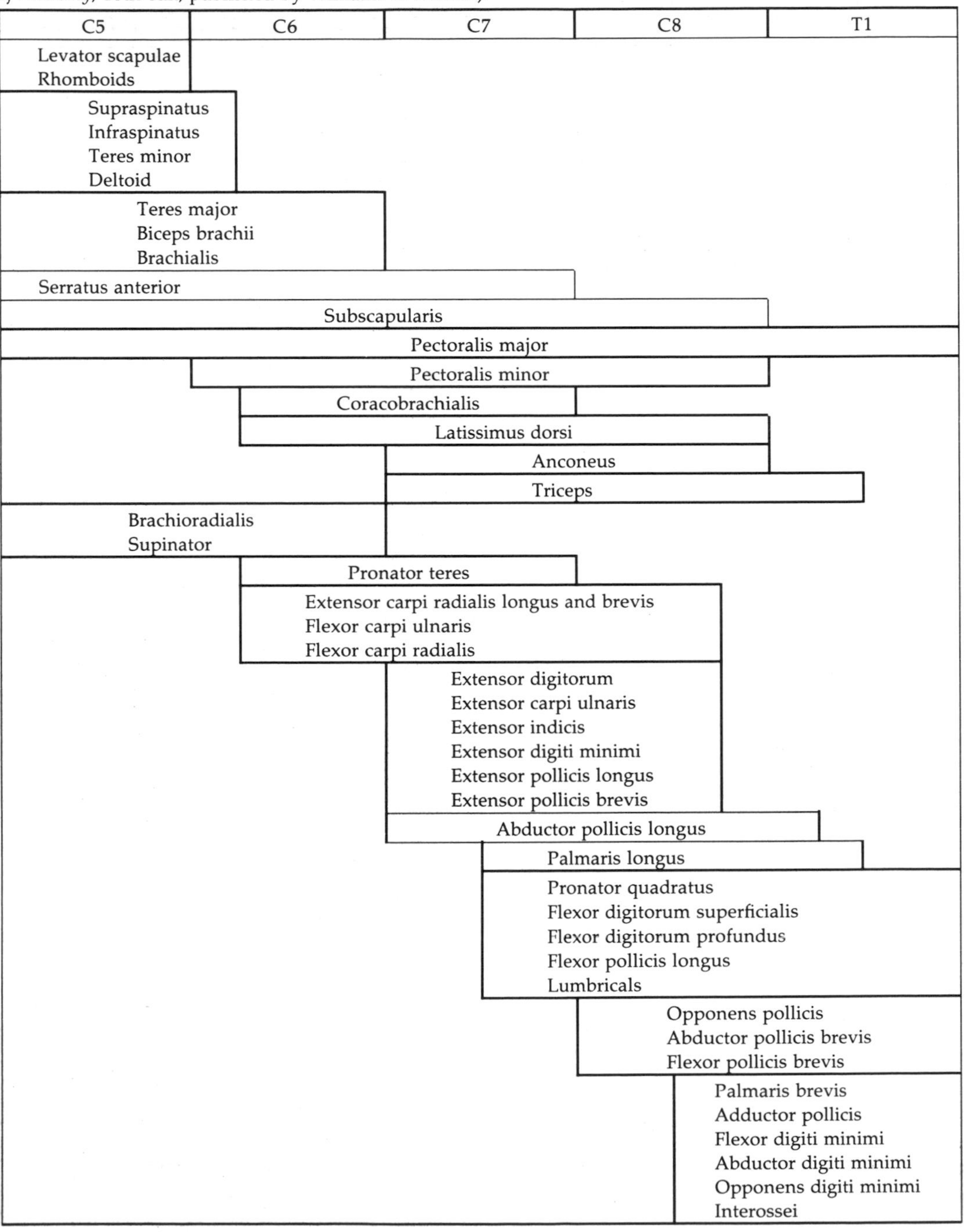

Muscles	C5	C6	C7	C8	T1
Levator scapulae Rhomboids	•				
Supraspinatus Infraspinatus Teres minor Deltoid	•	•			
Teres major Biceps brachii Brachialis	•	•			
Serratus anterior	•	•	•		
Subscapularis	•	•	•	•	
Pectoralis major	•	•	•	•	•
Pectoralis minor		•	•	•	
Coracobrachialis		•	•		
Latissimus dorsi		•	•	•	
Anconeus			•	•	
Triceps			•	•	•
Brachioradialis Supinator	•	•			
Pronator teres		•	•		
Extensor carpi radialis longus and brevis Flexor carpi ulnaris Flexor carpi radialis		•	•	•	
Extensor digitorum Extensor carpi ulnaris Extensor indicis Extensor digiti minimi Extensor pollicis longus Extensor pollicis brevis			•	•	
Abductor pollicis longus			•	•	•
Palmaris longus			•	•	•
Pronator quadratus Flexor digitorum superficialis Flexor digitorum profundus Flexor pollicis longus Lumbricals			•	•	•
Opponens pollicis Abductor pollicis brevis Flexor pollicis brevis				•	•
Palmaris brevis Adductor pollicis Flexor digiti minimi Abductor digiti minimi Opponens digiti minimi Interossei				•	•

Table A3.4. Segmental innervation of the muscles of the lower limb (after Basmajian, J. (ed) (1980) *Grant's Method of Anatomy*, 10th edn, published by Williams & Wilkins).

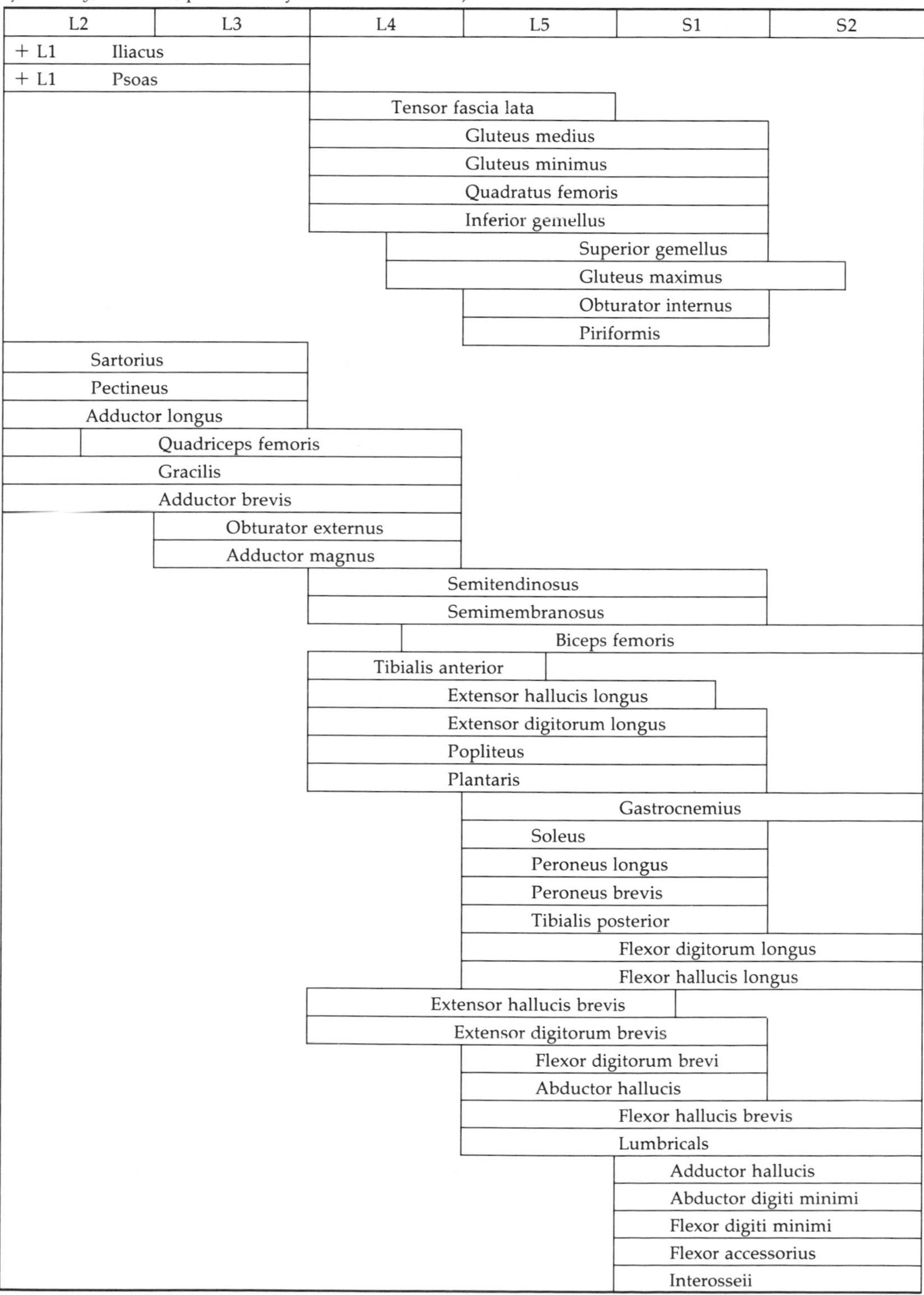

Muscle	L2	L3	L4	L5	S1	S2
+ L1 Iliacus	x	x				
+ L1 Psoas	x	x				
Tensor fascia lata			x	x		
Gluteus medius			x	x	x	
Gluteus minimus			x	x	x	
Quadratus femoris			x	x	x	
Inferior gemellus			x	x	x	
Superior gemellus			(x)	x	x	
Gluteus maximus			(x)	x	x	(x)
Obturator internus				x	x	
Piriformis				x	x	
Sartorius	x	x				
Pectineus	x	x				
Adductor longus	x	x				
Quadriceps femoris	x	x	x			
Gracilis	x	x	x			
Adductor brevis	x	x	x			
Obturator externus		x	x			
Adductor magnus		x	x			
Semitendinosus			x	x	x	
Semimembranosus			x	x	x	
Biceps femoris			(x)	x	x	x
Tibialis anterior			x	(x)		
Extensor hallucis longus			x	x	(x)	
Extensor digitorum longus			x	x	x	
Popliteus			x	x	x	
Plantaris			x	x	x	
Gastrocnemius				x	x	x
Soleus				x	x	
Peroneus longus				x	x	
Peroneus brevis				x	x	
Tibialis posterior				x	x	
Flexor digitorum longus				x	x	x
Flexor hallucis longus				x	x	x
Extensor hallucis brevis			x	x	(x)	
Extensor digitorum brevis			x	x	x	
Flexor digitorum brevi				x	x	
Abductor hallucis				x	x	
Flexor hallucis brevis				x	x	x
Lumbricals				x	x	x
Adductor hallucis					x	x
Abductor digiti minimi					x	x
Flexor digiti minimi					x	x
Flexor accessorius					x	x
Interosseii					x	x

Index